OECD CENTRE FOR CO-OPERATION WITH NON-ME

Review of Agricultural Policies

RUSSIAN FEDERATION

ORGANISATION FOR ECONOMIC CO-OPERATION AND DEVELOPMENT

ORGANISATION FOR ECONOMIC CO-OPERATION AND DEVELOPMENT

Pursuant to Article 1 of the Convention signed in Paris on 14th December 1960, and which came into force on 30th September 1961, the Organisation for Economic Co-operation and Development (OECD) shall promote policies designed:

- to achieve the highest sustainable economic growth and employment and a rising standard of living in Member countries, while maintaining financial stability, and thus to contribute to the development of the world economy;
- to contribute to sound economic expansion in Member as well as non-member countries in the process of economic development; and
- to contribute to the expansion of world trade on a multilateral, non-discriminatory basis in accordance with international obligations.

The original Member countries of the OECD are Austria, Belgium, Canada, Denmark, France, Germany, Greece, Iceland, Ireland, Italy, Luxembourg, the Netherlands, Norway, Portugal, Spain, Sweden, Switzerland, Turkey, the United Kingdom and the United States. The following countries became Members subsequently through accession at the dates indicated hereafter: Japan (28th April 1964), Finland (28th January 1969), Australia (7th June 1971), New Zealand (29th May 1973), Mexico (18th May 1994), the Czech Republic (21st December 1995), Hungary (7th May 1996), Poland (22nd November 1996) and Korea (12th December 1996). The Commission of the European Communities takes part in the work of the OECD (Article 13 of the OECD Convention).

OECD CENTRE FOR CO-OPERATION WITH NON-MEMBERS

The OECD Centre for Co-operation with Non-Members (CCNM) was established in January 1998 when the OECD's Centre for Co-operation with the Economies in Transition (CCET) was merged with the Liaison and Co-ordination Unit (LCU). The CCNM, in combining the functions of these two entities, serves as the focal point for the development and pursuit of co-operation between the OECD and non-member economies.

The CCNM manages thematic and country programmes. The thematic programmes, which are multi-country in focus, are linked to the core generic work areas of the Organisation (such as trade and investment, taxation, labour market and social policies, environment). The Emerging Market Economy Forum (EMEF) and the Transition Economy Programme (TEP) provide the framework for activities under the thematic programmes. The EMEF is a flexible forum in which non-members are invited to participate depending on the theme under discussion. The TEP is focused exclusively on transition economies. Country programmes, providing more focused dialogue and assistance, are now in place for Bulgaria, China, Romania, Russia, the Slovak Republic (a candidate for accession to the OECD), and Slovenia.

Publié en français sous le titre :

EXAMEN DES POLITIQUES AGRICOLES
FÉDÉRATION DE RUSSIE

FOREWORD

This review, which covers the period 1986-1997, evaluates the development and effects of agricultural policies in Russia against the benchmark of a competitive market economy. Although particular emphasis is given to the post-reform restructuring process of the agro-food sector, the review follows the general framework of the series of studies on National Policies and Agricultural Trade carried out for OECD countries.

The review gives detailed estimates of assistance to agriculture in Russia using the OECD's Producer and Consumer Subsidy Equivalent methodology (PSEs and CSEs). It includes an annex presenting the PSE and CSE calculations by product as well as a regional analysis of agricultural policies in Russia. Following the pattern in most transition economies, support levels to agriculture in Russia fell dramatically during the early period of transition. However, any trend towards increased market price support would give cause for concern. Although support levels to agriculture in Russia are below the OECD average, they are higher than those in most other countries in transition for which the PSE has been calculated.

As recognised in the review, the agricultural economy in Russia has vast potential and will continue to form a significant part of Russia's economic and social structure well into the 21st century. In order to fulfill this potential, Russia should resist pressures for increased market price support and seize the opportunity for removing market inefficiencies, establishing the necessary market infrastructure, and fostering transparency.

This study of Russia's agricultural policies was undertaken as part of the Special Country Programme for Russia within the framework of the programme of the OECD's Centre for Co-operation with Non-Members (CCNM).

The main author of this review was Andrzej Kwiecinski with contributions from Natacha Pescatore, Vaclav Vojtech and Peter Walkenhorst of the Directorate for Food, Agriculture and Fisheries. There was substantial input from Russian experts, notably Eugenia Serova and Olga Melyukhina of the Economy in Transition Institute, Victor Nazarenko of the All-Russian Institute of Information and Technical-Economic Research of the Agroindustrial Complex and Natalya Shagaida of the Agrarian Institute. The review benefited from co-operation with the Russian authorities.

The study was discussed in draft at a round table meeting with Russian officials in December 1997. It was subsequently examined by the OECD's Committee for Agriculture, meeting in an informal session in May 1998 which brought together policymakers from OECD Member countries and the Russian Federation. The final report was approved and declassified by the Committee for Agriculture and the Trade Committee in June 1998.

Kumiharu Shigehara
Deputy Secretary-General

TABLE OF CONTENTS

Tables

Figures

Boxes

SUMMARY AND CONCLUSIONS

The dissolution of the Soviet Union in 1991 marked the beginning of a long and laborious route to market-oriented reform in Russia. While significant progress has been made in liberalising the economy, the difficulties in applying a coherent reform strategy in the agro-food sector have resulted in the adoption of a series of interim policy measures which have inadequately addressed the structural impediments in agriculture and its upstream and downstream sectors.

The farm restructuring process needs to be completed, as only about ten per cent of large farms have been substantially restructured. The continuing uncertainty over land ownership rights is constraining the development of an operational land market and discouraging investment in the sector. Formal privatisation of the upstream and downstream food industries has taken place. However, privatisation alone in the absence of adequate corporate governance and enabling macro-economic conditions has failed to produce the necessary changes to increase productivity and return on capital and to inject the marketing and business skills on which the revitalisation of the sector depends. The absence of a supporting market infrastructure remains a serious constraint. The burden of carrying social assets and public utilities continues to weigh heavily on large agricultural enterprises, diverting their financial and managerial resources and hampering restructuring. While innovations in agricultural policies in certain Russian regions provide instructive examples, inconsistencies between regional and federal policies may create serious obstacles to agricultural reform and to specialisation based on comparative advantage.

A stable macro-economic and institutional framework is essential to agricultural reform in Russia and the recent macro-economic stabilisation is an encouraging sign in this respect. The levelling off of gross agricultural output, following a 36 per cent drop since the onset of reform, is also a positive development. So that the agro-food sector may benefit from these positive trends, policymakers should resist pressures for increased market price support and seize the opportunity for removing market inefficiencies, establishing the necessary market infrastructure and fostering transparency. Russia's economic potential is enormous and the rural economy will continue to form a significant part of Russia's economic and social structure well into the 21st century. In order to fulfil this potential, Russia needs to facilitate the development and implementation of effective market mechanisms and to eliminate the structural barriers that are impeding the emergence of a strong and competitive agro-food sector.

A. The economic and agricultural environment

1. Macroeconomic reform

Having implemented price liberalisation and mass privatisation...

Market-oriented economic reforms in Russia started during the last years of the Soviet Union and have accelerated since its dissolution in 1991. In January 1992, the government began the implementation of a series of reforms by **liberalising** almost all prices, floating the Rouble, slashing defence spending, cutting budget subsidies, eliminating the old centralised distribution system, and liberalising foreign trade. Between 1992-1994, the Russian **mass privatisation** programme was implemented. As a result, about 80 per cent of industrial enterprises passed, at least formally, into private hands. However, both inflation rates and the budget deficit remained very high in this period.

... the Russian government turned its attention to stabilisation policies

Toward 1995, to advance the **stabilisation** process the government formulated tight monetary and fiscal policies, which were associated with high real interest rates and an exchange rate anchor. The subsequent deceleration of inflation was significant, but institutional reforms and privatisation slowed down, partly due to political uncertainties surrounding the presidential elections held in mid-1996. Changes in the Russian Government in March 1997 gave new impetus to reforms.

At last, in 1997 GDP stabilised, with a small decline in the unemployment rate

Between 1990 and 1997 Russian **gross domestic product** (GDP) declined by about 40 per cent, according to official figures. The structure of GDP changed significantly with industry's and agriculture's shares in GDP falling, and that of the service sector increasing. In 1997 officially reported GDP stabilised for the first time since the beginning of the country's reforms. This positive development was also reflected by an apparent slight decline in the rate of **unemployment** to about 9 per cent at the end of the year.

Macro-economic stabilisation remains clouded by structural, institutional and legislative shortcomings

With the **inflation** rate falling to 11 per cent in 1997, a stable and predictable exchange rate, and lower real interest rates, Russia has achieved a considerable degree of macroeconomic stabilisation.* Progress in reform still needs to be consolidated, however. Remaining weaknesses are reflected in shrinking tax revenues, associated with high budget deficits and expenditure sequestration. Sequestration, in turn, contributes to escalating non-payment problems in the economy. Moreover, results of **institutional reforms** are mixed, with confusing and rapidly changing laws, weak law enforcement, lack of adequate corporate governance, crime and corruption remaining major obstacles to the development of efficient markets. So far, the privatisation process has meant very little for the internal restructuring of enterprises, with many firms having no clear owners apart from the managers and workers themselves. To speed up reforms, major legislation is still needed, including laws regulating taxation, foreign investments, natural monopolies and land markets.

2. *Agricultural in the economy*

Vast agricultural land has suffered from serious environmental problems

With 221 million hectares of agricultural land (about 1.5 hectare per capita), Russia has vast land resources which favour **extensive methods of farming**. However, on average the **soils** in Russia are of low quality, except for highly fertile soils in the southern half of European Russia (in particular Northern Caucasus and Central Black Soils Region), but also in the Southern Urals and the southern fringe of Siberia. Moreover, soil is being lost as a result of **severe water and wind erosion**, amplified by farming and livestock grazing practices that often ignore the need for soil conservation. Climate is extremely differentiated, but the continental climate predominates which favours occurrence of droughts that appear on average every three years and strongly effect production. The massive expansion

* In January 1998, the Rouble was redenominated with the rate of 1 new Rouble equal to 1 000 old Roubles. In this Report, all nominal Rouble amounts for the whole period under study are given in old Rouble, before the redenomination.

of irrigation systems in Russia in 1970s and 1980s aimed at alleviating vulnerability to weather conditions and increasing production potential, but caused serious environmental problems.

Agriculture's employment share has not matched the sharp decline in its GDP share

The **share of agriculture in the economy** has diminished considerably since the transition started. However, while the sector's contribution to GDP fell from 15.4 per cent in 1990 to 6.5 per cent in 1997, the proportion of the total working population in the sector grew from 13 per cent to about 14 per cent, mostly due to greater falls in employment in other sectors of the economy. The increased disparity between the agricultural contribution to GDP and its share in total employment indicates a sharp fall in output per worker since 1991 and suggests growing hidden unemployment in rural areas. In fact, while employment in agriculture fell by 7 per cent between 1990 and 1997, the volume of agricultural production fell by 36 per cent over the same period.

The drop in food consumption has been significant

Cuts in consumer subsidies and a fall in real incomes led to a significant decline in per capita **food consumption** in the 1990s. However, while demand for products with high income elasticities, such as meat and dairy products, decreased, demand for staple goods like potatoes and cereal products increased. Between 1990 and 1996, per capita consumption of meat and milk products fell by 31 and 38 per cent, respectively.

The severe cost-price squeeze...

The 36 per cent fall in agricultural output was due to the sharp **cost-price squeeze** which occurred after price liberalisation in 1992; while the decline in demand limited increases in farmgate prices, most input prices grew to the world market levels creating a sharp deterioration in the output/input price ratios for agricultural producers. The terms of trade for agriculture were further worsened by inefficient upstream and downstream enterprises trying to pass their high costs of production onto producers and/or consumers. As a result, between 1991 and 1993, input prices grew 2.3 times faster than agricultural output prices. In 1995, the terms of trade improved slightly, but worsened again in 1996. Worsening of the output/input price ratios occurred in all countries in transition, but it was the sharpest in those countries previously providing the highest levels of support for agriculture, such as Russia.

... has been exacerbated by a lack of access to financial resources

The cost-price squeeze combined with the lack of liquidity and rising indebtedness significantly reduced the ability of agricultural enterprises to purchase production inputs. A growing number of large farms has been reporting a **lack of working and investment capital**. Access to credit has been difficult because agriculture, due to its difficult financial situation, has been considered to be a high risk sector. This has been further exacerbated by tight monetary policy resulting in high real interest rates and by the poorly developed banking system in rural areas. Therefore, purchases of agricultural inputs fell sharply leading in turn to a fall in yields of both the crop and livestock sectors.

Like GDP, gross agricultural output stabilised at last in 1997...

As a result of all these factors, **gross agricultural output** (GAO) fell continuously between 1990 and 1996. In 1997, Russia's GAO stabilised, for the first time since the reforms started, which was partly due to favourable weather conditions for crop production, but might also indicate an end to the first and most difficult stage of transition for Russian agriculture.

... with the share of household plots nearly doubling from 1990...

Changes in agricultural output in Russia have been highly **differentiated** across sub-sectors, commodities and regions. While production by large-scale agricultural enterprises halved between 1990 and 1996, the output of **household plots** increased by 19 per cent over the same period. As a result, the latter's contribution to GAO increased from 26 per cent in 1990 to 46 per cent in 1996. The high share of household plots in total agricultural output can be partly attributed to their specialisation in high value products such as fruits, vegetables and animal products, but also to their reliance on various registered and unregistered produce and service transfers from large farms. In 1996, with the share of large farms in GAO at 52 per cent, the remaining 2 per cent of GAO was contributed by the emerging individual (family) farms.

... and grain production recovering

Following changes in consumer demand and relative changes in crop versus livestock prices, the share of **livestock production** in the total value of agricultural production dropped sharply from 64 per cent in 1990 to 31 per cent in 1992, but then rose again to 42 per cent in 1996. **Grain output** almost halved between 1990 and 1995, in part as a result of severe droughts in 1994 and 1995. It then recovered in 1996, rising to 69 million tonnes, and again in 1997, when it reached almost 89 million tonnes, mostly due to favourable weather conditions. Potatoes and vegetables, cultivated by household plots which resisted the general agricultural crisis, are among the few agricultural products for which production was rather stable or even increased during the reform period.

Certain regions persist in maintaining Soviet-style agro-food policies

Some **regional administrations** in particular at the initial stages of reform, using financial resources drawn for example from large oil and gas deposits located on their territories, have prevented a substantial decline in agricultural output in their regions. They have maintained Soviet style agro-food policies, such as large-scale state procurement and input subsidisation. However, such policies have proven to be very costly for taxpayers and consumers. Also, the experience from other transition countries indicates that postponing necessary systemic reforms makes the ultimate adjustments even more costly and painful.

B. Agro-food restructuring

1. Collective and state farm restructuring

The development of market-oriented, privately-owned farm structures...

The Russian agricultural **privatisation** programme, initiated at the end of the Soviet period and substantially further developed after the dissolution of the Soviet Union, was intended to make it possible to run privately owned farms for those who wished to do so, to transfer the land and non-land assets to the people who lived and worked on the large farms at the time of the reform, and to transform *kolkhozes* and *sovkhozes* into more market oriented legal entities. **No restitution** of land to heirs of pre-1917 landowners and to heirs of peasants collectivised in the 1930s was provided for.

... has been the goal of successive legislative measures

Between 1991 and 1996 the Russian government adopted a series of **measures** in the form of presidential decrees and federal regulations authorising the transfer of land and non-land assets to large-farm employees, pensioners and social workers. The legislation also provided for land and non-land assets to be allocated to large-farm members who decided to establish their own individual (family) farm. In March 1996 the Presidential Decree "On guarantees of constitutional rights of citizens to land" re-emphasised that land entitlements could be freely traded and required that all users of agricultural land conclude formal purchasing or leasing contracts with every individual land entitlement owner and that local authorities complete giving land certificates to land owners before the end of 1996.

Most land and non-land assets belong to collectives under share-based ownership...

Efforts to change land tenure fundamentally and to reform large-scale farm organisation and management, as well as legal ownership, clashed with attempts by the traditionalist and communist parties supported by many farm managers and much of the rural population to preserve the previous system. By the end of 1997, in the vast majority of large-scale enterprises, **reorganisation** has not proceeded beyond re-registration of the original farm under a new legal form. In practice, the reorganisation has changed very little in terms of the institutional structure of the farm, management practices, and agricultural techniques. Regardless of the variety of legal forms, the majority of farms has been converted into production co-operatives in which fixed assets belong to the collective under a form of share-based ownership. In this system, land and non-land assets are owned collectively by the enterprises, and the enterprises are in turn owned by shareholders, who are employees, pensioners and social workers entitled to participate in the distribution of land and asset shares. In these enterprises management is selected on the basis of the co-operative principles of "one member one vote" and profits distributed predominantly on the basis of work input rather than share contribution.

... with only about ten per cent of large farms substantially restructured

Only about 10 per cent of large farms underwent more substantial **restructuring** which took one of the following forms: the break-up of large farms into smaller technologically integrated production units; the concentration of large farm land and property entitlements in the hands of a limited number of owners by means of the purchase, exchange or leasing of entitlements; or the partition of large farms into household plots. The size of these household plots expanded through either formal or informal land take-overs from the large farms, whose functions thereafter were generally limited to providing inputs and services to household plots.

Some flexibility in the allocation of land use is apparent...

By October 1997, more than 90 per cent of land entitlement holders had received official certificates for their ownership rights. About 40 per cent of them contracted their entitlements to the users of land, mostly under **leasing** arrangements, with contracts registered officially, but the remaining 60 per cent did not sign contracts. Many managers of the large-scale farms were reluctant or unable to pay **rents** (usually paid in kind and/or in the form of services) for the use of land. Some of them expected that the new and long-discussed Land Code would give land users (*i.e.* large-scale enterprises) ownership rights to the non-contracted land. The vast majority of entitlements has been contracted to the original, but now re-registered farms. However, between 30 and 50 per cent of land entitlements are owned by pensioners and non-farm workers who have shown more flexibility in allocating their entitlements to other enterprises and some entitlement holders have preferred to lease land plots to neighbouring large enterprises or individual farms in the expectation of receiving higher rents. This is creating opportunities for the reallocation of land use to those enterprises and individual farms which are more efficient and are able to pay higher rents.

... but uncertain property rights constrain the development of an operational land market

To exploit the potential arising from the restructuring process of the agricultural sector, Russia needs to complete the reforms it has initiated. So far, the relations and respective rights in the triangle: collectives (collective owners of land and property), individuals (individual holders of land and property entitlements) and agricultural enterprises as legal entities (enterprises which are corporate users of land and property) are not sufficiently clear cut. **Uncertainty over land ownership rights**, as exemplified by the proposed Land Code which restricts land market operations, reduces interest in acquiring land and non-land assets and discourages investment. When property rights are uncertain, a practical alternative is long-term leasing. As experience shows in some other transition countries such as the Czech Republic, the Slovak Republic, Hungary, eastern Germany, and Poland, leasing contracts between land owners and land users, backed by an appropriate legal framework, leads to a more efficient use of factors of production in agriculture. To promote a similar transformation in Russia, long-term leasing contracts between the management of farms, on the one hand, and land and property entitlement holders, on the other, would help to promote investment and longer term business planning.

2. *Emerging farm structure*

Sixty-two per cent of agricultural land is privately owned, mainly collectively

As a result of the privatisation and reorganisation process, by November 1997, 137 million hectares (62 per cent) of the 221 million hectares of agricultural land in Russia, were considered **privately owned**, while the remaining 84 million hectares (38 per cent) were still owned by the State or local municipalities. The majority of "privately" owned land was in the form of collectively shared ownership. The rest of privately owned land was owned by individual farms and household plots. Of the 38 per cent of non-privatised land 9 per cent belonged to municipalities and 15 per cent to various types of agricultural and non-agricultural enterprises and institutions in which land was not privatised for various reasons.

i) Large-scale enterprises

... including about 27 000 large-scale enterprises...

In January 1997 there were about 27 000 large-scale agricultural enterprises operating on 134 million hectares of agricultural land and averaging about 4 950 hectares. While it is difficult to find strictly comparable data, it may be concluded that the number of such farms has increased and their average size has declined since 1990. In addition, at the beginning of 1997, there were over 14 000 other enterprises and institutions engaged in agricultural production operating on about 18 million hectares of land and averaging about 1 300 hectares. These enterprises were either not covered by the process of land privatisation and their reorganisation or reorganised, but without privatisation of land.

ii) Household plots

... and some 16 million household plots

In 1997, about 5.7 million hectares, about 3 per cent of total agricultural land, were divided among 16 million household plot owners, resulting in an average of just 0.4 hectares per household. Their share in total agricultural production has increased significantly since the transition started, to about 50 per cent of GAO in 1997, but so far production on household plots is mostly for **family use**. However, a growing part is for sale. Moreover, large farms, lacking liquidity, pay in kind for labour and sell products to workers at discount prices. In turn, households may resell the products received from the large farms to whatever buyer they may find. Owners of household plots pay a land tax, which is very low in Russia, but do not pay income tax on income earned from the plot. Following a Presidential Decree of March 1996, household plots are free to use their land entitlements to expand their scale of operation up to the upper limit fixed by the local administration (usually between 6 and 12 hectares), but few of them have used this opportunity and even fewer have shown any interest in developing into more independent, family-type farms.

iii) Individual farms

Interest in private family-type farms remains limited

The family farm sector, operating on about 6 per cent of agricultural land in Russia in 1997, has remained of rather **minor importance**. Due to lack of capital, legislative and political uncertainty, difficult macroeconomic conditions, difficult access to information, credits and markets, and a lack of tradition and experience with individual farming in Russia, only a small proportion of farm workers has decided to establish private farms. Moreover, potential individual farmers are afraid of losing access to production infrastructure (storage, repair service, grain drying facilities etc.) located on large farms and to the social infrastructure provided through employment contracts with large farms.

Although their average size has recently expanded through leasing...

The number of individual farms stabilised at about 280 000 between 1995 and the beginning of 1997, but then declined by about 2 per cent by January 1998. However, the total amount of land in use by individual farms increased in 1997 by 7 per cent, mostly due to the enlargement of farms through **leasing** contracts made with land entitlement holders. As a result the average size of individual farms increased from 44 hectares at the beginning of 1997 to 48 hectares at the beginning of 1998.

... such farms cannot replace viable, restructured large-scale farm enterprises

While in some regions the development of individual farming may be complementing or even stimulating the restructuring of large-scale farms to some extent, it will not obviate the need for a more radical conversion of the large-scale enterprises into viable, business oriented, large-scale farms based on corporate principles.

3. *Privatisation of upstream and downstream industries*

Agents throughout the food chain need to be privatised and commercially linked

The privatisation of agricultural upstream and downstream enterprises combined with the creation of competitive commercial relationships between all agents in the food chain are essential for the revitalisation of agriculture and for securing its sustainable development. The economic justification for privatisation is to ensure transfer of productive assets to those who have an active interest in using them most efficiently and who are able to improve these assets through new investment. In turn, **competitive marketing links** are essential for creating a stimulating environment for economic agents, forcing them to cut processing and handling costs to the benefit of agricultural producers and consumers alike. Despite significant progress made since the reforms started, none of these processes has been completed in Russia.

De jure privatisation has failed to produce the necessary behavioural changes...

By the end of 1997, privatisation of agricultural upstream and downstream sectors was almost finalised. However, as in the case of formally privatised agricultural enterprises, the impact of the change in property rights on input-producing enterprises and food processing companies has been rather weak. So far most of them remain **inefficient** and try to pass their high production costs onto agricultural producers and consumers. Their slow adaptation to market conditions is partly due to the **methods** of their privatisation, which gave strong preferences to insiders and, in the case of about one fourth of upstream and downstream enterprises, to agricultural producers. In total the two groups (in particular employees) hold well over half of the shares in the vast majority of medium and large upstream and downstream enterprises. Such a structure of ownership may lead to a policy maximising employee benefits: awarding wages and salaries not related to productivity, giving an excessive preference for job security, avoiding layoffs, etc.; where enterprises are dominated by agricultural producers, the temptation may be to increase producer returns leaving **capital** without adequate reward. This in turn would deter potential external (domestic and international) providers of the capital as well as the management, financial and marketing know-how, that the enterprises so urgently need.

... and to inject necessary marketing and business skills

Both employees and agricultural producers lack not only capital but also **marketing and business skills**, which hampers or at least significantly delays the restructuring of upstream and downstream enterprises and, thus, contributes to their low efficiency. This may in turn retard the revitalisation of the agricultural sector which depends on high quality, reasonably priced and timely supplied inputs and on reliable, financially solvent, efficient purchasers and processors able to produce and sell food products at quality and prices attractive to domestic and foreign consumers.

C. Agricultural policies

1. Policies in the Soviet period

Consumer subsidies in the Soviet period rose to unsustainable levels...

Agricultural policy during the Soviet period was aimed at ensuring social stability and guaranteeing the supply of cheap food to the population. The Soviet policy of keeping retail food prices at stable and depressed levels, combined with an increase in nominal wages and a lack of spending opportunities for other goods and services, led to a high demand for food and shortages in the state retail stores. The increasing gap between rising procurement prices for producers and stagnant retail prices during the 1980s resulted in **unsustainable** levels of consumer subsidies.

... in a framework of misallocated resources and production-distorting agricultural policies

In the agro-food sector the policies pursued under central planning led to a huge misallocation of resources and practically eliminated any incentive to lower costs of production. Although retail price controls and accompanying subsidies were the most damaging to the budget, the rigid system of procurement prices and state orders were the most **distorting** for agricultural production.

2. Main post-reform agricultural policy objectives

Early* ad hoc *measures failed to tackle the roots of the problem...

During the first years of the transition there was no clear concept of coherent agricultural policy. Agricultural policy was limited to *ad hoc* measures addressing the most immediate problems of the agricultural sector. In response to the worsening output/input price ratios for agricultural producers, the government introduced livestock **subsidies** and input cost subsidies for farmers. The lack of financial resources in agriculture was addressed by credit subsidies. However, most of the underlying problems were consequences of macroeconomic instability and inadequate institutional reform, so that these agricultural policy measures addressed only the **symptoms** but not the causes of the problems.

... and proposed input subsidies and market intervention would mark a further step backwards

The programme set for agriculture for the period 1996-2000 aims to increase Russian **food self-sufficiency** and to reduce dependence on food imports. The main tools envisaged to achieve these objectives are various input subsidies, and market intervention. These measures, if applied, would lead Russian agriculture back to a dependence on state intervention at great costs to consumers and taxpayers and to serious misallocation of resources. Moreover, they would re-introduce trade distorting subsidies at a time when WTO members are disciplining these sorts of policies and detract from advances made through attracting investment and securing capital.

3. New policy instruments

i) Trade measures

Tariffs are aggravated by de facto trade barriers including lack of policy transparency

In the first years of reform, domestic markets were protected against imports by the undervalued Rouble. In this period trade policies were intended to prevent large outflows of agro-food products induced by high world market prices expressed in Roubles. The protection provided by a lowly valued currency was eroded by the subsequent appreciation of the Rouble and protectionist tendencies started to develop from 1993. Tariffs on agricultural products currently applied in Russia, at between 10 per cent and 30 per cent, are relatively low compared with some OECD countries which have tariffied their import restricting measures. However, their combination with a multiplicity of legal acts issued at different levels (including federal laws, presidential decrees, government decisions or resolutions as well as regulations introduced by regional and local authorities) with complex and sometimes arbitrary modalities of customs valuation and with bureaucratic, time consuming and expensive certification, creates important trade barriers. Moreover, frequent changes to specific requirements and regulations, often introduced on an *ad hoc* basis, make trade policy untransparent for both domestic and foreign traders.

Increasing border protection would be costly to consumers and complicate WTO negotiations

The government is under constant **pressure** from producer interest groups to increase border protection further against agricultural and food imports. However, any further tariff increases will have an adverse impact on domestic price levels, will increase transfers from consumers, and will complicate Russia's negotiations with the World Trade Organisation (WTO). Moreover, trade restrictive measures, including import tariff barriers, rarely produce the expected effects, in particular when institutions function poorly and regulations are unclear. Rather, they encourage corruption and rent-seeking behaviour. This is also the case for the inter-regional trade barriers erected by local administrations within Russia to a range of agro-food products.

ii) Price support and market regulation

State intervention has evolved through several stages...

In 1992 the Soviet state procurement system was phased out but the state retained some control on prices for agricultural products delivered to government reserves. The federal state procurement system was used to ensure a supply of food to the large cities (Moscow and St. Petersburg), northern regions in Russia, the army, prisoners, etc. State procurement at the federal level served also to maintain strategic reserves. Since 1995, the Federal Foodstuff Corporation (FFC) has handled the procurement for federal food stocks and in each region the FFC has its branch or representative company for procurement of regional stocks. In 1997 the federal government established a Federal Agency for the Regulation of the Food Market which is to replace the FFC. The main declared objective of **state intervention** is to stabilise agricultural prices on the domestic market.

... and carries distortive and costly risks

The operation of any **price support** (stabilisation) programme requires accurate information on domestic and foreign market conditions. In setting a support price, the responsible agency must be able to anticipate the supply response to the announced price and must have a reasonably accurate forecast of demand. Market intervention policies based on inaccurate information can be more destabilising to the market than no policy at all. Moreover, experience with market intervention in OECD countries shows that policies originally designed to stabilise markets and to ease adjustment problems in agriculture invariably result in growing support and protection for agriculture as a result of interest group pressure. Once adopted, it has proved politically very difficult to reduce such support. The negative consequences are well known: extra burdens on taxpayers and consumers, delayed structural adjustment and misallocation of resources, and reduced economic efficiency and international competitiveness in agriculture and food processing industries.

No effective, integrated marketing system exists and regional price disparities are considerable...

The agro-food market in Russia still lacks both horizontal and vertical coherence and an effective private marketing system has yet to emerge. Internal links between major production and consumption areas, as well as vertical price transmission across the agro-food chain are weak. Different approaches to controlling producer and retail prices at the regional level contributed to market segmentation, particularly at the beginning of the reforms, and slowed down the process of market integration necessary for a well-functioning Russian agro-food market. Price differences between regions are still considerable and not explicable by transportation costs alone. In order to solve these problems the Russian government should aim at creating a better **institutional framework** for an effective agro-food market, within which contracts will be strictly enforced and the **rule of law** will be upheld. If second-best market regulation and price support policies are adopted as temporary measures they must be properly co-ordinated at the federal level. Effective and enforceable federal legislation to restrict efforts by regional governments to control prices on their own would be needed. With such overall co-ordination in place, regions could be granted a certain degree of autonomy in applying less market distorting types of support to agriculture such as infrastructure investment and incentives for more environmentally-friendly farming.

iii) Direct payments

... and aggravated by the regional subsidy systems

During the Soviet period consumer subsidies were greatest for livestock products, so their discontinuation in 1992 resulted in a sharp decline in demand and excess supply of livestock products appeared on the domestic market. This depressed prices for livestock products. To counter declining farm revenues, the government introduced direct payments to livestock producers in March 1992. **Livestock subsidies** were one of the most important programmes supporting agriculture between 1992 and 1996. In 1993, the subsidies were transferred to the regional level and their payment became dependent on the market intervention schemes in operation in different regions and on the availability of finance in the regional budgets. The differentiated treatment of livestock subsidies strengthened the incentives of local governments to limit inter-regional flows of food products and thus impeded the creation of a well-functioning market for livestock products in Russia.

iv) Reduction of input costs

State intervention has undermined the development of agricultural finance...

Up to 1996 large budget resources were engaged under various programmes to compensate for the lack of finance in agriculture. Many of these programmes supplied financial support in the form of loans from the state budget; however, the rate of repayment of these resources was very low and, until 1995, resulted in debt **write-offs**. The tradition of write-offs, combined with the extensive system of preferential credits subsidised by the state, has hampered the development of financial discipline among agricultural producers, increased the risk of lending to agricultural producers, and crowded out commercial banking activities in rural areas. As a result, the majority of credit activities depend strongly on state involvement. Moreover, at the macroeconomic level, the massive postponements of repayments and write-offs increase domestic debt and contribute to inflationary pressures in Russia.

... although recent changes mark an improvement

In 1997 the preferential credit system in Russia was substantially reformed with the most distorting "commodity credit" programme replaced by a new one financed and distributed in a much less distorting way. The main source of finance, in addition to the federal budget, has been the receipts from the sale of the 1996 commodity credit debts, restructured into bonds and sold at public auctions. Since 1997, these credits have been allocated by commercial banks which, since 1988, have been selected through tender. Also, the government programme supporting purchases of agricultural machinery, the so-called leasing system, was reformed: *Rosagrosnab* lost its monopoly position as machinery supplier and private leasing companies were selected by open tenders. These are positive moves in that they allow commercial firms and competition to play a greater role in the allocation of subsidised credit resources to agriculture.

4. General services

i) Research, education and training

Research, education and training lack funding, co-ordination and relevant skills

Agricultural research, education and training institutions have had to face considerable financial difficulties in recent years because of the reduction in public funding. In this new context many institutions started to offer research services, courses and training often on a fee-per-service basis. However, there is still a substantial lack of **co-ordination** between research, education and extension activities. Also, the education programmes are still highly concentrated on **production skills** as most schools and universities are short of staff able to teach economics, management and marketing. Changes in the curricula are being introduced with technical assistance from international programmes such as Tempus and TACIS.

ii) Marketing and promotion

Market infrastructure is badly needed to link producers and consumers nationwide

Appropriate market infrastructure has been very slow to develop in Russia. There is a lack of commodity exchanges, wholesale markets and auctions that would improve market transparency and provide clear market signals to producers. Barter trade remains important, and a significant part of production is used for payments in kind for workers on farms. In 1994 the Ministry of Agriculture and Food (MAF) adopted a national programme for the development of agro-food **wholesale markets**. Mainly due to budget constraints the programme was implemented on a very limited scale. Only some regional and local governments supported a few activities to develop such markets. At the end of 1997, the MAF started to prepare a new, two-stage programme with the aim of creating a system of wholesale markets for the whole country by the year 2005. Although rather late, this would be a positive step towards linking producers and consumers in the widely separated and isolated agricultural markets in Russia.

5. Rural development

The rapid formulation of an integrated rural development policy is essential

The rural economy is still to a large extent **dependent on the agricultural sector** which employs almost 50 per cent of the active population in rural areas. Rural areas have been under strong economic and social pressures in recent years, due to the general process of structural adjustment and the specific problems related to the reorganisation of large-scale agricultural enterprises. Another particular problem is the high number of low-skilled workers in rural areas, which hinders occupational mobility within the labour force. As non-agricultural activities have been slow to emerge in rural areas, reorganised enterprises have felt obliged to keep employees despite the fact that agricultural production has fallen sharply. That policy has helped to ease short-term social tensions in rural areas, but has hindered the effective restructuring of enterprises and contributed to hidden unemployment. However, social tensions may be simply postponed if the lack of effective restructuring brings large scale enterprises to bankruptcy. The Russian government has so far not formulated an integrated policy, embracing economic, social and environmental aspects of rural development. Support for the rural population is provided almost exclusively through agricultural policy measures.

6. Social measures

Carrying social assets and public utilities continues to weaken agricultural enterprises

Apart from lower contributions to the Pension Fund, the same social policy measures are applied to agriculture and rural areas as to the other sectors of the economy. However, despite legislation dating back to 1991/1992 and providing for the **transfer of social assets and public utilities** to local authorities, large agricultural enterprises are still charged with the provision of public services. It has been estimated that at the beginning of 1997 these agricultural enterprises continued to be responsible for about 70 per cent of their pre-transition stock of social assets and public utilities. This situation diverts financial and management resources from the commercial functions of the enterprise; hampers the restructuring process of agricultural enterprises; and keeps village inhabitants dependent on services provided by the enterprise making them less interested in reallocating their land and non-land entitlements to other, possibly more efficient, enterprises and/or family farms.

D. International trade relations

1. Agro-food trade

The recent decrease in Russia's traditional agro-food trade deficit...

Since the 1960s Russia has been a net importer of food and agricultural products. In the 1990s, agricultural and food trade has represented a substantial part of Russia's total **imports** but only a small proportion of its exports. In 1996 agro-food products accounted for 25 per cent of total imports and 4 per cent of total exports. The agro-food trade deficit amounted to US$8.2 billion, down from US$10.5 billion in 1995. This fall may indicate a weakening of the initial surge in imports of selected food products that arose from the drop in output and competitiveness of domestic producers, the appreciation of the Rouble, and the increased availability of a wide range of imported products whose appeal to consumers was amplified by aggressive marketing and export subsidies applied by some exporting countries.

... is confirmed by the declining ratio of net agro-food imports to the total value of agricultural production

Such a conclusion seems to be confirmed by the declining **ratio of net agro-food imports** (valued at the annual average exchange rate) to the total value of agricultural production (valued at current prices). This ratio fell from 24 per cent in 1994 to 15 per cent in 1996. While the rate of the decline is somewhat overestimated because of the effect of the appreciation of the Rouble on the different valuation methods, a real decline is confirmed by changes in the shares of net imports of essential individual agricultural products in total domestic use of these products (except for meat and meat products), calculated on the basis of agricultural product balances provided by Russian Goskomstat.

Sources of agro-food imports have changed in the 1990s...

Nevertheless, Russia remains a net importer of agro-food products *vis-à-vis* all its trading partners. The geographical structure of Russian imports has changed significantly in the post-reform period: while agro-food imports from OECD countries, especially the European Union (EU), increased, imports from traditional suppliers, such as New Independent States (NIS) and central and eastern European countries (CEEC), declined. Since 1994, the share of agro-food imports from the NIS has been increasing and in 1996 exceeded the share from the EU. In 1996, Russia's main food suppliers were Ukraine, the US, Kazakhstan, Germany and the Netherlands.

... and their composition has shifted

Since the transition started, there has been a striking shift in the **composition** of Russia's agro-food imports: imports of raw agricultural products have fallen sharply while those of processed foods have increased. More specifically, grain imports declined from about 30 million tonnes in 1992 to about 4 million annually between 1994 and 1996, while imports of meat and meat products rose from 0.5 million tonnes in 1992 to about 2 million tonnes in 1995, but then declined to 1.7 million tonnes in 1996.

Russia's position as a net importer is no argument for inefficient policies

It may be that **comparative advantage** will dictate that Russia will be a net importer of food and agricultural products for some time to come. That should not be a cause for concern nor an argument for attempts to restrict imports and support exports. Such policies would only reduce overall economic efficiency by penalising other sectors. At the same time it does not mean that every effort should not be made to increase efficiency all along the agro-food chain.

There is potential for grain exports if structural impediments are tackled

In the 1997/1998 season, following a sharp rise in grain output in 1997, Russia became a **net exporter** of feed grains. However, this new situation revealed many impediments to realising Russia's full export potential, such as: low quality of grains, high transaction and transportation costs in Russia, internal barriers on grain flows, and a lack of adequate market information. Domestic deficiencies disrupting linkages between Russian grain producers and foreign buyers need to be tackled if Russia wants to develop its possible comparative advantage and competitiveness on international grain markets. The medium term and long term perspectives seem to be promising. In the medium term, crop production in Russia will most probably still fluctuate quite sharply at the lower end of its historical output range, depending on weather conditions. However, due to lower livestock numbers and more efficient use of animal feedstuff, the volume of grain used for feed will be lower, enabling Russia to become a net exporter of grains in years with relatively good weather conditions. In the longer term, Russian agriculture can be expected to capitalise on its potential and on the positive results of restructuring. Those gains, combined with more market-oriented agricultural policies strengthened by Russia's prospective WTO membership and more efficient management of large farms, could enable Russia to develop competitive advantage in grain production, even though it may remain dependent on meat imports. Both internal reallocation of agricultural resources and changes in trade flows in recent years seem to support such developments.

2. International trade agreements

Russia's reintegration into the world economy includes trading agreements with traditional partners...

After the collapse of the Council for Mutual Economic Assistance (CMEA) and the disintegration of the Soviet Union, Russia needed not only to negotiate new trading arrangements with each of its traditional trading partners, but also actively to seek to **reintegrate** itself into the world economy on new market terms. Several new trade agreements have been concluded, in particular the Partnership and Co-operation Agreement (PCA) with the EU and various agreements with the NIS.

... and new ones like the European Union

The **PCA** was signed in June 1994 and came into force in December 1997. However, an Interim Trade Agreement, the trade part of the PCA, came into force in February 1996 and put bilateral relations in a new GATT/WTO-type framework. Despite the PCA, the EU left Russia on its list of **non-market economies**, which might in some cases be harmful to Russian interests in trade disputes, in particular in antidumping investigations against Russian producers. In January 1998, the European Commission proposed to remove Russia, along with China, from the list. The proposal has yet to be approved by EU governments.

Various CIS trade arrangements have yielded few concrete results

Russia continues to develop bilateral and multilateral initiatives to strengthen its economic and political links with the rest of the **NIS**. As early as December 1991, the loose Commonwealth of

Independent States (CIS) was set up to co-ordinate some of the activities of countries previously forming the Soviet Union (except for the Baltic States). Russia has signed Free Trade Agreements (FTA) with all of the CIS countries and a number of agreements on regional economic co-operation. In January 1995 an agreement was signed between Russia, Belarus and Kazakhstan to create a customs union (subsequently joined by Kyrgyzstan in 1996). One of the latest initiatives was an agreement on the creation of the common CIS agricultural market (CAM) signed in October 1997 by the CIS governments (except for Azerbaijan and Uzbekistan). To a large extent, the CAM is modelled on the EU's Common Agricultural Policy before its reforms of 1992. However, on the whole, all these efforts have not yielded many concrete results. While countries within the CIS normally have no tariffs on each other's goods, they apply varying tariff rates and other import measures against goods coming from outside. So far it has not proved possible to harmonise CIS trade policies towards third countries.

3. WTO negotiations

The WTO membership negotiation process...

Russia formally applied for WTO membership in December 1994. Since then, during the meetings of the relevant WTO working party, Russia's trade regime, economic policies and laws have been reviewed to determine their compliance with WTO rules and to develop the terms of accession. Selected problems related to Russian **agricultural policy** were discussed such as market access, internal support, export subsidies, sanitary and phytosanitary measures, and technical barriers to trade. Russia argues that the country's still unstable economic situation makes it difficult to accept definite commitments. To preserve some room for manoeuvre in the future, Russia wants a level of agricultural support and tariff protection "comparable with the levels in other WTO-members".

... revealed several areas of concern

Selected **areas of concern** have been identified such as: Russia's highly complicated and bureaucratic systems of trade regulations which contrasts with relatively low tariffs applied on agricultural imports; trade relations with the NIS countries, which lack transparency; the tendency to regionalise some agricultural policy measures which may contradict trade concessions negotiated with the federal government; and the base period for the estimation of the Aggregate Measure of Support (AMS). As for the latter, Russia advocates that the 1989-1991 period of "average, normal conditions", and not the "crisis period of Russian agriculture" of 1993-1995, should be used as the base for further reductions in support. However, such a proposal may not comply with the WTO practice of "normally using the average of the most recent three-year period" (WT/ACC/4) as the base period for acceding countries.

WTO membership will provide MFN status and an institutional foundation for reforms

WTO membership, while giving Russia most-favoured-nation trade status *vis-à-vis* all other WTO members and assisting the overall reform process in Russia by providing an institutional foundation ensuring the continuation and consolidation of reforms, will require Russia to abide by the provisions of the 1994 **Uruguay Round Agreement on Agriculture** as well as other Uruguay Round agreements which would have a bearing on trade in agriculture (*e.g.* GATT Article XVII on State Trading Enterprises). The commitments made will constitute a framework for future agricultural policy in quantitative and qualitative terms. While the former will fix the maximum amount of support allowed, both aggregate and, to some extent, on a product-specific basis (export subsidies), the latter will determine the choice among various policy measures with some of them being allowed and others, such as quantitative trade restrictions, prohibited.

E. Assistance to agriculture

The level of support to Russian agricultural measured by OECD's PSE...

The level of support to Russian agriculture has been estimated using the concept of **Producer Subsidy Equivalent (PSE)**. This measure of assistance has been applied to all OECD countries and more recently to several central and eastern European countries. The PSE measures the money value of transfers from consumers and taxpayers to agricultural producers arising from government policies. The percentage PSE gives an indication of the proportion of total farm revenues originating from support, whether that support comes through domestic prices higher than on world markets or more directly from government budgets. Such direct transfers include subsidies paid directly on outputs, subsidies on the use of inputs, and more general subsidies that lower the costs of production. The Consumer Subsidy Equivalent (CSE) measures the implicit tax paid (or subsidy received) by consumers as a result of higher (lower) domestic prices maintained by market price support measures net of consumer subsidies. The results of the analysis should be interpreted with caution in view of the major macro-economic changes that took place in Russia during the period analysed.

... can be divided into three periods:

Between 1986 and 1997, **support** policies to Russian agriculture, as measured by the percentage PSE, calculated at the official exchange rate, can be divided into three periods:

– support to agriculture in the Soviet period was extremely high

– In the **Soviet era**, support was very high, averaging 90 per cent between 1986 and 1990. The whole agro-food economy was under strong government control with high subsidies paid both to producers and consumers. Most of the support provided for producers was in the form of market price support as a result of the high administered prices (relative to world reference prices) maintained in the central planning framework. Budgetary support to reduce input costs and provide soft credits and capital grants to the agricultural sector was also very high during the Soviet period.

– support fell dramatically to negative levels in the early transition period

– In the **1992-93 period**, when major macroeconomic changes and rather chaotic adjustments took place, support to Russian agriculture fell sharply, with the PSE falling from 61 per cent in 1991 to minus 105 per cent in 1992 and minus 26 per cent in 1993, meaning that agricultural producers were implicitly taxed in this period. The fall in the level of support resulted mostly from general macroeconomic developments, rather than from specific agricultural measures and, to some extent, reflected the major shift from a planned to a more market-oriented economy. However, the decline was strongly exacerbated by the inefficiencies in the food chain ("systemic failure") and by restrictions imposed on agricultural exports that kept producer prices much lower than they would be otherwise.

– and, since 1994, the level of support has been increasing

– **Since 1994**, strong real appreciation of the Rouble in 1994, combined with the introduction of border protection against imports of many products between 1994 and 1996, contributed to the increase in the level of measured support as domestic prices for most commodities moved closer to world reference prices or even exceeded them. In 1995, the level of support to Russian agriculture turned positive and was 21 per cent. In 1996, it increased to 32 per cent, and then, according to provisional data, fell to 26 per cent in 1997 which reflects the fall in prices of several agricultural commodities, the decline in the budgetary support provided to agriculture, the government's resistance to demands for increased border protection and the relatively stable real exchange rate of the Rouble. The 1997 level of protection in Russia is lower than the OECD average of 35 per cent. Nonetheless, it exceeds the level of support in some OECD countries and in most other countries in transition for which OECD has calculated PSEs.

Support to consumers has been measured for the same three periods

In the 1986-1991 period, direct budgetary subsidies to consumers were used to reduce the impact of high prices paid to producers. However the subsidies were smaller than the overall taxation effect on consumers. As a result although CSEs were substantially lower than the PSEs, they nevertheless were on average minus 50 per cent, indicating implicit taxes on consumers. During the early years of the reform, consumer subsidies were substantially reduced and were mostly granted to milk and cereal products. Between 1992 and 1994, CSEs mirrored the high implicit taxation of agricultural producers and showed implicit subsidies to consumers. However, the level of consumer subsidies declined from 172 per cent in 1992 to about zero per cent in 1995, in 1996 CSE became negative at minus 18 per cent, and in 1997 was again negative at minus 20 per cent, indicating an implicit tax on consumption.

Livestock producers have been those most affected by shifts in support

Between 1986 and 1991, support to livestock products dominated support, accounting for 75 per cent of the total, which was due to the high share of livestock in agricultural production. There was less difference in net percentage PSE for livestock and crop products during that period, which averaged 90 per cent between 1986 and 1990. In 1992, the PSE fell sharply for both livestock and crop products leading to implicit taxation of producers, but the fall was steeper for livestock than for crops. In the following years, the rise in the level of support was more rapid for livestock commodities than for crops and in 1996 aggregate PSEs were positive at 16 per cent and 39 per cent, respectively. In 1997, the numbers declined slightly to 14 per cent and 32 per cent, respectively, with potatoes and sugarbeet the most highly supported crops, and poultry and eggs the most highly supported livestock products. The **difference in the level of support** provided to the two groups of products is a reflection of increasing border protection provided for animal products and special direct subsidies paid to animal producers by the Russian government. Moreover, a steep decline in animal production at the beginning of the transition period to below the level of demand, combined with a large increase in meat imports meant that domestic prices started to increase to the levels determined by the new domestic market equilibrium and high border protection.

PSE/CSE calculations are highly sensitive to the exchange rate applied

Since the main component of support to Russian agriculture is market price support, PSE estimates are very sensitive to the exchange rate applied. The basic set of PSEs/CSEs presented above is calculated at official exchange rates, on the assumption that these rates reflect the economic conditions in which the economic agents made their decisions. However, a second set of PSEs/CSEs was calculated with the **shadow (adjusted) exchange rate** that more closely represented the real effective exchange rate over the period under study. The greatest disparity in the results obtained took place in 1992 when the Rouble depreciation was four times faster than the increase in prices. Subsequent very high inflation rates combined with much lower depreciation in 1993 and 1994 led to a real appreciation of the currency which returned to a more market-related equilibrium. In effect, the exchange rate adjustment deferred the impact of the policy induced depreciation of 1992, spreading it out over several years when the adjusted exchange rate depreciated in line with inflation.

F. Conclusion

Large-scale restructuring and market re-orientation have not yet been completed

Considering the long period of Soviet rule in Russia, the difficulties of restructuring and reorienting agriculture towards an efficient, more market-oriented mode of operation have inevitably been immense. Progress achieved so far has been limited. In the process of privatisation, the vast majority of state and collective farms have re-registered as private enterprises and have formally transferred ownership of collectively held land and non-land assets to workers and pensioners. However, the reorganisation undergone so far has not given a sense of ownership to farm workers and has done little to improve the organisational structure, size, management and economic behaviour of the farms. Large-scale enterprises of more than 100 hectares operate about 90 per cent of agricultural land in Russia. This could be a positive factor that may create favourable conditions for productivity growth and increased international competitiveness of Russian agriculture, provided that the large farms are adequately organised and managed.

The inconsistencies between regional and federal policies may be an obstacle

While innovations in agricultural policies in certain Russian regions provide instructive examples of resolving selected structural issues, such as land reform, inconsistencies between regional and federal policies with unequal levels of farm support across regions and inter-regional trade barriers may create obstacles to agricultural reform. Farm operators do not engage in those activities that would be most suitable for their particular location, but rather in those that receive the highest support from their respective regional governments. Such a situation inhibits the process of specialisation based on comparative advantage and, thus, makes the allocation of resources within the Russian agricultural sector sub-optimal.

The trend towards increased market price support and regulation is worrisome

The progress made by Russian agriculture towards developing a market oriented agricultural policy framework can be measured partly by the trend in the PSE. The level of government support, including its most distorting form, market price support, fell sharply at the beginning of the transition, but has increased in more recent years. The sharp fall and subsequent increase in support were predominantly due to macroeconomic factors, especially the massive depreciation of the Rouble followed by rapid appreciation, and less a result of agricultural policies. However, this evolution was also influenced by the government's policy of taxing agricultural exports at the beginning of the transition, further intensified by the inefficient downstream sector, and then protecting domestic agricultural markets by increasing import barriers. The policy should focus attention on removing inefficiencies in the food chain (through improved market infrastructure and increased market competition), greater market transparency (through better market monitoring and information systems), and on the provision of training, education, research results and advice to producers rather than on market regulatory measures which are distorting market signals for producers and are detrimental to consumers. In addition, the environmental dimensions of agricultural policies development and implementation deserve more careful consideration in Russia.

A stable macroeconomic and institutional framework and respect of the rule of law are indispensable

Even the best designed agricultural policies will not be sufficient to assist the development of Russian agriculture in the absence of a more stable macroeconomic environment and well-functioning institutional framework. These should provide lower interest rates, easier access to finance, stable and predictable exchange rates, a stable, transparent and simplified tax system, a well developed banking sector in rural areas, well established commercial links between farming enterprises and upstream and downstream sectors, and a business environment in which the honouring of contracts is the norm and recourse to law is affordable and accessible when needed. These conditions are also critical for the development of non-agricultural employment opportunities in rural areas, which would ease the flow of labour from restructuring agricultural enterprises to other activities. A well defined local tax base and tax revenue transfers between local and federal budgets would enable local authorities to take over some responsibility for the provision of social services which to date have been financed to a large extent by agricultural enterprises. Many of these services could also be provided by an invigorated private sector.

Within this framework, Russian agriculture is potentially a major international player

Russia's economic potential, like the country, is enormous. Agriculture and food production will be a significant part of Russia's economic and social structure into the next century. That potential can be dissipated by pursuing policies, modelled on those failed policies of some OECD countries which have, in the past, been overly influenced by sectoral interest. Russia can omit this wasteful diversion and take advantage of the opportunity to put in place a set of agricultural and related policies that will make its economy, with agriculture in its rightful place, a force to be reckoned with.

ECONOMIC AND AGRICULTURAL ENVIRONMENT

A. GENERAL ASPECTS

1. Background data

The Russian Federation with its **territory** of 17.1 million square kilometres (12.6 per cent of the earth's land surface, and 76 per cent of the former USSR's territory) is by far the largest country in the world, slightly more than 1.7 times larger than Canada, the second largest country, and five times larger than EU-15. The distance from the eastern to the western border is above 9 000 kilometres and from north to south above 4 000 kilometres. Russia neighbours 14 countries. Out of its 20.1 thousand kilometres of land boundaries, about half border Kazakhstan and China. Due to its size, Russia has a great variety of landscape, climate, soils and wildlife but is rather unfavourably located in relation to major sea routes of the world and much of the country lacks favourable soils and climates (either too cold or too dry) for agriculture.

About 75 per cent of the territory is flat and characterised by broad plains with low hills west of the Urals to vast taiga forests and tundra in Siberia. Uplands and mountains dominate along southern border regions with the highest point, Elbrus, on the border with Georgia, at 5 642 metres. About 45 per cent of Russia's territory is forested, 4 per cent is covered by water, about 13 per cent is **agricultural land**, 19 per cent, mostly in the tundra area, is used as pasture for deer and the remaining 19 per cent for other purposes. Almost 40 per cent of the territory is under permafrost. In 1995, about 221 million hectares of utilised agricultural land (UAA) were broken down into arable land (130 million, 59 per cent), pastures and meadows (about 40 per cent) and permanent crops (about one per cent). About 2.8 per cent of the UAA is under irrigation. On average, the soils in Russia are of low fertility, ranging from acidic soils, containing few natural nutrients in the northern regions to highly fertile soils in the southern half of European Russia, in particular Northern Caucasus and Central Black Soils Region, but also in the Southern Urals and the southern fringe of Siberia.

Climate is extremely differentiated although the continental climate with cold, windy and snowy winters and hot and dry summers predominates. Such climate favours the occurrence of droughts that appear on average every three years and strongly affect agricultural production, particularly in the regions with the most fertile soils. Average temperatures in January range from –50 °C in north-east Siberia to about –5 °C in the western part of Russia, and in July from 1 °C on the northern Siberian border to 24-25 °C in the Caspian Lowlands. The coldest point of the northern hemisphere where temperatures fall to –70 °C, is located in Siberia. Average annual precipitation amounts to 500 mm, but ranges from 250 mm in the Caspian Lowlands to 1 000 mm in southern Siberia and the far East.

Russia disposes of a vast **natural resource** base including major deposits of oil, natural gas, coal, timber and many minerals of strategic importance. However, enormous obstacles of climate, terrain, and distance hinder exploitation.

At the beginning of 1997 Russia's **population** amounted to 147.5 million (6th in the world) with an average density of 8.7 people per square kilometre, ranging from 1.2 in the far East to 50 in the northern Caucasus and 325 in the Moscow region. Since 1992 the rate of natural population increase has been negative every year and was –0.3 per cent in 1996. The negative rates of natural population increase have been largely compensated by significant net immigration of mostly Russian nationals from other Newly Independent States (NIS) and Baltic countries after the collapse of the Soviet Union in 1991.

Between 1990 and the beginning of 1997 the total net immigration to Russia was about 2.8 million. As a result, total population did not change much in the 1990s, but there has been a slight tendency toward a decline in more recent years. About 73 per cent of the population live in towns and 27 per cent in rural areas. More than 10 per cent of the population reside in one of the three largest cities (Moscow 8.7 million in 1996; St. Petersburg 4.8 million; Nizhny Novgorod 1.4 million). Russia is a multinational country with 83 per cent of inhabitants declaring Russian nationality (according to the last micro-census of 1994), 3.8 per cent Tatar, 2.3 per cent Ukrainian, 1.2 per cent Chuvash, 0.9 per cent Bashkir, 0.7 per cent Belorusian, 0.6 per cent Moldavian and 7.5 per cent other. In some regions, strong minorities with their own identities and interests have used the weakened position of the federal government during the early years of the transition to push for greater independence from the centre. In most cases it was possible to reconcile conflicting interests between the federal state and constituent regional entities through negotiations. The war over Chechenyan independence represents the tragic exception.

2. Administrative framework

The Russian Federation consists of 89 components (regions): forty-nine *oblasts*, one autonomous *oblast*, twenty-one republics, six *krays*, ten *okrugs*, and two major metropolitan centres (Moscow and St. Petersburg). The regions are further subdivided into *rayons* and these into municipalities and rural administrations. Moreover, major regional cities, including regional capitals have their own administration structures, parallel to the *rayon* administrations (Figure I.1).

Republics are inhabited to a large extent by minorities, such as Tatars and Chuvash, and they enjoy a high degree of autonomy. They have the right to their own constitution and legislation and to elect their own president. O*krugs* are ethnic subdivisions of *oblasts* or *krays*. The Federation Treaty of 1992 and the Russian Constitution of 1993 define the division of legislative authority between the Federation and its components. Article 71 of the Constitution reserves exclusive rights for federal authorities in areas

◆ Figure I.1. ***The administrative and territorial structure of the Russian Federation***

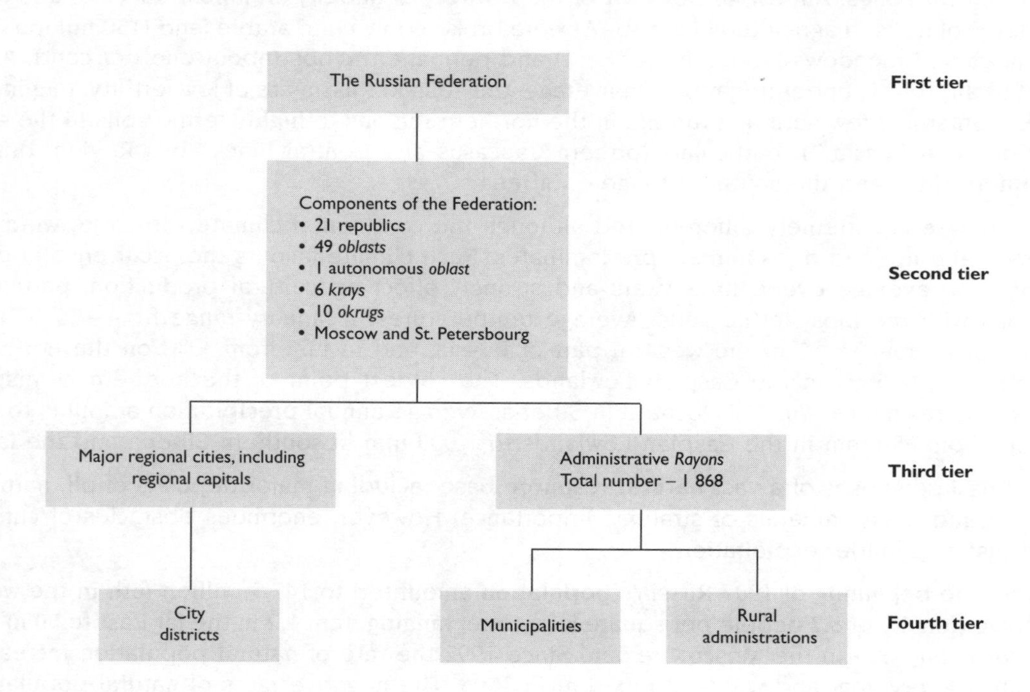

Source: OECD.

such as foreign and military policy, monetary affairs, and energy, transportation, and communication policies. Article 72 provides for joint federal-regional jurisdiction in a number of spheres, including tax administration, selection of judicial and law enforcement officials, and ownership and use of land and natural resources. All policy areas in which federal involvement is not explicitly mentioned in these two Articles are left to regional jurisdiction (Article 73). Although all the entities of the Russian Federation are supposedly equal in their relationship to the federal authorities (Article 5), considerable differences in the delimitation of responsibilities and powers between the Federation and individual regions exist. These differences are the result of bilateral agreements between the central government and some regional authorities, which have created an "asymmetric federalism" of unequal relations and overlapping jurisdictions.

3. Political system

The State power in the Russian Federation is exercised on the basis of the separation of executive, legislative, and judicial powers (Figure I.2). Their competencies are fixed in the Constitution approved by referendum on 12 December 1993.

The executive powers have been given to the **President** and the **Government**. The President is elected for a four-year term by a popular vote, and must receive a majority. In case one candidate does not receive more than 50 per cent of the vote, a run-off election is held. As Head of State, the President disposes of broad powers which essentially make the Russian Federation a presidential republic. The President forms and heads the Defence Council and is in charge of the country's foreign policy. The Prime Minister is selected by the President, but must be approved by the State Duma. However, if the President's three consecutive candidates are rejected, the chamber can be dissolved by the President and new elections called. The President together with the Prime Minister choose the cabinet. The heads of the three "power Ministries": the ministers of defence, interior, and foreign affairs report to the President. The President can draft laws to submit to the Federal Assembly and has veto rights. The veto can be overridden by a two-thirds majority of both chambers of Parliament. In addition, the President may pass decrees and orders without the Assembly's approval, with some exceptions specified by the Constitution. The President schedules elections to the State Duma and has the power to dissolve the chamber. The President is also the Commander in Chief of the armed forces, can appoint or dismiss top military commanders, and may declare martial law or a state of emergency.

The Federal Assembly consists of two chambers: the upper Federation Council and the lower State Duma. The two houses meet separately, but may hold joint meetings. The **State Duma** consists of 450 deputies elected for a term of four years. The deputies are elected through two types of mandate. The first is the party-list vote, whereby 225 seats are divided among those parties which receive more than 5 per cent of the vote. The other 225 seats are direct mandates distributed through single-member constituencies. The Duma's main responsibility is passing federal laws. A law is adopted by a majority vote unless otherwise provided for by the Constitution. It is then passed to the Federation Council, which has 14 days to take a vote on the issue. If it is rejected, it is returned to the Duma, which then can only pass it if two-thirds of the Duma deputies vote for it again. The Duma's responsibilities also include approving the President's choice of Prime Minister, holding a confidence vote on the government, approving or dismissing the Head of the Central Bank.

The **Federation Council** has 178 deputies, two from each of Russia's 89 components: one is the locally elected executive head of the region and the other is the head of the regional legislature, selected by the regional deputies. The Federation Council has the power to confirm border changes within the federation, approve the introduction of martial law or a state of emergency by the President and vote on the deployment of Russian armed forces outside of its borders. Federal laws passed to the Federal Council are deemed to have been approved by the Council if more than half of the total number of members have voted for them or if they have not been examined by the Council within fourteen days. In case of the rejection of a law by the Council the two chambers may set up a conciliation commission to overcome differences, and the law may go back to the Duma for another vote.

◆ Figure I.2. ***The political system of the Russian Federation***

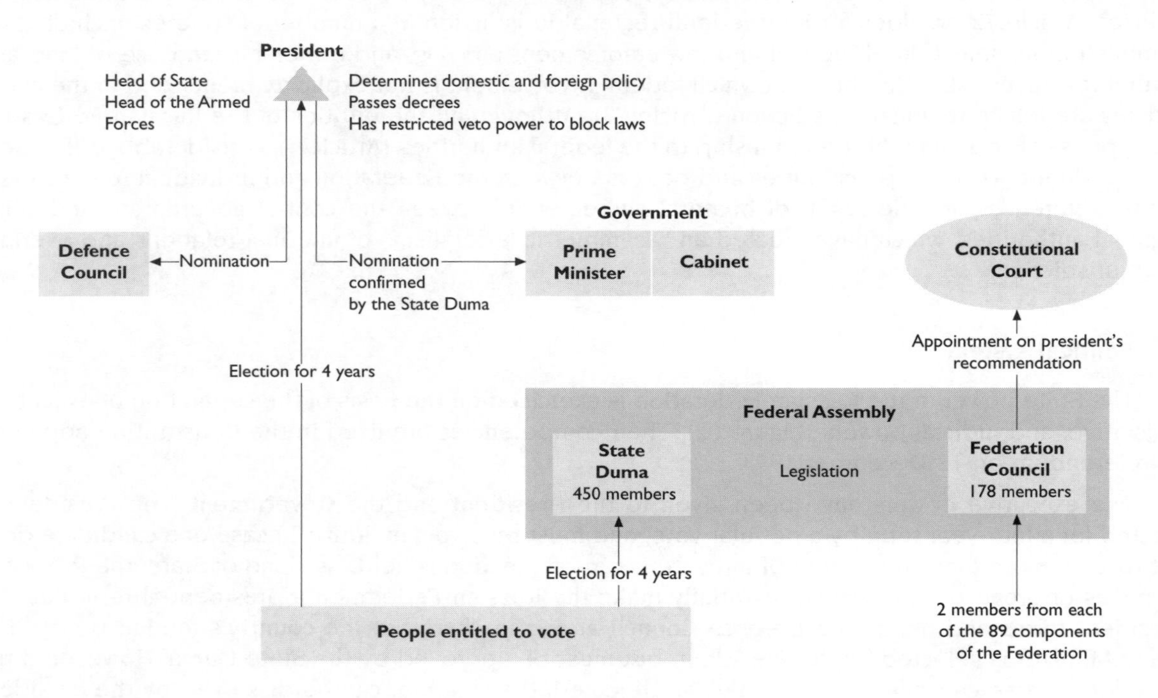

Source: Wichern (1997); Constitution of the Russian Federation of 12 December, 1993.

Box I.1. **Results of elections held in December 1995**

Parties	Number of Duma seats
Communist Party of the Russian Federation (led by G. Zyuganov)	157
Independents	78
Our Home is Russia (pro-government, democratic party led by V. Chernomyrdin)	55
Liberal Democratic Party of Russia (nationalist party led by V. Zhirinovskiy)	51
Yabloko Bloc (pro-market, democratic party led by G. Yavlinskiy)	45
Agrarian Party of Russia (traditionalist and nationalist party led by M. Lapshin)	20
Russia's Democratic Choice Party (pro-market, liberal, democratic party led by Y. Gaidar)	9
Power to the People (traditionalist party led by N. Ryzhkov and S. Baburin)	9
Congress of Russian Communities (centrist party led by Y. Skokov)	5
Forward, Russia! (pro-market, democratic party led by B. Fedorov)	3
Women of Russia (centrist party led by A. Fedulova and Y. Liakhova)	3
Other parties	15

Box I.2. **Types of basic legal acts in the Russian Federation**

The precedence of major forms of legislation, from highest to lowest, is as follows:

Federal Constitution of Russian Federation and Federal Constitutional Laws (constitutional amendments): the supreme acts; not "directly acting", *i.e.* for their implementation other legal acts are required.

Federal Laws (*zakon*) and **Codes** (*kodeks*): passed by the Federal Assembly and signed by the President; the President has a veto power, which may be overridden by the Federal Assembly; take effect on publication; codes are a system of laws concerning specific issues, such as the Land Code.

Presidential decrees (*ukaz*): issued by the President on his own authority; with some exceptions not subject to confirmation or rejection by the Federal Assembly; may be directly acting (not requiring other legal acts for implementation) or require specification through further presidential orders, cabinet decrees or ministerial regulations; take effect on the date specified in the document.

Presidential orders (*rasporiazhenie*): issued by the President on his own authority; specify matters already covered by decrees; directly acting documents.

Cabinet decrees (*postanovlenie*): issued by the Prime Minister on his own authority in the name of the cabinet; serve as the implementation of laws and presidential decrees; take effect on the date specified in the document.

Cabinet orders (*rasporiazhenie*): issued by the Prime Minister on his own authority; directly acting documents further specifying matters already covered by decrees.

Administrative rules (*podzakonnye akty*): issued by individual administrative agencies; explanatory documents circulated within the administrative agency and its regional sub-units; establish procedures for implementing laws.

Source: OECD, 1997*b*.

The **judicial** powers are distributed to the Constitutional Court, Supreme Court (highest court for criminal, civil, and administrative cases) and Superior Court of Arbitration (highest court that resolves economic disputes). The judges for all these courts are appointed by the Federation Council on the recommendation of the President.

4. Recent macroeconomic developments[1]

i) Main economic reforms

The Soviet economy operated on the basis of state ownership and administrative planning, and was heavily militarised. Economic structures, built during almost seven decades of central planning, were characterised by the domination of heavy and defence industries; underdevelopment of the service and, in general, consumer sectors relative to industry; energy and material-intensive production; an uneven but, by international standards, generally low technological level; a structure of production that bore little relation to competitiveness on world markets; practically no foreign capital involvement; an extensive enterprise-based social security system; and, to a large extent, a devastated environment, particularly in areas with a heavy concentration of industry.

In the second half of the 1980s, Gorbachev's half-way reforms, despite their inconsistencies, facilitated the first emergence of new entrepreneurship and provided the first legal framework for non-state enterprises through the Law "On Co-operation in the USSR" (1988). Making use of the newly defined legal status of the co-operative, a wide diversity of non-state enterprises emerged, mostly small and medium sized ones. The leasing of state enterprises by employees and management became one of the preferred forms of lessening state control over enterprises. At that time, new entrepreneurs profited from many market niches left by the dominating state sector but also from low wages, cheap credits and a lack of competition from imported goods. However, the process was still at the margins of the overall economy. Entrepreneurial co-operatives were forced to operate in a generally hostile economic and political environment during this period.

Significant movement towards market reforms began only in 1991 when the right to private property was established and 1992, when most prices were liberalised. The reforms were accelerated by the break-up of the USSR into 15 successor states in late 1991. This break-up, on the one hand, destroyed major economic links between previous Soviet republics, but on the other, gave Russia more flexibility to conduct its own economic policy unconstrained by considerations of the USSR's territorial integrity. In January 1992, under the presidency of Boris Yeltsin, the government led by Yegor Gaidar began the implementation of a series of "Big Bang" reforms which were intended to bring market forces more fully to bear on the economic activity of the country. This was achieved by freeing nearly all prices, slashing defence spending, cutting budget subsidies, eliminating the old centralised distribution system, and liberalising foreign trade. Between 1992-1994, the Russian mass privatisation programme was implemented, including buy-outs of small-scale enterprises, "corporatisation" of Russian state-owned enterprises into joint stock companies, and a voucher programme giving 144 million participating Russians the possibility of using their vouchers to buy shares of enterprises undergoing privatisation (Part III). The programme was considered "rapid, extensive, and unprecedented in world history" (Blasi, 1997). As a result, about 80 per cent of industrial enterprises passed, at least formally, into private hands. However, for many reasons, including the dominant role of insiders (workers and managers) in privatisation, the process has so far meant very little for the internal restructuring of most enterprises.

At the beginning of 1995, in light of the previous three years' inconclusive macroeconomic stabilisation efforts with inflation rates remaining extremely high and the budget deficit even rising (Table I.1), the Russian Government decided to advance the stabilisation process significantly with the key objective of rapid deceleration in inflation rates (World Bank, 1996a). To achieve this goal tight monetary and fiscal policies were associated with high real interest rates and an exchange rate anchor, corresponding to a band introduced in July 1995 (see below). The deceleration of inflation rates was remarkable (Table I.1), but institutional reforms and privatisation slowed down, partly due to political uncertainties preceding presidential elections held in mid-1996 and "wait and see" attitudes caused by President Yeltsin's health problems. In 1996, federal budget revenues from privatisation were only one-fifth of planned levels.

The President's return to office and changes in the Russian Government in March 1997 gave new impetus to institutional reforms with the government's declared determination to implement privatisation on the basis of more open and competitive principles; push through new laws to protect the rights

Table I.1. **Selected macroeconomic indicators, 1991-1997**

Indicator	Units	1991	1992	1993	1994	1995	1996	1997p
Real change in GDP	Per cent	–5.0	–14.5	–8.7	–12.7	–4.0	–5.0	0.4
GDP in US$ terms	US$ billion	n.a.	98.7	184.0	286.3	357.3	440.4	462.5
Exchange rate (annual average)	Rb/US$	n.a.	192.5	932.0	2 204.0	4 554.0	5 124.0	5 784.4
Consumer price index (December-December)	Per cent	160.0	2 509.0	840.0	215.1	131.3	21.8	11.0
Unemployment rate	Per cent	n.a.	4.8 [2]	5.5 [2]	7.5	8.9	9.3	9.0
Federal Budget balance	Per cent of GDP	n.a.	–16.7	–7.2	–10.6	–5.3	–7.8	–7.0
Merchandise exports [1]	US$ billion	n.a.	42.4	44.3	66.9	80.1	87.0	86.7
Merchandise imports [1]	US$ billion	n.a.	37.0	26.8	38.7	46.6	45.4	n.a.
Current account	Per cent of GDP	9 [2]	5 [2]	4 [2]	4	3	2	1
External public debt (end-year)	US$ billion	96.8	107.7	112.7	119.9	120.4	125.0	130.8

p: provisional.
n.a.: not available.
1. Non-CIS plus CIS trade, but excluding unrecorded trade.
2. Figures for 1991-1993 may not be comparable with figures for 1994-1996 due to methodological changes.
Source: OECD 1997*b*; Goskomstat; World Bank (1996*b*).

of foreign investors; simplify and rationalise the tax system (new Tax Code); eliminate the ability of banks and their associated financial-industrial groups (FIGs) to profit from special relations with state administration; improve the regulation of natural monopolies; reform the housing sector and utilities; and better target social assistance. However, the results achieved so far are mixed with confusing and rapidly changing laws, weak law enforcement, corporate governance, crime and corruption remaining major obstacles to the development of efficient market relations.

ii) Output

Between 1990 and 1997 Russian GDP declined by 40 per cent, according to official figures (Table I.1). However, the uncertainties concerning official data, in particular problems with the measurement of shadow economy activities, make any clear-cut assessment of aggregate activity rather difficult (OECD, 1997*b*). The registered fall was caused by many factors, including the inevitable squeeze of sectors favoured under central planning; the break-up of the Soviet economy itself, which separated different parts of what was once a unified economy; and dramatic changes in macroeconomic policies which ended much of the state support for enterprises provided under the previous system. In particular the dismantling of central planning had a strong impact on overall activity and the structure of Russian economy. For example, the 68 per cent reduction in government orders for defence equipment in 1992 had a shocking impact on the military-industrial sector as such, which accounted for between 20 and 35 per cent of the Soviet economy, as well as on supplier enterprises (OECD, 1995). While the 1992 liberalisation shock, combined with the strong devaluation of the rouble, left a large part of enterprises unadjusted, tight monetary and fiscal policies in 1995-97, and the discontinuation of implicit subsidisation through the commercial banking system forced enterprises to cope with a fundamentally new incentive system. However, considerable adjustment efforts are still needed. In 1997 officially reported GDP stabilised for the first time since the beginning of the country's reforms, growing by 0.4 per cent. It seems plausible that the economy is at the turning point with modest real growth projected for 1998 (OECD, 1997*b*).

The composition of GDP changed significantly between 1990 and 1996 with industry's and agriculture's shares in GDP falling, and that of the service sector increasing. More specifically, industry's contribution to GDP declined from about 35 per cent to about 32 per cent, and agriculture's more than halved from above 15 per cent to below 7 per cent, whereas the share of services rose from about 33 per cent to about 49 per cent (Table I.2). Value added in the production of goods shrank by around 50 per cent. At a more disaggregated level, some subsectors showed dramatic falls (military output), while in some others production remained almost unchanged (natural gas output). The decline in the provision of services was much smaller at about 22 per cent which, combined with an increase in their relative prices, contributed to a significant increase in services' contribution to GDP. Some services, especially financial, boomed. Similar differences are visible across regions with the magnitude of the industrial fall ranging from 15 to 87 per cent between 1990 and 1996 (OECD, 1997*b*).

Table I.2. **The structure of GDP at current prices, 1990-1997**

Per cent

Sector	1990	1991	1992	1993	1994	1995	1996	1997p
Industry	35.1	36.7	34.5	32.4	31.5	31.4	32.1	n.a.
Agriculture	15.4	13.7	7.2	7.4	6.0	6.9	6.7	6.5
Services	32.5	35.9	52.7	46.3	48.8	46.0	48.5	n.a.
Other[1]	17.0	13.7	5.6	13.9	13.7	15.7	12.7	n.a.
Total	100	100	100	100	100	100	100	100

p: provisional.
n.a.: not available.
1. Including construction and other goods producing branches of economy, imputed financial intermediary services and net taxes on goods and imports.
Source: Goskomstat.

iii) Employment

The shifts in the level and structure of production have been accompanied by significant movements in employment. Total employment increased slightly in the second half of the 1980s to 75.3 million in 1990, but then shrank to 65.3 million in 1997, *i.e.* by about 13 per cent. While the fall in employment has been much slower than that of real GDP, reflecting significant over-employment in many branches of the economy and reduced labour productivity at the beginning of transition, the structure of employment evolved more directly in line with the changes in the composition of output. Employment in industry fell by 25 per cent during the first half of the 1990s, and by another 8 per cent in 1996. Employment stagnated in agriculture (see below), but rose rapidly in trade, banking, insurance, and public administration. Employment in larger enterprises steadily declined, whereas small business employment increased substantially and in 1995 it accounted for 14 per cent of total employment. However, the share was still very low by international standards.

Registered unemployment is relatively low, at 3.4 per cent as of late 1996 but unemployment, according to the broader International Labour Organisation (ILO) definition was higher at 9.3 per cent. By the end of 1997, the rates had declined slightly to 2.8 and 9.0 per cent, respectively. Hidden unemployment, in the form of forced leaves and shortened work hours, continues to be widespread and affected about 8 per cent of the total labour force in late 1996, particularly in industry. Moreover, there are significant regional differences in the rate of unemployment with the ILO-type rate ranging from less than 5 per cent in Moscow city to 23 per cent in the Republic of Dagestan. Interregional labour mobility continues to be hampered by deficiencies in the housing market and high registration fees for residency.

Table I.3. **Structure of employment, 1990-1997**

Per cent

Sector	1990	1991	1992	1993	1994	1995	1996	1997p
Industry	30.3	30.4	29.5	29.4	27.1	25.9	24.7	n.a.
Agriculture	12.9	13.2	14.0	14.3	15.0	14.7	14.0	13.9
Construction	12.0	11.5	10.9	10.1	9.9	9.3	9.7	n.a.
Transport and communication	7.7	7.8	7.8	7.6	7.8	7.9	8.0	n.a.
Services and other	37.1	37.1	37.8	38.6	40.2	42.2	43.6	n.a.
Total	100	100	100	100	100	100	100	100

p: provisional.
n.a.: not available.
Source: Goskomstat.

iv) Inflation

Price liberalisation announced by the federal government on 2 January 1992 covered about 80 per cent of producer prices and 90 per cent of retail prices. However, direct controls over retail prices for basic consumer goods such as milk, baby food, bread, salt, sugar, and vodka, as well as electricity, housing rents, medicines, and public transportation prices, were not lifted until 7 March 1992 and even afterwards prices of many of these goods continued to be regulated, most often through cost mark-ups (for more details on food prices see Part IV). Moreover, actual liberalisation of prices at the regional level has been even more gradual and strongly differentiated across regions. As of mid-1996, some 18 per cent of consumer prices were still controlled, most of them at the local level (OECD, 1997*b*).

Following price liberalisation, a one time adjustment in consumer prices (CPI) was 245 per cent in January 1992 (296 per cent based on the "urban CPI") and then the rate of inflation strongly fluctuated to average 16 per cent per month in 1992-94 . The end-year inflation rate increased from 160 per cent in 1991 to 2 509 per cent in 1992 and fell to 840 per cent in 1993. Much tighter monetary and fiscal policies, resulted in an impressive decline in the rate to 22 per cent by 1996 and to 11 per cent in 1997 (Table I.1.). In part to signal that the days of high inflation were in Russia gone for good, the government redenominated the rouble by removing three zeros at the beginning of 1998.

v) Monetary and exchange rate policy

In the 1980s, the official exchange rate was fixed at a level of Rb 0.6 per US$. Particularly in the period of Gorbachev's partial reforms, growing repressed inflation resulted in a rising wedge between the official and black market exchange rates. In 1987, so-called differentiated foreign exchange coefficients (DVKs) were introduced which differed by commodity, but also by country of origin and destination. In November 1990, a commercial exchange rate was introduced at a depreciated level compared with the official exchange rate and was applied to most trade transactions. The rate was established to ensure that 90 per cent of exports were profitable. At the same time the DVKs were eliminated. A special rate for tourist transactions was retained.

In mid-1992, a unified exchange rate was introduced and the authorities largely allowed the rate to float which led to a dramatic depreciation of the rouble. The rate was determined on the Moscow Interbank Currency Exchange (MICEX) on the basis of daily trades in the hard-currency interbank market. Until mid-1995, exporters were required to convert 50 per cent of foreign exchange earnings into roubles. At various times the Central Bank of Russia (CBR) intervened heavily in favour of the rouble. In July 1995, in a major policy turning point, the CBR introduced a fixed Rb/$ 4 300-4 900 exchange rate corridor. The policy allowed the CBR to set upper and lower rates for rouble fluctuation over certain periods, and to announce daily buying and selling rates for the rouble within the band. In mid-1996, the system was modified and the authorities committed themselves to a crawling band which made the nominal exchange rate even more predictable. Combined with high interest rates, the exchange rate anchor had a visibly stabilising impact on expectations and contributed to the dramatic fall in inflation rates. In June 1996, Russia accepted the obligations of Article VIII of the IMF charter. This article commits the authorities to refrain from imposing restrictions on payments for current account transactions or operating multiple exchange rate regimes. In November 1997, as a reaction to turbulences on international financial markets causing falls on the Russian stock market and an outflow of capital, the Central Bank of Russia increased interest rates, tightened reserve requirements for foreign currency deposits, and announced a more flexible exchange rate policy for 1998. At the beginning of February 1998, the measures were tightened again with the refinancing rate rising from 28 per cent to 42 per cent. A new wave of turbulence on Russian financial markets in May 1998 caused the CBR to raise the refinancing rates again to 50 per cent in mid-May and to triple the rates to a dramatic 150 per cent on 27 May.

After strong depreciation in real terms in 1992, the rouble appreciated sharply in real terms between 1993 and 1995. During 1996 and 1997, the gradual nominal depreciation against US$ has continued at a rate close to the CPI, meaning a relative stabilisation in real terms.

The 1993 Constitution granted the CBR full autonomy from the Duma and the government. A new law passed in April 1995 confirmed the independence of the CBR and explicitly forbade interference in the bank's activities by the President, government or Parliament. The CBR determines the rate of interest on credit issued to commercial banks (the discount, Lombard and refinancing rates), but there are no limits or restrictions imposed by the monetary authorities on the rates banks charge or offer their clients.

vi) Government budget

In 1992, the Federal Budget deficit of Russia was 16.7 per cent of GDP. Since then it declined to 5.3 per cent in 1995, but increased again to between 7 per cent and 8 per cent in 1996 and 1997 (Table I.1.).

On the revenue side several major factors contribute to low efficiency of tax collection: tax evasion resulting partly from the complicated tax system with a large number of taxes cumulating to very high rates; widespread tax exemptions; corruption; reluctance of some regional authorities to remit the full amount of revenue due to the federal budget; and inexperience of the staff of the State Tax Service. Moreover, enterprises delay tax payments which, due to high inflation, particularly up to 1995, meant a significant fall in real terms. Identified arrears on taxes and social security contributions reached 10 per cent of the 1996 GDP. The successive versions of the draft Tax Code that have been discussed since 1995 aim at cutting the number of taxes and at restricting the scope for tax evasion, but the 1997 version had not been adopted by the Duma by the end of that year. To meet its budget deficit targets, the government is relying heavily on expenditure sequestration meaning cuts in budgetary expenditures and long delays in payments to salaried workers and pensioners.

Up to 1994 budget deficits were almost entirely financed by the CBR with strong inflationary implications. Since 1995 direct CBR money creation has been replaced by the issuance of treasury bills (GKOs) and bonds (OFZs). In 1995, annualised yields on GKO were generally above 100 per cent, sometimes 150 per cent. This created a crowding-out effect, with an increasing portion of banks' assets taken up by GKOs.

Growing liquidity problems of enterprises and mounting fiscal problems at all levels of government, combined with a spiral of tax, budgetary, inter-enterprise and wage arrears, led to an explosion in issues of bills of exchange and promissory notes (*veksels*) by local governments, commercial banks and enterprises.[2] In recent years these bills served to ease liquidity constraints in the face of restrictive monetary and fiscal policies. Moreover, various types of highly inefficient barter deals have developed, reflecting among other factors an attempt to evade taxes. Barter deals may account for more than 40 per cent of industrial sales.

vii) Foreign trade

Foreign trade expanded rapidly in recent years, contrasting with the continuing decline in domestic output. In dollar terms, registered merchandise exports more than doubled between 1992 and 1996 reaching US$87 billion in 1996 (Table I.1), *i.e.* almost 20 per cent of GDP. Energy exports, including crude oil, natural gas and oil products, continue to play a dominant role and accounted for 45 per cent of total exports in 1996. Other major exports include metals, machinery and equipment and precious stones and metals. In general, Russian exports are characterised by rather low value-added products. Registered merchandise imports grew by about 23 per cent between 1992 and 1996, reaching US$45.4 billion in 1996. However, non-registered imports, both so called shuttle trade and smuggled imports, account for a large share of total imports. It is estimated that shuttle trade alone accounted for about one fourth of total imports in 1996. The resulting trade surplus, partly adjusted for non-registered trade, was about 5 per cent of GDP in 1996, one of the largest in the world.

viii) Foreign debt

Russia assumed former Soviet Union (FSU) debts to creditors in the Paris and London clubs and to other countries of the former Council for Mutual Economic Assistance (Comecon). At the end of 1991 the debts amounted to US$96.8 billion. Together with the debt accumulated by the Russian Federation, the total amount of debt was US$131 billion at the end of 1997, representing 28 per cent of 1997 GDP. Most of the debt has been covered by comprehensive long-term restructuring schemes concluded with the Paris and London clubs in 1996 and 1997. Moreover, agreements with Bulgaria, Poland and Hungary roughly halved the debt owed to the previous Comecon members and agreements with other creditors foresee that around one third of the remainder is to be repaid in cash and two thirds in kind.

Russia also inherited FSU loans totalling 96.4 billion of old roubles, which enabled Russia to claim a Paris Club membership which was finally granted in September 1997. However the hard currency valuation of these loans, largely granted on concessional terms to such countries as Cuba, Mongolia, Vietnam and India, remains highly controversial and, therefore, the size of Russia's net external debt is unclear. The rescheduling schemes made the Russian debt burden rather manageable, but debt service obligations are increasing and are already exerting strong pressures on the budget (OECD, 1997*b*).

ix) Capital inflows

Foreign investment is still low in Russia, but increased sharply in 1995-96. Foreign direct investment (FDI) increased from US$0.6 billion in 1994 to US$2.5 billion in 1996, and portfolio investment from virtually zero in 1994-95 to US$7.5 billion in 1996 as restrictions on purchases of state securities (GKOs and OFZs) by non-residents were gradually relaxed. Borrowing from the IMF increased sharply as well, with new credits rising from US$1.5 billion in 1994 to US$5.5 billion in 1995 and US$2.9 billion in 1996 (first tranches of the US$10.1 billion extended IMF facility). In addition, France and Germany provided loans of a total value of US$2.4 billion in March 1996.

The stock of FDI was only about $8 billion at the end 1996, or $55 per capita, compared with $140 per capita in Poland and over $1 300 per capita in Hungary. Political uncertainties, taxation, unfriendly legislation and practices, potential conflicts between federal and regional regulations, weak law enforcement, and deficiencies of domestic infrastructure and input supplies are still major barriers.

x) *Social policy issues*

The social protection system in the Soviet period was based on three pillars: the "right to work"; a general system of social benefits such as free education, free health care and diverse types of pensions administered by the state; and a system of administered, subsidised prices which made "necessities" (food, housing, public transport, etc.) affordable (OECD, 1995). The right to work, guaranteed by permanent labour shortages in the central-planning framework, entailed an access to a wide range of social facilities (kindergartens, vacations spots, housing etc.) provided to employees at low or even without any charges by state-owned enterprises. Since the reforms started, employment, however, is no longer "guaranteed"; the link between state-provided benefits and the workplace for the most part has been broken; and the provision of free or subsidised access to goods and services is diminishing. While the dismantling of the former system was occurring, hardship and insecurity were increasing.

Living standards deteriorated after 1991 in a large part as a consequence of the sharp declines in output and employment, the erosion of real wages and benefits, and the increase in unemployment. According to official statistics, between 1991 and 1995 real wages declined by 56 per cent. During 1993-94 only 40 per cent of the workforce was being paid in full and on time (Klugman, 1996). The Gini coefficient, measuring income inequality, increased from 0.26 in 1991 to 0.41 in 1994, but narrowed slightly thereafter.[3] Most sources and methods measuring poverty indicate that particularly vulnerable are single-headed families and households with at least two children or at least one unemployed. Another indicator of living standards, life expectancy at birth, dropped dramatically from 70 years in 1988 to 64 years in 1994, with the sharpest decline among men from 65 to 58 years, but it has since recovered slightly (Goskomstat, 1996 and 1998).

In 1995 about one quarter of the population had monetary incomes below the official minimum subsistence standard. In 1996, the share declined to 21.6 per cent, but the incidence of poverty varied across regions from 13.3 per cent (Tyumen) to 74.6 per cent (Republic of Tyva). Moscow stands apart as the region with the highest per capita income, lowest rate of unemployment, and below-average poverty rate. Household surveys also confirm that households with access to a private plot of land and/or engaged in shadow economic activities are less affected by poverty.

The reforms of the social security system in Russia are relatively slow and, often, *ad hoc*. There is a large array of cash and in-kind benefits which often are overly complicated, inadequately administered, and poorly targeted. Moreover, there is a gap between *de jure* and *de facto* benefits depending on the federal and local budgets' financial situation: in addition to persistent problems of pension and unemployment benefit arrears, 30 per cent of the poor who were entitled to assistance did not receive any benefits in late 1995. Some of these inadequacies have been recognised by the government, and several proposals, such as gradual phasing out of housing and utilities subsidies to households, were submitted in 1997 to strengthen targeting and reduce cumulation of different benefits with other sources of income (OECD, 1997*b*).

B. AGRICULTURAL SITUATION

1. Agriculture and the food sector in the economy

The share of agriculture in the Russian economy has diminished considerably since the transition started. The sector's contribution to GDP fell from 15.4 per cent in 1990 to 6.5 per cent in 1997 (Table I.4). That drop was partly due to a fall in the volume of agro-food production, but was more the consequence of a relative decline in agricultural prices compared to the prices of non-agricultural products after price controls were largely removed in 1992.

Despite the fall of agriculture's contribution to GDP, the proportion of the total labour force employed in agriculture increased from about 13 per cent in 1990 to 15 per cent in 1994 and fell slightly to 14 per cent in 1997. The increased disparity between the agricultural contribution to GDP and its share in total employment indicates a sharp fall in labour productivity since 1991.

Box I.3. **The major economic, political and social organisations in the agro-food sector**

In the USSR only one political party was allowed, therefore no agricultural or other political organisation existed as of the end of the 1980s. The then existing Union of Kolkhozes combined all *kolkhozes* of the country and their representatives gathered regularly for congresses, but its role was negligible.

During the *perestroika* period a spontaneous process of party formation started and several agricultural organisations of different political orientations emerged. One of the first and most important was the **Agrarian Party** led by Mikhail Lapshin, then a Moscow-area *sovkhoz* director. The party has branches in practically all regions of the Russian Federation and after the last parliamentary elections of 1995 formed the fifth largest fraction in the State Duma (Box I.1). It stands for limited private property rights for agricultural land, massive governmental support for agricultural producers and strong protection of the domestic agro-food markets against imports. Members of the Agrarian Party take up most of the seats on the parliamentary Committee on Agriculture. In the State Duma the party usually votes with the Communist Party.

At the end of the 1980s the Association of the Farmers and Farm Co-operatives of Russia (AKKOR) was formed. Initially, it was designed as a social organisation to provide support for emerging private farmers and, in this respect, was in strong opposition to the Agrarian Party. At a later stage, **AKKOR** moved gradually to more economic activities: it became involved in governmental programmes for distributing inputs and finance to individual farmers; established several banks and an insurance company servicing mostly individual farmers; organises the individual farmers' fair in St. Petersburg; supports education programmes for farmers; and promotes development of marketing and supply co-operatives servicing individual farmers. Politically, AKKOR demands financial support for agriculture and protection against imports. AKKOR has subsidiaries in all regions of the country. Its leader, Mr. Bashmachnikov, is a member of the State Duma, elected as a member of the pro-governmental party "Our Home is Russia".

There are several agricultural organisations without political purposes in their charters. One of the most influential is the Agrarian Union renamed later as the **Agro-Industrial Union**. The union is led by Mr. Starodubtcev, one of the leaders of the failed August 1991 coup organised by communist hard-liners and, currently, a governor of the Tula region. The Union represents the interests of large farms. It co-operates with the Agrarian Party and the Communist Party but is even more conservative on some issues, for instance it opposes any form of land market.

The agricultural **co-operative movement** has been very weak in Russia. Large farms do not see the advantages of economic co-operation and individual farmers are not willing to co-operate due to a long period of bad experiences with "co-operation" in the *kolkhoz-sovkhoz* system. Another reason is the inadequate co-operative legislation. The Agricultural Co-operative Law, adopted only in December 1995, is oriented mostly toward the production co-operatives. Existing vertical farmers' co-operatives are few, weak, and badly managed.

In a number of food sub-sectors non-governmental **business associations** have begun to emerge. The first (Grain Union) was established in 1994 by grain producers, processors and traders. Another was established in 1996 by sugar producers and traders (Sugar Union) and united all 96 sugar plants of the Russian Federation. Recently, the creation of a tea processors' and traders' association was announced. The major objectives of these associations are to elaborate and promote federal support policies on the corresponding markets, to facilitate the establishment of trade institutions (such as futures markets and exchanges), and to stimulate market studies and monitoring of domestic and foreign markets. While some of these activities may stimulate the development of market institutions, the major efforts of most of the associations are concentrated on lobbying for more protective domestic policies, and there is a danger that monopolistic structures may be strengthened. The recent (1997) introduction of a new white sugar import tariff for example was induced to a certain extent by the Sugar Union.

Table I.4. **Share of agro-food sector in the economy, 1990-1997**

	1990	1991	1992	1993	1994	1995	1996	1997p
Share of agriculture in GDP	15.4	13.7	7.2	7.4	6.0	6.9	6.7	6.5
Share of food industry in GDP	n.a.	n.a.	n.a.	n.a.	0.9	1.1	1.8	n.a.
Share of agriculture in employment	12.9	13.2	14.0	14.3	15.0	14.7	14.0	13.9
Share of food industry in employment	2.1	2.1	2.1	2.2	2.3	2.3	2.2	n.a.
Total capital investments in agro-industrial complex (billion roubles, 1991 prices)	70.4	65.8	26.9	17.6	9.3	6.5	4.7	4.3
of which: agriculture	39.5	37.4	12.9	8.2	4.0	2.4	1.8	1.5
Share of capital investments in agro-industrial complex in total capital investments (%)	28.3	31.3	22.6	16.9	11.5	9.4	7.9	7.6

p: provisional.
n.a.: not available.
Source: Goskomstat; Nazarenko (1997).

The share of investments in the agro-food complex in total capital investments has also fallen significantly from 28 per cent in 1990 to below 8 per cent in 1997, including in agriculture from 16 per cent to 2.5 per cent, respectively. The food industry's contribution to total GDP increased from 0.9 per cent to 1.8 per cent between 1994 and 1996 while its share in total employment decreased slightly from 2.3 per cent to 2.2 per cent over the same period (Table I.4).

2. Evolution of market conditions: prices and costs

Under the previous system of central planning, prices in the agro-food sector were strictly controlled. Agricultural producer prices were fixed by the state, but differentiated by production zones according to the cost-plus concept, which was designed to provide "normal levels of profits" for agricultural enterprises producing under average conditions. Input prices were controlled as well and retail food prices were kept stable and relatively low due to significant consumer subsidies. The system was partly deregulated in 1991 and prices were largely liberalised in 1992 (Part IV).

While in the early years of the economic reform process (1992-1993) the sharp decline in demand following cuts in consumer subsidies and a fall in real incomes limited inflationary increases in farmgate prices, input prices often soared to world market levels resulting in a sharp cost-price squeeze for agricultural producers. The output/input price ratios were worsened by inefficient upstream and downstream enterprises trying to pass their high costs of production onto producers and/or consumers. Between 1991 and 1993, input prices soared by a factor of 173, while agricultural output prices increased only by a factor of 76. The largest increases in input prices were registered for agricultural machinery and energy inputs. For example the price of tractors rose by a factor of 240 and the price of fuel and lubricants by a factor of 390 over the period 1991-1993. In 1995, the price conditions for farmers improved slightly, when agricultural input and output prices rose by 222 and 235 per cent respectively. The cost-price squeeze tightened again in 1996, with agricultural input prices rising by 75 per cent and output prices by 40 per cent. In overall terms, between 1990 and 1996, agricultural input prices increased 4.6 times faster than agricultural output prices (Table I.5. and Graph I.1). It should be noted that the sharp worsening of the output/input prices occurred in all countries in transition at the beginning of the reform process, but the extent depends more on the level of support provided under the previous system than on policies that are applied after the liberalisation of prices. It seems that in countries providing the strongest support for agriculture under the previous system, such as Russia and other countries covered by Soviet policy, the fall was sharpest (Graph I.2).

The retail food price index and consumer price index (CPI) have followed almost the same pattern since 1990, with the CPI increasing only 1.04 times faster than the retail food price index between 1990 and 1996 (Table I.5 and Graph I.1).

Table I.5. **Price indexes, 1991-1996**

1990 = 100

	1991	1992	1993	1994	1995	1996
Agricultural output price index[1]	160	1 504	12 211	37 120	124 353	174 094
Agricultural input price index[1]	193	3 129	33 457	140 897	454 078	793 327
Grain combines	206	3 460	43 247	232 153	658 849	1 426 640
Tractors	229	4 997	54 863	212 013	651 342	1 222 576
Fertilisers	175	2 232	23 941	155 882	523 920	848 227
Mixed feed	213	3 814	32 820	121 828	320 774	699 608
Electrical energy for production purposes	152	2 009	43 341	282 757	807 836	1 244 877
Fuel and lubrificants	131	4 523	51 051	175 259	680 181	1 021 632
Retail food price index[2]	259	6 913	64 841	216 181	481 218	544 258
Consumer price index[2]	260	6 783	63 764	200 920	464 729	566 039

Note: The numbers contained in the above table, particularly for input and output prices, have to be treated as an illustration of tendencies rather than a precise measure of relative price changes.

1. Average annual change.
2. End-year change.

Source: Goskomstat; OECD 1997*b*; Nazarenko (1997).

◆ Graph I.1. ***Price relationships, 1991-1996***

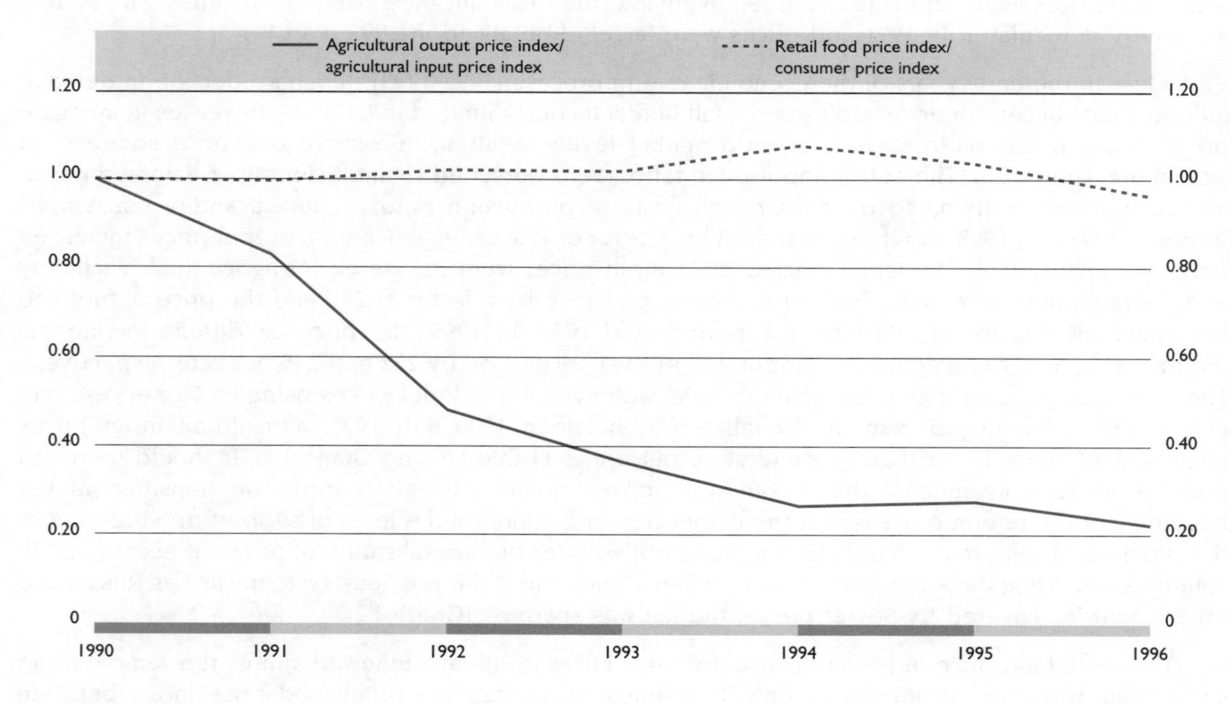

Source: Table I.5.

◆ Graph I.2. ***Agricultural output/input price ratio, 1990-1996***
1990 = 100

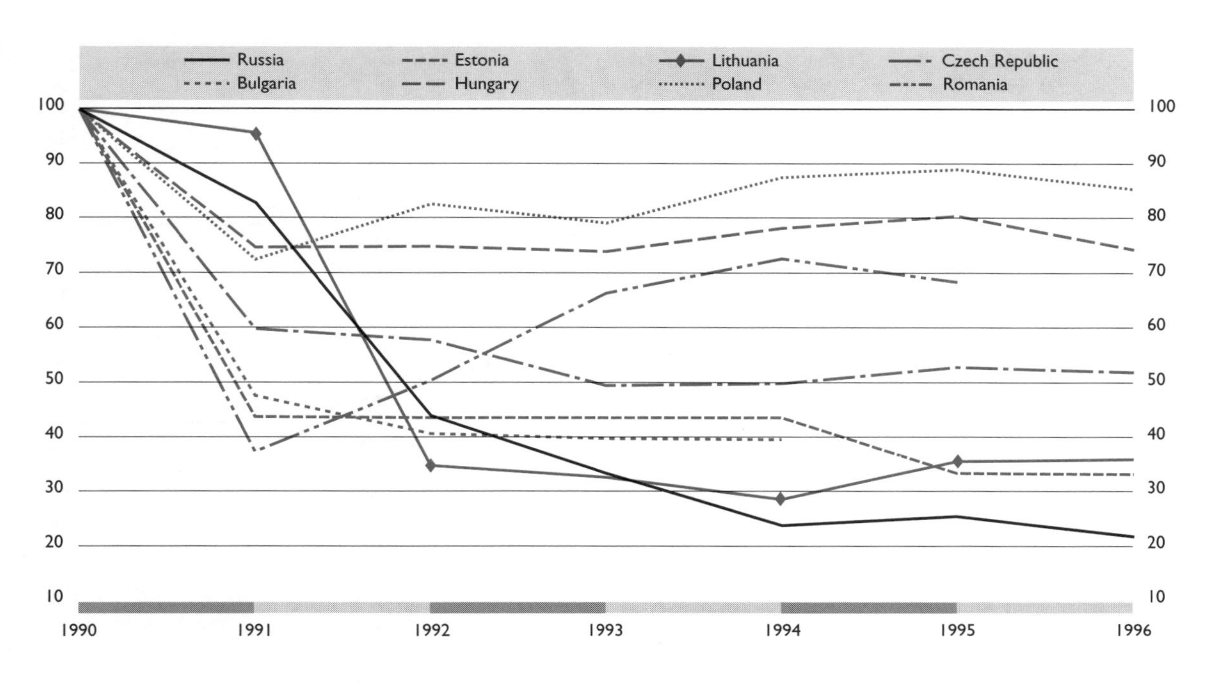

Source: Goskomstat, OECD database on CEECs and NIS.

3. Sectoral performances

i) Output

Between 1990 and 1996 the Gross Agricultural Output (GAO) fell continuously, reflecting the adjustment to new market conditions. This decline was in part due to the severe cost-price squeeze in agriculture which resulted in a significant decline in purchases of agricultural inputs and to the negative demand-side effect of reducing consumers' real income which decreased demand for foodstuffs, particularly for livestock products which have a relatively high income elasticity of demand (see below). Moreover, the break-up of the state monopoly for trade has given Russian agriculture the additional shock of integration into the world agricultural economy. Since world prices for certain foodstuffs, mainly livestock and other high value processed foods, are lower than Russian prices, imports of these products have increased in recent years (see Part II).

In 1996, agricultural output was 64 per cent of the 1990 level. However, compared to other countries in transition, particularly to those emerging from the previous Soviet Union, the fall in Russia was not particularly sharp (Graph I.3). Smaller falls are registered in countries with relatively good macroeconomic performance and less affected by the farm restructuring process (for example Poland and the Czech Republic) or in which agriculture was taxed under the previous system and reforms eased some previously existing constraints (Romania). In 1997, the GAO stabilised in Russia, for the first time since the reforms started, which may indicate an end to the first and most difficult stage of transition for Russian agriculture (Table I.6).

Changes in the volume of agricultural production in Russia have been highly differentiated across sectors. While production by large-scale enterprises fell by more than half between 1990 and 1997, the output of household plots increased by 19 per cent over the same period (Table I.6). As a result, the former's contribution to total agricultural output fell from 74 per cent in 1990 to 52 per cent in 1997,

◆ Graph I.3. ***Gross Agricultural Output (GAO), 1990-1997***
1990 = 100

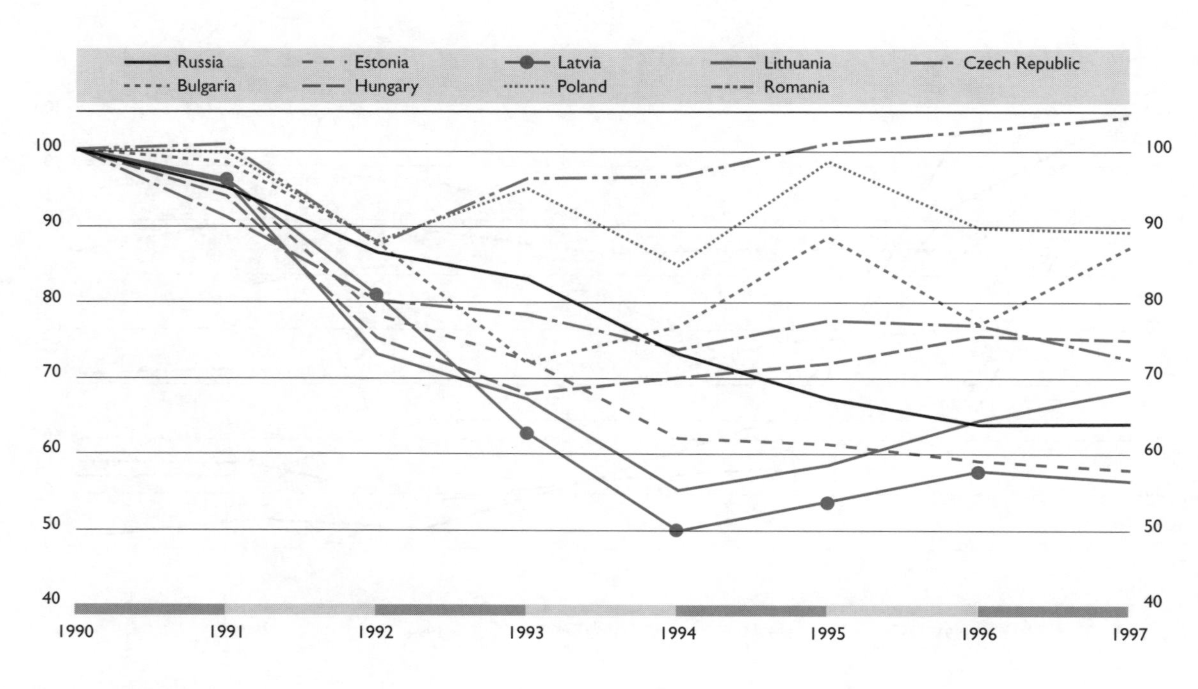

Source: Goskomstat, OECD.

while that of the latter increased from 26 per cent to 46 per cent. The high share of household plots in total agricultural output can be partly attributed to their specialisation in relatively labour intensive and high value products such as fruits, vegetables and animal products, but also to various registered and unregistered transfers of inputs and agricultural products from large scale farms to household plots (Part III). Newly emerging individual (family type) farms accounted for 2 per cent of agricultural output in 1997.

The transition period also brought significant changes in the structure of agricultural production in Russia. Following changes in consumer demand (see below) and the relative increase in crop prices compared to livestock prices, the share of livestock production in total value of agricultural production dropped sharply from 64 per cent in 1990 to 31 per cent in 1992. However, in more recent years price trends reversed with livestock market prices rising more than crop prices. As a result, the share of livestock output rose to 42 per cent in 1996.[4]

Output changes have been highly differentiated across commodities. With the reduced application of fertilisers (17 kg per hectare in 1996, down from 88 kg per hectare in 1990) and plant protection chemicals, as well as the general deterioration in the technologies used, crop production became more dependent on weather conditions. However, the percentage decline in use of material inputs has been much stronger than the percentage drop in crop output. This means that the productivity of those inputs being used has risen, indicating that with fewer inputs available, farms are being motivated to use them more productively (Sedik *et al.*, 1996).

Grain output almost halved between 1990 and 1995, being strongly affected by severe drought conditions in 1994 and 1995 and the massive contraction of livestock production reducing demand for feed grain. It then recovered in 1996, rising by 9 per cent to 69.3 million tonnes and, mostly due to

Table I.6. **Change in Gross Agricultural Output, 1990-1997**

	1990	1991	1992	1993	1994	1995	1996	1997p
Total agricultural sector								
% change		–5	–9	–4	–12	–8	–5	0.1
index 1990 = 100	100	95	86	83	73	67	64	64
Large-scale farms								
% change		–9	–17	–9	–16	–15	–10	0.0
index 1990 = 100	100	91	76	68	57	49	44	44
Household plots								
% change		9	8	3	–5	3	0.4	0.0
index 1990 = 100	100	109	118	121	115	119	119	119
Individual farms								
% change				67	–14	–3	0	7.7
index 1990 = 100								

p: provisional.
n.a.: not available.
Source: Goskomstat.

Table I.7. **Crop sown areas, 1985-1997**

Thousand hectares

		1985	1990	1991	1992	1993	1994	1995	1996	1997	1997/1990
Grains [1]		68 138	63 068	61 783	61 939	60 939	56 280	54 705	53 388	53 611	–15%
of which:	Wheat	25 284	24 244	23 152	24 284	24 666	22 190	23 909	25 707	26 026	7%
	Barley	16 144	13 723	15 282	14 564	15 478	16 404	14 710	11 793	12 493	–9%
	Rye	7 214	8 008	6 480	7 592	6 000	3 903	3 247	4 147	4 005	–50%
	Oats	10 981	9 100	9 032	8 540	8 402	8 333	7 928	6 904	6 436	–29%
	Maize	1 080	869	733	810	800	524	643	622	918	6%
Sugar beet		1 492	1 460	1 399	1 439	1 333	1 104	1 085	1 060	935	–36%
Sunflower		2 320	2 739	2 576	2 889	2 923	3 133	4 127	3 874	3 585	31%
Potatoes		3 538	3 124	3 187	3 404	3 548	3 337	3 409	3 404	3 350	7%
Vegetables		n.a.	618	662	682	684	704	758	737	749	21%

n.a.: not available.
1. "All grains" are wheat, barley, rye, oats, maize, buckwheat, millet, unmilled rice and pulses.
Source: OECD database on CEECs and NIS.

Table I.8. **Production of major crops, 1985-1997**

Thousand tonnes

		1985	1990	1991	1992	1993	1994	1995	1996	1997	1997/1990
Grains [1]		98 564	116 676	89 094	106 855	99 094	81 297	63 406	69 341	88 505	–24%
of which:	Wheat	38 362	49 596	38 899	46 166	43 547	32 129	30 119	34 917	44 188	–11%
	Barley	23 013	27 235	22 174	26 988	26 843	27 054	15 768	15 933	20 771	–24%
	Rye	10 691	16 431	10 639	13 887	9 166	5 989	4 098	5 934	7 480	–54%
	Oats	15 201	12 326	10 372	11 241	11 556	10 757	8 562	8 346	9 381	–24%
	Maize	3 020	2 451	1 969	2 155	2 441	892	1 738	1 088	2 671	9%
Sugar beet		31 450	32 327	24 280	25 548	25 468	13 946	19 072	16 166	13 841	–57%
Sunflower		2 621	3 427	2 896	3 110	2 765	2 553	4 200	2 765	2 824	–18%
Potatoes		33 840	30 848	34 329	38 330	37 650	33 828	39 909	38 652	37 015	20%
Vegetables		11 131	10 328	10 425	10 018	9 827	9 621	11 275	10 731	11 085	7%

1. "All grains" are wheat, barley, rye, oats, maize, buckwheat, millet, unmilled rice and pulses.
Source: OECD database on CEECs and NIS.

favourable weather conditions, reached as much as 88.5 million tonnes in 1997. During the 1990-1997 period, there was an evident shift in land allocation and production from feed grains (barley, oats and maize) to wheat (Table I.7) in response to changes in domestic demand. The area sown to wheat was more or less stable between 1990 and 1995 but increased by about 9 per cent between 1995 and 1997 and wheat production rose by 47 per cent. Potatoes and vegetables are among the few agricultural products for which production was more or less stable or even increased in the reform period, mostly due to the fact that these crop products are cultivated in particular by household plots, which have resisted the general agricultural crisis. In 1997 potato production was 20 per cent higher than in 1990 and accounted for 31 per cent of the total value of crops. The high profitability of sunflower production at the beginning of the transition led to a significant expansion of the area allocated to sunflower, which contributed to a rise of 23 per cent in production between 1990 and 1995. However adverse weather conditions and inappropriate agro-technology (lack of crop rotation, low fertiliser and other chemicals use) led to a reduction in yields and a drop in output by 34 per cent in 1996. Sunflower production increased then slightly in 1997 by 3 per cent, due to improved yields (Table I.8).

The fall in demand for animal products was most severe in the case of pigmeat (consumption of which dropped by 45 per cent between 1990 and 1996), beef (down by 44 per cent) and milk and milk products (down by 37 per cent). The livestock sector and meat processing industries have also to face tough competition from strongly increased meat imports (Part II). Between 1990 and 1998, total inventories of cattle decreased steadily by 46 per cent, pigs by 57 per cent and poultry by 45 per cent. The decline in cow numbers was less dramatic as the cumulative fall since 1990 was 30 per cent in 1998 (Table I.9). Animal numbers have dropped on all types of farms, but by appreciably more on large farms than on household plots and individual farms. Total meat production in 1997 was 52 per cent below the 1990 level, with poultrymeat falling by 65 per cent, pigmeat by 55 per cent and beef and veal by 46 per cent (Table I.10). Milk production fell less sharply than meat production and was 39 per cent down in 1997 compared to 1990, but milk yields per cow decreased significantly, amounting to 2 330 kg per cow in 1997 in comparison with 2 710 kg in 1990.

Table I.9. **Animal numbers, 1985-1998**

Thousand heads, as of 1 January

	1985	1990	1991	1992	1993	1994	1995	1996	1997	1998	1998/1990
Cattle	60 044	58 841	57 043	54 677	52 226	48 914	43 297	39 696	35 103	31 719	–46%
of which: Cows	22 000	20 760	20 557	20 564	20 243	19 831	18 398	17 436	16 874	14 620	–30%
Pigs	38 732	39 982	38 314	35 384	31 520	28 557	24 859	22 631	19 115	17 292	–57%
Poultry	616 600	653 640	659 808	652 211	568 278	566 812	490 966	422 601	371 866	360 442	–45%

Source: OECD database on CEECs and NIS.

Table I.10. **Production of basic animal products, 1985-1997**

Thousand tonnes

	1985	1990	1991	1992	1993	1994	1995	1996	1997	1997/1990
Total meat (carcass weight)	8 487	10 112	9 375	8 260	7 513	6 803	5 796	5 336	4 811	–52%
of which: Beef	3 572	4 329	3 989	3 632	3 359	3 240	2 733	2 630	2 338	–46%
Pigmeat	2 960	3 480	3 190	2 784	2 432	2 103	1 865	1 705	1 565	–55%
Poultry	1 527	1 801	1 751	1 428	1 277	1 068	859	690	632	–65%
Milk	50 169	55 715	51 886	47 236	46 524	42 174	39 241	35 819	34 066	–39%
Eggs (million pieces)	44 277	47 470	46 875	42 902	40 297	37 473	33 830	31 902	31 884	–33%

Source: OECD database on CEECs and NIS.

Table I.11. **Population in rural areas and population engaged in agriculture, 1980-1997**

		1980	1985	1990	1991	1992	1993	1994	1995	1996	1997p
Rural population	Thousands	41 530	39 838	38 332	39 117	39 837	39 987	40 051	39 938	39 790	39 520
Total population	Thousands	138 839	143 528	148 164	148 326	148 295	147 997	147 938	147 609	147 137	146 737
Share of rural population in total population	Share (%)	30%	28%	26%	26%	27%	27%	27%	27%	27%	27%
All agriculture employment	Thousands	10 718	10 405	9 728	9 736	10 101	10 104	10 278	9 744	9 243	9 070
	Share in total employment	14.6%	13.9%	12.9%	13.2%	14.0%	14.3%	15.0%	14.7%	14.0%	13.9%
Large farms only	Thousands	9 689	9 277	8 341	n.a.	8 062	n.a.	7 333	6 678	n.a.	n.a.
	Share in total agriculture	90.4%	89.2%	85.7%	n.a.	79.8%	n.a.	71.3%	68.5%	n.a.	n.a.
Total APK[1] (annual average)	Thousands	n.a.	17 835	16 974	16 618	16 580	16 417	n.a.	n.a.	n.a.	n.a.
Total APK[1]	Share in total employment	n.a.	23.8%	22.5%	22.6%	23.0%	23.2%	n.a.	n.a.	n.a.	n.a.
Total civilian employment	Thousands	73 275	74 937	75 325	73 548	72 071	70 852	68 484	66 441	65 950	65 400

p: provisional.
n.a.: not available.
1. "APK": "Agro-industrial Complex", standard Soviet-type statistical construct approximately equivalent to "Agro-food sector", however excluding retail trade in food.
Source: OECD database on CEECs and NIS.

ii) Employment in agriculture

While the overall population in Russia has declined slightly since 1990, the number of rural dwellers increased from 38.3 million in 1990 to 40.1 million in 1994, partly due to net migration from urban to rural areas, immigration from other republics of the former Soviet Union and a change in the classification of rural areas which turned some small towns into villages at the beginning of the 1990s. By 1997, the number had declined to 39.5 million, but was still higher than in 1990. About 27 per cent of the Russian population lived in rural areas in 1996.

As indicated above, the share of agriculture in total employment increased from about 13 per cent in 1990 to about 14 per cent in 1997. However, the main reason for the increase was a significant decline in non-agricultural employment as the total number of people employed in agriculture fell from 9.7 million in 1990 to 9.1 million in 1997. Employment in agriculture is still dominated by large-scale farms although numbers of those employed on large farms declined from 8.3 million in 1990 to 6.7 million in 1995 and the share of large farms in total agricultural employment fell from 86 per cent to 69 per cent (Table I.11.).

In general, registered unemployment rates in rural areas are low, but there are some regional differences and the rates varied between 1 to 5 per cent in 1995 (Table I.12). In the Central Black Soil region the rate was the lowest at 1 per cent while in the North, the North-West and the Far East regions the rates were the highest at 4-5 per cent in 1995. The share of rural unemployment in total unemployment was the highest in the North Caucasus, Siberia and Far East regions. However, these numbers do not include unregistered unemployment in rural areas which on average would probably double the unemployment rates in Table I.12.

Table I.12. **Registered unemployment rates in rural areas by regions, 1995**

Per cent

	Rural unemployment	Share of rural unemployment in total unemployment
Russia	3	29
North	5	29
North-West	4	17
Center	3	18
Volga-Viatka	3	22
Central Black Soil	1	22
Volga	2	27
North Caucasus	3	55
Ural	3	26
West Siberia	3	42
East Siberia	3	37
Far East	4	34

Source: MAF (1996).

iii) Agricultural income

In 1990, wages in the agricultural sector were close to the national average. Since 1991, average nominal wages in agriculture have increased but at a lower rate than in other sectors of the economy, with the result that agriculture has become the sector with the lowest wages. While in 1990, the average monthly wage in agriculture amounted to 95 per cent of the national average, by 1997, it had fallen to 44 per cent (Table I.13). In 1997, the average monthly wage on large and middle size farms was about Rb 423 000 (US$73). By the end of 1996, wage arrears in agriculture were equal to 3.3 months' wages, while arrears in the economy as a whole averaged 2.7 months' wages. However, wages paid by large-scale farms are a declining part of rural families' incomes. The shift in the structure of agricultural production from large-scale enterprises to household plots indicates that a growing proportion of the

Table I.13. **Nominal wages in agriculture and in industry, 1980-1997**

Roubles

	1980	1990	1991	1992	1993	1994	1995	1996[1]	1997p[1]
Agriculture									
Average monthly wages	142	289	459	3 984	36 019	111 266	236 707	382 042	423 248
Share in national average (%)	82	95	84	66	61	50	50	48	44
Industry									
Average monthly wages	191	311	606	7 064	63 447	228 528	528 829	868 823	1 136 896
Share in national average (%)	110	103	111	118	108	104	112	110	118
National economy									
Average monthly wages	174	303	548	5 995	58 663	220 351	472 392	790 210	964 507

p: provisional.
1. Large and middle size enterprises.
Source: Goskomstat.

incomes of rural households comes from their own production which is used for self-consumption but also sold in an increasing proportion on local markets. The results of household budget surveys, which take account of households' money incomes and the market value of food produced and consumed by the family, indicate that in 1995 monthly income per head in the rural areas of three Russian regions surveyed was higher than the per capita income of the urban population in the same regions (Braun, 1997). While the value of other goods and services produced and consumed by families were not covered by the survey and the results are not necessarily representative of the country as a whole, it might indicate that aggregate incomes of the rural population are not as low as shown by the data in Table I.13.

According to official statistics, the profitability of large-scale farms, measured as a ratio of net profits to costs of production, also fell significantly and the total percentage of loss-making farms increased from 5 per cent in 1992 to 81 per cent in 1997 (Table I.14). However, there are numerous methodological problems related to the measurement of "profitability" derived from the accounts of large-scale farms. For example, there is a significant methodological confusion in the estimation of depreciation charges using antiquated accounting practices where fixed assets, such as buildings and machinery, are valued at their book value disregarding the fact that they may be totally unusable, obsolete and have negligible opportunity cost as a result of the huge changes that have taken place in farm structure and in the economic environment. In most countries in transition, including Russia, there is a great temptation to inflate these "costs" and thereby present an overly pessimistic measurement of farm economic performance. Also, what is termed a farm in these calculations is rather different from the Western perception of a farm as it includes many activities which are not farming activities *per se* but are related to processing, transport, provision of services and inputs to agriculture as well as social services for the rural population. Private (individual) farms and household plots are not included in the calculations, even though the latter account for about half of total agricultural output. Household plots are heavily dependent on various forms of inputs and services obtained from large enterprises and as expenditures for these inputs are included in the costs of large farms, these transfers have an

Table I.14. **Profitability of Russian farms, 1990-1997**

	1990	1991	1992	1993	1994	1995	1996	1997
Profitable farms (as % of all farms)	97	95	95	90	41	43	21	19
Loss-making farms (as % of all farms)	3	5	5	10	59	57	79	81
Net profits per farm (million roubles)	1.1	1.8	21.2	117.7	–9.6	53.9	–864.5	–839

Source: Agricultural Accounting Report of Agricultural Enterprises, Russian Agriculture, various years.

Table I.15. **Production and delivery of inputs, 1980-1997**

			1980	1985	1990	1991	1992	1993	1994	1995	1996	1997p
Tractors	Produced	000 units	249	261	214	178	137	89	29	21	14	12.6
	Delivered	000 units	177	187	144	131	65	40	22	10	9	n.a.
	Inventory[2]	000 units	1 472	1 592	1 520	1 500	1 444	1 381	1 148	1 052	966	916
Grain combines	Produced	000 units	117	112	66	55	42	33	12	6	3	2
	Delivered	000 units	75	70	38	32	17	14	9	4	3	n.a.
	Inventory[2]	000 units	448	511	408	394	371	347	317	292	264	248
Mineral fertilizer[1]	Produced	000 tonnes	11 772	17 304	15 979	15 042	12 300	9 917	8 266	9 639	9 076	9 532
	Delivered	000 tonnes	8 911	12 674	11 051	10 102	5 510	3 721	1 398	1 601	1 579	1 587
	Applied	000 tonnes	7 480	9 790	9 923	10 100	8 600	4 295	2 091	1 487	1 473	1 500
Compound feed[3]	Produced	000 tonnes	32 464	37 896	40 976	37 405	27 426	25 218	18 137	14 300	9 600	6 700
Gasoline	Delivered	000 tonnes	18 300	18 500	11 264	10 633	9 456	6 223	3 670	3 345	2 945	n.a.
Diesel fuel	Delivered	000 tonnes	21 600	23 200	20 032	19 424	16 522	12 767	7 846	7 105	6 212	n.a.

p: provisional.
n.a.: not available.
1. Mineral fertilizer amounts are recalculated to a nominal 100 per cent of active ingredients.
2. "Inventory" is amount on hand in agricultural enterprises at end of period.
3. Compound feed *(kombikorma)* produced at state feed mills.
Source: OECD database on CEECs and NIS.

increasingly negative impact on larger farms' incomes. The symbiotic relationship between the large-scale farms and the small plots is therefore not adequately dealt with by the above mentioned accounting system (see OECD, 1997*a*).

iv) Performance of the upstream sector

Huge indirect and direct input subsidies were applied during the Soviet period, mainly to the production of agricultural machinery and fertilisers, as an important means of lowering agricultural production costs. This led to an inefficient input use and significant wastage. Very often, the application of fertilisers and pesticides took place too late in the season because of insufficient or broken machinery. As indicated above, price liberalisation led to a much faster rise in input prices compared to output prices. This sharp rise in relative prices led to strong declines in input purchases by the farming sector, followed by a dramatic fall in input production and deliveries (Table I.15.). Between 1990 and 1997, the production of tractors and agricultural machinery collapsed, with production of tractors falling by 94 per cent and output of grain combines by 97 per cent. In most cases, capacity utilisation rates were below 10 per cent in recent years. While output of agricultural machinery fell drastically, inventories of tractors and grain combines did not decline as quickly and in 1997 they had fallen by about 23 and 28 per cent respectively compared to 1990.

Application of mineral fertilisers plummeted by 85 per cent between 1990 and 1996, while total fertiliser production in Russia has fallen by 43 per cent since 1990. In effect, thanks to substantial exports of nitrogenous and potash fertilisers, the fertiliser industry has been less affected than other upstream industries by the crisis in Russian agriculture. In 1996, about 80 per cent of total fertiliser production was exported. The decline in the application of lime and mineral and organic fertilisers on agricultural land resulted in increasing problems of soil depletion. The proportion of total sown area treated with mineral fertilisers fell from 66 per cent in 1990 to just 25 per cent in 1996, and the share of sown area treated with organic fertilisers was 2.9 per cent in 1996 down from 7.4 per cent in 1990 (Table I.16.).

Table I.16. **Chemical use by large-scale enterprises**

	Units	1970	1980	1990	1993	1994	1995	1996	1997
Mineral fertilisers	kg/ha	28	62	88	46	24	17	15	15
Share of sown area treated with mineral fertilisers	per cent	36	58	66	45	29	25	25	n.a.
Organic fertilisers	tonnes/ha	1.7	3.1	3.5	2.6	1.8	1.4	1.2	1.1
Share of sown area treated with organic fertilisers	per cent	n.a.	9.0	7.4	5.2	3.9	3.2	2.9	n.a.
Lime	tonnes/ha	3.9	5.9	6.7	6.5	6.4	6.6	n.a.	n.a.

n.a.: not available.
Source: Goskomstat.

v) Performance of the food industry

Partly in line with the decline in food consumption, the registered output of the food industry has fallen continuously since 1990; by 1997 the overall change ranged from a 84 per cent decline in wine production to a 0.1 per cent rise in sugar production. In general, there was a greater drop in animal production (meat and milk processing) than in crop processing which reflects changes in the consumption structure in recent years. In 1997, output of meat and milk products amounted to 23 per cent of their 1990 levels (Table I.17.). It should be noted that the decline in food consumption (see below) has been much smaller than the registered decline in amounts supplied by domestic food industries. The gap can be explained by much stronger reliance of households on unprocessed agricultural products

Table I.17. **Food industry output, 1985-1997**

Thousand tonnes

	1985	1990	1991	1992	1993	1994	1995	1996	1997p	1997/1990
Meat, including category I subproducts[1]	5 334	6 629	5 815	4 784	4 099	3 282	2 416	1 937	1 464	–78%
Whole milk products (in fluid milk equivalent, million tonnes)	18	21	19	10	8	7	6	5	4.8	–77%
Total canned goods (million standard cans)	7 057	8 206	6 944	5 353	4 517	2 817	2 428	2 158	2 162	–74%
of which:										
Dairy products	769	980	767	629	682	582	527	544	569	–42%
Meat	547	545	478	558	488	352	348	380	287	–47%
Fish and other seafood	2 204	2 335	1 982	1 332	918	591	574	471	467	–80%
Fruits and vegetables	n.a.	3 060	2 712	2 222	1 913	834	645	560	564	–82%
Total granulated sugar	3 642	3 758	3 426	3 923	3 918	2 736	3 155	3 284	3 765	0%
of which: from sugar beet	2 569	2 630	2 052	2 248	2 497	1 655	2 064	1 711	1 337	–49%
Bread and bakery products (million tonnes)	19	18	19	17	15	12	11	10	8.9	–51%
Margarine	822	808	627	560	438	278	198	200	222	–73%
Vegetable oil	775	1 159	1 165	994	1 127	909	802	879	687	–41%
Confectionery goods	2 268	2 869	2 641	1 829	1 746	1 530	1 372	1 262	1 347	–53%
Vodka and liqueurs, million decalitres	161	138	154	152	157	125	123	71	87	–37%
Wine, million decalitres	143	76	65	40	25	21	15	11	12	–84%

p: provisional.
n.a.: not available.
1. "Category I subproducts" are tongue, liver, lungs, and pigs' heads. These data "include an allowance for production by small and joint enterprises, as well as processing enterprises that are part of non-industrial enterprises".
Source: OECD database on CEECs and NIS Goskomstat (1998).

originating from household plots, and by increased imports of processed food (meat products in particular) at the beginning of transition (Part II).

Capacity utilisation rates in most branches of the industry are extremely low, ranging from 12 per cent to 45 per cent in the meat, dairy, bakery and confectionery branches, for instance. Output trends in 1995-1997 show that in a few industries (canned dairy products, sugar and margarine) the drop in production bottomed out in 1995 and output rose slightly in 1996 and, according to provisional data, continued to rise in 1997. This may be a sign that the food industry sector is reacting better to market signals and is demonstrating some ability to withstand international competition. The food industry is becoming more attractive to foreign investors, who are becoming involved in the production of high value products such as ice cream, baby food and confectionery.

4. Food consumption

Food consumption in Russia was relatively high in the 1980s, even by Western standards and especially considering the country's relatively low per capita GDP. Low prices of highly subsidised foodstuffs resulted in an increased demand for food, in particular for high-value meat and milk products. In the reform period, the liberalisation of prices and the discontinuation of high consumer subsidies combined with a fall in real incomes, led to a sharp decline in demand.

Table I.18. shows the change in the structure of food consumption. In the 1990s per capita consumption of products with high income elasticities, such as meat and dairy products decreased, while demand for staple goods like potatoes and cereal products, increased. In 1997, per capita consumption of meat and milk products amounted to 51 kg and 235 kg respectively, a decline of 31 and 38 per cent respectively compared with 1990. The consumption of potatoes and bread and bakery products increased at the beginning of the transition but then stabilised at 1993 levels. Consumption of such products as sugar, fruits and vegetables, and vegetable oil declined at the beginning of the transition, but then stabilised at 1992/1993 levels.

However, the major concern is not the average level of food consumption but rather its differentiation. In effect, price liberalisation has led to a growing income differentiation among the Russian population. According to the Goskomstat, the 10 per cent of the population with the highest incomes received 34 per cent of total money income in 1996, while the 10 per cent of the population with lowest incomes received only 2.6 per cent. The resulting income ratio between the highest and lowest income brackets was 13:1 in 1996. As the data do not take into account payments in kind and food produced by households themselves, the actual income differentiation may be somewhat lower. However, it may be assumed that a large part of urban pensioners and other groups of the urban population, who have to

Table I.18. **Food consumption pattern, 1980-1997**

Kg per person per year

	1980	1985	1990	1991	1992	1993	1994	1995	1996	1997p
Meat and meat products	62	67	75	69	60	59	57	55	52	51
Milk and dairy products[1]	328	344	386	347	281	294	278	253	240	235
Eggs (pieces)	279	299	297	288	263	250	236	214	200	200
Sugar	47	45	47	38	30	31	31	32	34	32
Vegetable oil	9	10	10	8	7	7	7	7	8	n.a.
Potatoes	118	109	106	112	118	127	122	124	124	126
Vegetables and melons	94	98	89	86	77	71	68	76	75	74
Fruits and berries	37	46	35	35	32	29	28	29	28	28
Cereal products[2]	126	119	119	120	125	124	124	124	127	128

p: provisional.
n.a.: not available.
1. In fluid milk equivalent.
2. "Cereal products" include flour, groats, and pulses. Bread and macaroni in flour equivalent.
Source: OECD database on CEECs and NIS.

live on very low incomes and have less scope for producing food themselves than do rural dwellers, have inadequate diets in terms of both quantity and nutrient content. In 1996, the average share of household incomes spent on food was high at 42.6 per cent (up from 39.1 per cent in 1994), while the share of expenditure on food for low income households was between 70 and 80 per cent.

Another area of concern relates to consumption imbalances between regions due to a lack of an efficient institutional and physical infrastructure linking production and consumer regions. These barriers are sometimes administratively reinforced by local governments which fear local food shortages (Part III and Annex III). Such developments provoke excessive price differentiation between regions, implicitly taxing producers in surplus regions and consumers in deficit regions.

5. The agro-environmental situation

Constant pressure for increased production led Soviet authorities to expand area under cultivation. During the New Lands Programme of 1954-58, dry, marginal areas of the southern Urals and southern Siberia (along with northern Kazakhstan) were brought under cultivation. Also, large areas of permanent grassland were turned into arable area. As a result, the total **agricultural area** in Russia grew from 146.3 million hectares in 1950 to about 230 million hectares in 1980. In the 1980s some decline in total agricultural land was registered and the process was accelerated in the 1990s with cuts on subsidies and the drop of zonal pricing (Part IV) leading to some shifts away from marginal and fragile land. As a result, total agricultural land shrank to 221 million hectares in 1995, but with above 1.4 hectare of agricultural land per inhabitant, Russia has still vast agricultural land resources which favour extensive farming practices.

The average natural quality of soils in Russia is rather poor, compared to most western European countries, and part of the soils is further damaged by human activity. Soil is being lost as a result of severe **water and wind erosion**, amplified by farming and livestock grazing practices that ignore the need for soil conservation. The risk of water erosion is highest in the zone across the country, which is north of the line of latitude 50° N (Libert, 1995). Wind erosion is particularly widespread in southern, semi-desert areas where strong, dry winds are removing top layers of soil. Studies show that during last 15-25 years annual loss of humus in the soils amounted to 0.62 tonnes per hectare (Roskomzem, 1995). The plowing of steep slopes, overgrazing, fertility-exhausting cropping patterns, and removal of vegetation have been particularly damaging in the forest zone of the European part of Russia. Moreover, soil has been compacted by massive, heavy farming machines.

Seventeen regions of the Russian Federation face the problem of **desertification**. In the Kalmykia, Dagestan, Astrakhan, Volgograd and Rostov regions the process affects more than 50 per cent of their territories and in Altay almost 40 per cent. The Stavropol and Krasnodar regions are also concerned by this problem (Roskomzem, 1995).

Agriculture is one of the largest **water** users in Russia, including for irrigation, but its share in total water use in Russia fell from 24 per cent in 1980 to 19 per cent in 1996. Its share in total water pollution was estimated at 10 per cent in 1995 (Goskomstat, 1996). The massive expansion of **irrigation** systems in Russia in the 1970s and 1980s increased the production potential of agriculture, but has also caused serious environmental problems, such as concentration of pollutants, a disturbed salt/fresh water balance, lower water levels in the delta lands, higher groundwater levels turning parts of irrigated land into swamps (Libert, 1995). In 1990, total irrigated land in Russia was 6.2 million hectares, *i.e.* 2.8 per cent of total agricultural land. By 1995, this area had fallen to 5.3 million mainly due to a sharp fall in maintenance and a lack of cash for new equipment.

It is estimated that almost 8 per cent of total farm land in Russia suffers from **excessive wetness**, resulting both from natural processes and human activity (Roskomzem, 1995). In 1995, the total amount of **drained** land was 4.8 million hectares, compared to 5.1 million hectares in 1991.

One third of the farm land in Russia is excessively **acid** and requires large applications of lime. Also, about one third of soils are highly deficient in phosphorus and between 10 and 30 per cent of soils, depending on the region, are deficient in potassium. Nitrogen deficiency is widespread. In the Soviet period highly subsidised prices of fertilisers and pesticides, combined with inadequate spreading machinery and bad management, resulted in inefficient use of chemicals. The strong input price increase relative to the increase in output prices (cost-price squeeze, see above) resulted in a significant fall in chemical use during the reform period (Table I.16). These developments to some extent eased environmental pressures, but further deteriorated soil fertility and made yields even more dependent on weather conditions. Moreover, soil pollution by residual pesticide elements still strongly exceeds environmental norms in a number of the regions, particularly those producing rice and cotton and those with intensive orchard production.

In some areas, surface and ground waters have also been polluted by manure from highly **concentrated livestock units** (livestock complexes) created particularly in the 1970s and 1980s all around Russia (as in other parts of the USSR). In the 1990s livestock production halved (see above) and the concentration of animals per farm fell, thus easing its negative impact on the environment.

A 1994 survey showed that 15 per cent of the European part of the Russian territory (19 regions) was affected by the Chernobyl accident. There are also **radioactively** polluted agricultural lands in the south Ural area and Tomsk region (West Siberia).

In the Soviet period large mining and other industries severely damaged agricultural land. In more recent years the annual amount of **industrially damaged** land was lower than the annual amount of restored land. However, by the end of 1995, 1.1 million hectares of agricultural land remained damaged.

NOTES

1. Any analysis of the dramatic changes in the Russian agricultural sector must be situated in the context of the macroeconomic and structural changes that shaped them. Therefore, in this section, some major developments in reforming Russian economy are presented, mostly on the basis of the OECD economic surveys of the Russian Federation (OECD, 1995; OECD, 1997*b*).
2. For a detailed description of the reasons and consequences of the development of money surrogates in Russia, see OECD (1997*b*), pages 113-119 and 178-184.
3. The Gini coefficient ranges from 0 (each quintile of the population receives the same share of total income) to 1 (total income is concentrated within a population quintile).
4. The percentage shares are based on the OECD Secretariat data. The official Goskomstat data give different shares with livestock production accounting for 63 per cent of total agricultural production in 1990, 53 per cent in 1992 and 45 per cent in 1996. The difference results mostly from different prices used for calculations. In effect, the Goskomstat uses normative costs of production as a proxy for the value of production used on large scale farms as inputs – the normative costs being much higher than market prices. The production on households (both for family use and for sales) is valued at market prices received by household plots; these market prices are based on household surveys and are, in most cases, higher than prices received by large scale producers. So for instance, in 1995, for some products, such as beef and veal, the value of production estimated in this way was more than twice as high as the value estimated on the basis of market prices received by large scale producers.

AGRO-FOOD FOREIGN TRADE

A. TRADE FLOWS

1. Pre-reform trade flows

Since the 1960s Russia has been a net importer of agricultural and food products from all major groups of countries, including other republics of the former USSR, former members of the Council for Mutual Economic Assistance (CMEA) and OECD countries. In 1988 and 1989, agro-food imports accounted for 23 per cent of total Russian (Russian Soviet Federative Socialist Republic – RSFSR) imports and agro-food exports for about 4 per cent of total Russian exports. The range of imported agro-food products was rather narrow and was dominated by purchases of raw products (in particular grains) for feed and further processing.

Statistics on agricultural trade between the RSFSR and other republics inside the former USSR as well as between the RSFSR and other countries are incomplete, but it can be estimated that at the end of the 1980s former republics supplied about half of total food and agricultural products originating from outside the RSFSR and absorbed about 70 per cent of all food and agricultural products "exported" by the RSFSR.[1] The main products "imported" by the RSFSR from other republics were meat and meat products, grains and fruits and vegetables. Russia "exported" a few food products such as milk products, potatoes, sugar, flour, and fish. Moreover, the RSFSR was a major producer and "exporter" of agricultural inputs such as grain combines, tractors, fertilisers, feed additives and pesticides.

The CMEA countries were the second largest exporters of agro-food products to Russia. Meat and meat products came from Hungary, fresh and tinned fruits from Hungary and Bulgaria, raw sugar from Cuba and tobacco products from Bulgaria. The main agro-food imports from Poland included potatoes, onions, and apples.

From the 1970s until 1990, the USSR was one of the largest grain importers (including maize) in the world. These imports were partly used in other USSR republics, but at the same time Russia "imported" large amounts of grains from Ukraine and Kazakhstan. Between 1986 and 1989, USSR grain imports increased from 27 to 37 million tonnes but then declined to 32 million in 1990. Between 75 per cent and 90 per cent of grain imports came from developed countries, in particular the United States. Developed countries were also major suppliers of butter and poultry meat (France and the Netherlands), as well as citrus fruits (Spain and Greece). The USSR's exports to developed countries mainly consisted of a small range of products such as canned crab, caviar and alcoholic beverages. Exports of fish and marine products exceeded one million tonnes in 1990 of which approximately 10 per cent was exported to Japan.

2. Post-reform trade flows

Economic reforms in Russia coupled with the disintegration of the USSR and CMEA had a significant effect on Russian agro-food trade, both its direction and composition. Russian agro-food imports from countries other than NIS ("far abroad" in Russian trade statistics) fell from US$16.6 billion to US$6 billion between 1990 and 1993. However, total trade was equally adversely affected by the changing economic environment, and the share of agro-food imports in total imports remained high at above 20 per cent (Table II.1). In 1992, imports of some food products such as grain, raw sugar and macaroni were subsidised making these products affordable for domestic buyers and, thus, competitive

Table II.1. **Russian Federation: agricultural trade, 1990-1996**[1]

	1990	1991	1992	1993	1994	1995	1996
Exports ($billion)							
NIS	n.a.	n.a.	n.a.	n.a.	0.5	0.4	0.5
All other countries	1.5	1.3	1.6	1.6	2.3	2.3	2.7
Total	n.a.	n.a.	n.a.	n.a.	2.8	2.7	3.2
Imports ($billion)							
NIS	n.a.	n.a.	n.a.	n.a.	2.1	3.5	3.6
All other countries	16.6	12.4	9.6	5.9	8.6	9.7	7.8
Total	n.a.	n.a.	n.a.	n.a.	10.7	13.2	11.4
Balance ($billion)							
NIS	n.a.	n.a.	n.a.	n.a.	–1.6	–3.1	3.1
All other countries	–15.5	–11.1	–8	–4.3	–6.3	–7.4	–5.1
Total	n.a.	n.a.	n.a.	n.a.	–7.9	–10.5	–8.2
The share of agro-food trade in total trade:							
exports (per cent)	2.1[2]	2.6[2]	3.9[2]	3.8[2]	4.2	3.4	3.8
imports (per cent)	20.3[2]	27.9[2]	26.0[2]	22.2[2]	27.7	28.3	25.1

1. It should be noted that this table is based on data provided by Goskomstat which differ from data supplied by the Russian Federation Customs Committee, particularly on agro-food exports. To compare, Customs Committee shows agro-food exports for the first 24 two-digit categories in the Harmonised System at $1.3 billion in 1995 and $1.7 billion in 1996. The corresponding figures for imports of $13 billion and 11.2 billion are close to the Goskomstat figures above. Customs Office Trade Statistics of the Russian Federation, Russian Federation Customs Committee, Moscow, 1997.
2. Between 1990 and 1993, the shares relate to trade with the non-NIS ("far abroad") only.

Source: Goskomstat.

Table II.2. **The share of net agro-food imports in total value of agricultural production in Russia, 1994-1996**

	1994	1995	1996
Gross Agricultural Output (GAO), current prices; trillion Roubles	73.7	260	282
Net agro-food imports, US$ billion	7.9	10.5	8.2
Exchange rate (annual average) Rb/US$	2204	4554	5124
Net agro-food imports, trillion Roubles	17.4	47.8	42
Share of net agro-food imports in GAO; per cent	23.6	18.4	14.9

Source: OECD Secretariat's calculations based on Goskomstat data.

Table II.3. **The share of net imports of main agricultural products in total domestic use in Russia,**[1] **1986-1997**

Per cent

	1986	1987	1988	1989	1990	1991	1992	1993	1994	1995	1996	1997p
Grains	8.7	11.3	15.0	16.2	11.4	18.6	22.0	10.7	3.2	2.5	4.1	1.4
Meat and meat products	13.6	13.3	13.3	13.0	12.7	13.2	13.8	15.5	19.1	27.8	28.0	33.2
Milk	12.0	12.8	14.8	14.5	12.2	11.4	5.9	10.9	9.9	13.1	10.1	13.2
Potatoes	2.1	2.3	2.3	2.5	2.3	2.5	0.4	0.2	0.2	0.0	0.1	n.a.
Sugar	40.4	44.8	41.1	38.1	45.5	32.0	25.8	33.3	29.7	36.6	27.0	n.a.
Eggs	5.5	5.1	3.6	2.9	2.8	1.1	–0.6	–0.7	–0.2	0.2	0.7	n.a.
Vegetables	n.a.	n.a.	n.a.	n.a.	18.6	22.3	21.3	12.6	14.4	9.2	n.a.	n.a.

p: provisional.
n.a.: not available.
1. Imports minus exports divided by total domestic use including personal consumption, processing, losses and changes in stocks (ending stocks minus beginning stocks).

Source: OECD Secretariat's calculations based on the Goskomstat data.

on domestic markets. Partly due to the subsidies, grain imports increased from 20.8 million tonnes in 1991 to 30.1 million tonnes in 1992. Import subsidies were discontinued in 1993, but to increase the availability of food products on domestic markets and protect consumers, between 1992 and mid-1996 the government applied export duties on selected agricultural products. In 1992 and 1993 (until September) export duties on grains and meat products were at a prohibitive level of 70 per cent, rendering exports of these products unprofitable. In October 1993 and September 1995 the duties were lowered or phased out for some products, although for some others such as durum wheat, flour and oilseeds export duties of 10 per cent were introduced. By mid-1996 all export taxes were abolished (Part IV).

In 1994 and 1995 Russian agricultural imports increased again due to several factors: re-established trade ties with the NIS ("near abroad" in Russian trade statistics) and "far abroad" countries; strong appreciation of the Russian rouble; reduced domestic food output and competitiveness of domestic producers; aggressive marketing policies by foreign suppliers quite often supported by Western export subsidies; and a lack of market infrastructure to link agricultural surplus with deficit regions. As a result, agro-food imports increased to US$13.2 billion in 1995 (28 per cent of all imports), including imports of US$3.5 billion from the NIS and US$9.7 billion from other countries. The reaction of the government to the increased food imports was to introduce border protection from March 1994, representing a change in government policy from consumer to producer protection (Parts IV and V).

Partly as a result of the increased border protection, but also due to a better adaptation of the Russian agro-food industry to the new economic environment, in 1996 agro-food imports declined to US$11.4 billion, *i.e.* to 25 per cent of total imports. Since exports were valued at US$3.2 billion (4 per cent of total exports), the resulting agro-food trade deficit amounted to US$8.2 billion, down from US$10.5 billion in 1995 (Table II.1). According to preliminary data, in 1997 imports of agro-food products were slightly above the 1996 level and the share in total imports was at about 26 per cent.

Despite the sharp decline in agricultural production in Russia, significant rises in imports of selected food products and the striking visibility of imported food in shops, some indicators show that Russia is becoming less dependent on food imports in most recent years. The ratio of net agro-food imports valued at the annual average exchange rate to total value of agricultural production valued at current prices declined from 23.6 per cent in 1994 to 14.9 per cent in 1996 (Table II.2; see also Goskomstat, 1997*b*). While the rate of the decline is somewhat overestimated due to the appreciation of the Rouble, the decline as such seems to be confirmed by changes in the shares of net imports of essential agricultural products in total domestic use of these products, calculated on the basis of agricultural product balances provided by Russian Goskomstat. Between 1990 and 1996, the shares declined to less than 5 per cent for grains (after an initial increase between 1990 and 1992), and nearly to zero for potatoes and eggs; more than halved to about 9 per cent for vegetables (in 1995); fluctuated between about 6 and 13 per cent for milk (milk products are expressed in milk equivalents); and remained high, at about 30 per cent for sugar. The only category for which the share increased significantly is meat and meat products with the share at 33 per cent in 1997 compared to 13 per cent in 1990 (Table II.3).

Some other indicators applied by the Goskomstat show Russia's dependence on food imports as much higher. For example, in the last quarter of 1997, the share of imports in total retail turnover of selected food products varied between 22 per cent for red meat to 57 per cent for poultry meat. The shares were also high for such products as vegetable oil (56 per cent), cheeses (50 per cent), and pasta (41 per cent), but they declined slightly compared to the first quarter of 1996 (Goskomstat, 1997 and 1998). However, this indicator seems to disregard flows of agricultural and food products from domestic producers to consumers through local markets and other unregistered channels, which overestimates the measured share of imports in total retail turnover.

i) Regional breakdown of agro-food trade

Russia remains a net importer of agro-food products *vis-à-vis* all its trading partners. However, the geographical structure of Russian imports has changed significantly in the post-reform period: while

◆ Graph II.1. ***Russia's agro-food imports by region, 1996***

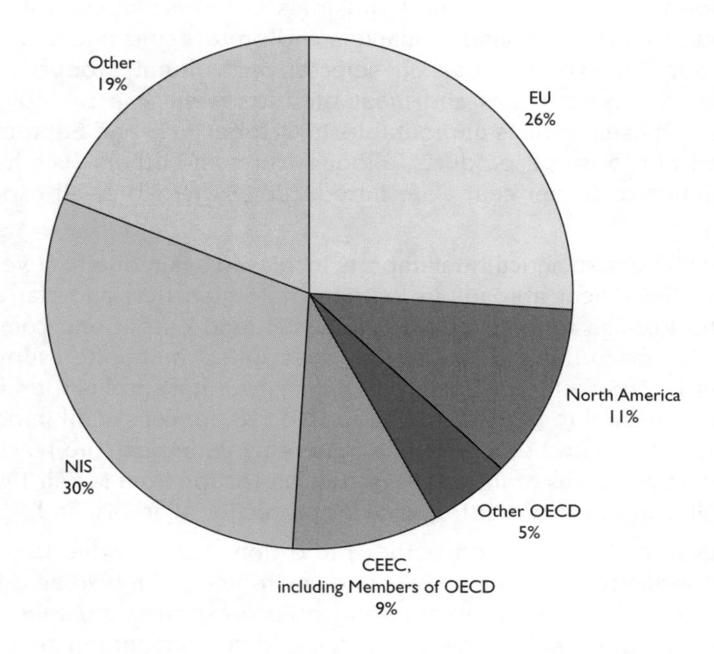

Source: OECD.

◆ Graph II.2. ***Main suppliers of agro-food products to Russia, 1996***

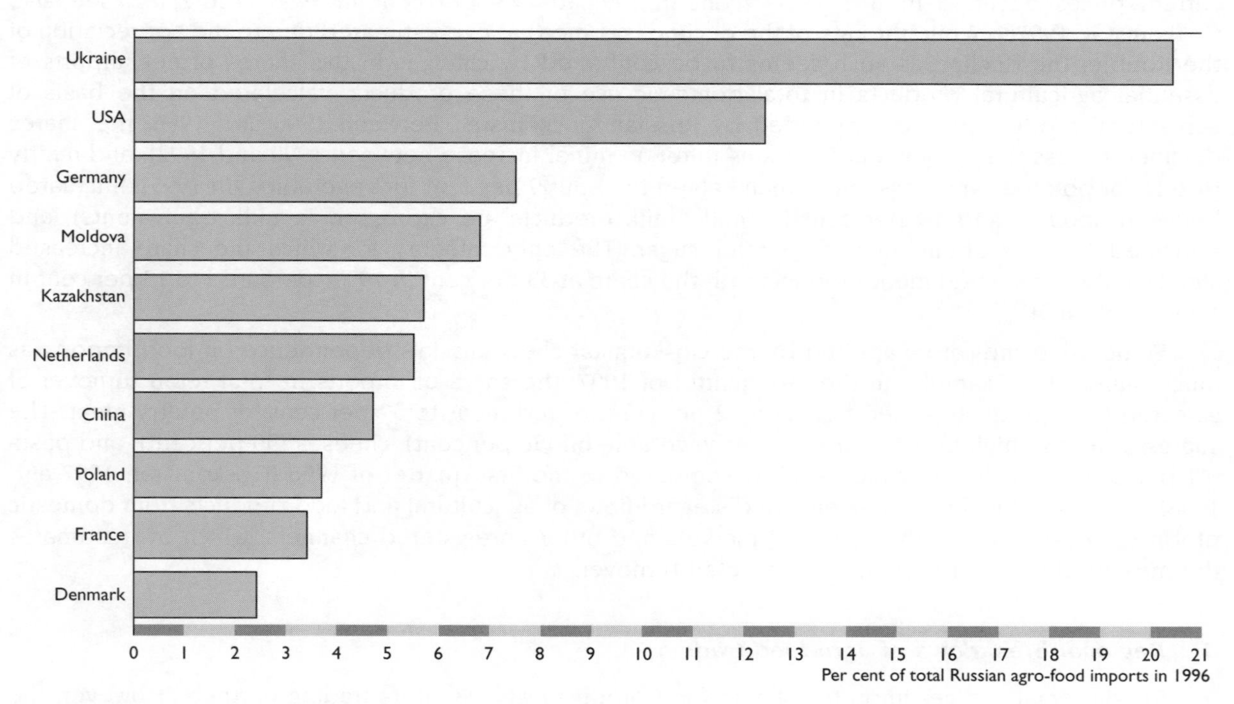

Source: Goskomstat.

agro-food imports from OECD countries, especially the EU, increased, imports from traditional suppliers, such as NIS and CEECs, declined.[2] Since 1994 the share of agro-food imports from the NIS has been increasing and in 1996 exceeded the share from the EU (Graph II.1).

With respect to the OECD area, between 1994 and 1996 the EU was the source of 25-30 per cent of Russia's total agro-food imports. Major EU suppliers were Germany, the Netherlands, France and Denmark (Graph II.2). In 1996, the main products imported from the EU were frozen beef, frozen fish, chocolate, apples, poultry and cheeses. The EU is also an important market for selected Russian agro-food exports such as sunflower seeds and alcoholic beverages. In 1996, the EU accounted for about a fourth of total Russian agro-food exports.

In 1996, the US was the second largest agro-food supplier to Russia accounting for about 12 per cent of total Russian agro-food imports (Graph II.2). The commodity structure of Russian agro-food imports from the US changed dramatically from feed grains in the pre-reform period to meat and meat products in recent years. In 1995 and 1996 poultry meat accounted for about a third of total Russian agro-food imports from the US.

At the beginning of the reform period, the NIS share in total Russian agro-food trade fell sharply from about 50 per cent in 1991 to less than 20 per cent in 1993. The decline resulted partly from payment problems related to the Rouble zone still existing during this period. Economic agents were unwilling or unable to use the banking system to pay for imported goods and services.[3] The situation started to improve with the July 1993 currency reform in Russia, followed by the introduction of national currencies by NIS countries in the second half of 1993 and lower inflation rates in 1995 and 1996. Moreover, commodity prices in the NIS countries are generally lower than those in Russia and free trade arrangements between Russia and the NIS countries eased access of NIS products to the Russian market. Hence, the NIS share in total Russian agro-food imports increased to 30 per cent in 1996, mostly due to increased supplies of sugar, frozen beef, wheat and wheat flour.[4] Between 50 and 70 per cent of agro-food imports from the NIS come from Ukraine which is the single most important supplier of agro-food products to Russia accounting for about 20 per cent of total Russian agro-food imports in 1996 (Graph II.2). Major products imported from Ukraine include sugar (US$578 million in 1996), frozen beef and wheat flour. The NIS is also the major export market for Russian agro-food products, accounting for 30 per cent of total Russian agro-food exports in 1996 (Graph II.3). The main products exported to these countries are refined sugar and alcoholic beverages.

The share of agro-food products imported from the CEECs was about 9 per cent in 1996 (including imports from new OECD members, *i.e.* Poland, Hungary and the Czech Republic), still below pre-reform levels, mainly because of the decline in agricultural production in the CEECs, but also due to the reorientation of their exports to EU markets. However, agro-food imports from Poland and Hungary increased sharply in recent years with Poland becoming the eighth largest exporter of agro-food products to Russia in 1996 (Graph II.2).

Russian post-transition imports from China have grown, primarily those of meat and rice imports delivered to the Siberian and Far Eastern regional markets. Russian agricultural and food imports from China totalled US$0.6 billion in 1994, but dropped to US$0.4 billion in 1996. Trade with Cuba shrank with raw sugar imports declining from 3.3 million tonnes in 1990 to 1.2 million in 1996.

ii) Commodity structure of agro-food trade

Imports

Since 1990 there has been a significant change in the structure of Russia's agro-food imports: imports of raw agricultural products have fallen sharply while those of processed foods have increased. More specifically, massive feed grain imports in the Soviet period have been replaced by significant imports of meat and meat products.

In the post-reform period Russian imports of grain declined from 30.1 million tonnes in 1992 to about 4 million tonnes annually between 1994 and 1996. The structure of imported grains changed as well: while in the Soviet period the major part of imported grains was feed grains, in 1995 and

◆ Graph II.3. ***Russia's agro-food exports by region, 1996***

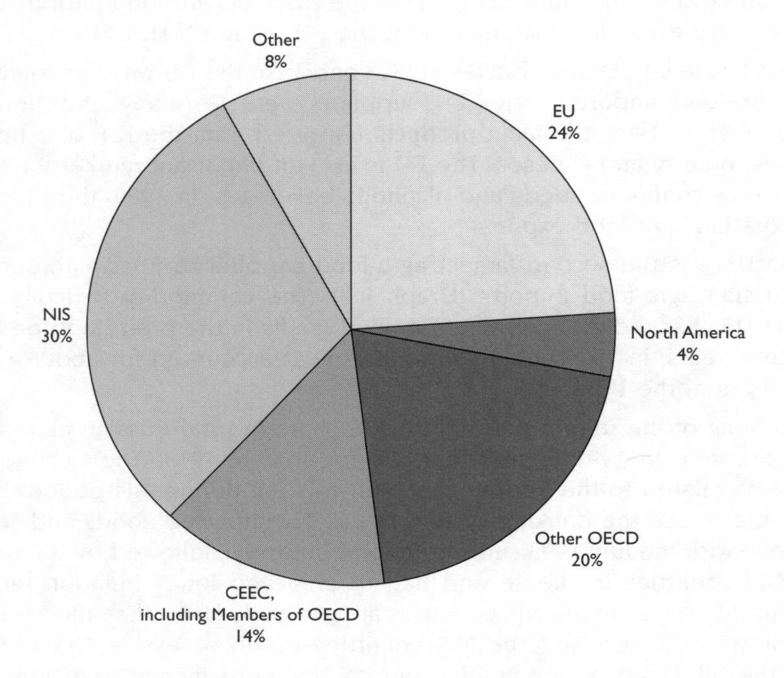

Source: OECD.

◆ Graph II.4. ***Russia's agro-food imports by products, 1996***

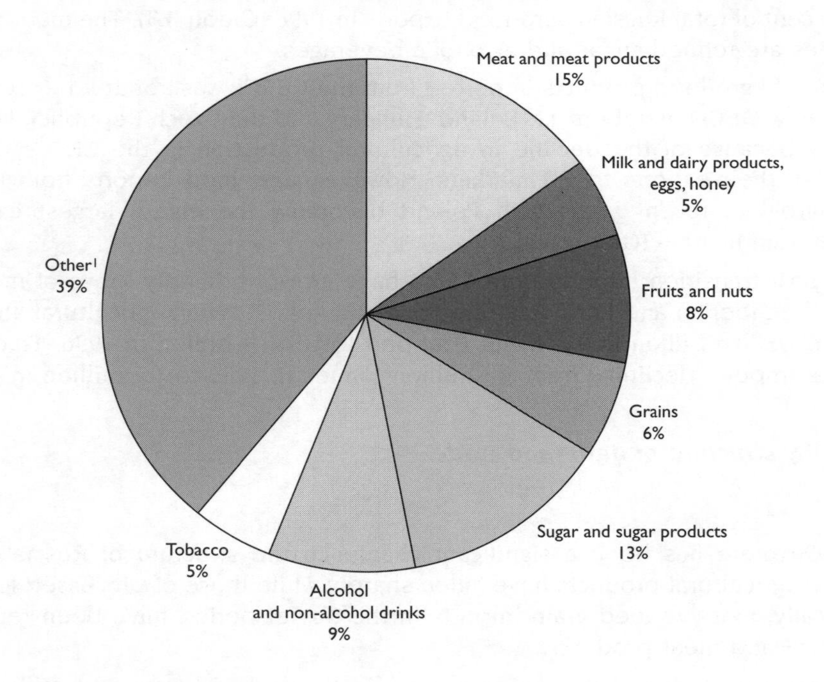

1. Groups of products with a share below 5 per cent each of the total.
Source: OECD.

1996 grains for human consumption accounted for more than 70 per cent of the total. The decline in feed imports is mostly due to the reduction in animal numbers in Russia by about half between 1990 and 1996. More than 50 per cent of Russian grain imports came from Kazakhstan in 1996 and the share of grains in total Russian agro-food imports was about 6 per cent (Graph II.4).

While grain imports dropped, imports of meat and meat products rose from 0.5 million tonnes in 1992 to about 2 million tonnes in 1995, but declined to 1.7 million tonnes in 1996. Total meat imports were valued at $1.7 billion in 1996 and accounted for 15 per cent of total Russian agro-food imports (Graph II.4). Major suppliers were the US, the EU and the NIS with shares at 26 per cent, 26 per cent and 21 per cent, respectively, in 1996. However, while imports of beef came mostly from the EU (from Ireland and Germany) and NIS (from Ukraine), about three -quarters of poultry meat originated from the US.[5] Meat and meat products were not traded with China during the pre-reform period, but in 1996 China's share in total Russian meat imports amounted to about 13 per cent, including more than 40 per cent share in pigmeat imports. Polish exports of meat and meat products (particularly sausages) to Russia also strongly increased after the liberalisation.

The volume of Russian sugar imports has remained relatively constant in the post-reform period at around 3 million tonnes. Traditionally, Russia imported raw sugar from Cuba and white sugar from other Soviet republics. New trading arrangements with Cuba in the 1990s, based on more commercial terms than in the past, resulted in a 50 per cent decline in imports of raw sugar from this country. Moreover, the collapse of traditional production ties between the FSU republics at the beginning of the 1990s led to a five-fold reduction in Russian imports of white sugar from the NIS and a corresponding jump in Russian imports from non-NIS countries. By 1995 the NIS market had recovered and accounted for nearly 80 per cent of Russian white sugar imports (from 13 per cent in 1992). In 1996, the share increased to 92 per cent, including Ukraine's share at 85 per cent. Purchases of sugar abroad are dominated by the leading commercial structures such as Menatep-Impex, Sucden, Alpha-Eco, Russian Sugar and Soyuzcontract.

Up to 1996, sugar imports from Ukraine were stimulated by low prices and duty free access to the Russian market, but a 25 import duty imposed by Russia in May 1997 adversely affected the competitiveness of Ukrainian sugar on the Russian market. In November 1997, an agreement was reached between the Russian and Ukrainian governments on the annual customs quota for tariff-free import of 600 000 tonnes of Ukrainian sugar into Russia. Sugar bought above and beyond the 600 000 tonne limit would be subject to 25 per cent duty. According to the Russian government resolution, the customs quota is to be divided into lots and distributed between the state and private enterprises on a tender basis.

Russia has become increasingly self-sufficient in potatoes over the last few years, and at present only early potatoes are imported in small quantities, for consumption in the larger cities. Fresh and processed fruits and vegetables are imported from the NIS and the CEECs though the share of the latter, traditional suppliers of these products to Russia, has fallen considerably in the post-reform period.

Exports

In 1996, the value of Russian agro-food exports increased to US$3.2 billion (3.8 per cent of total exports) from US$2.7 billion (3.4 per cent) in 1995, though the value of agro-food exports remains significantly lower than of agro-food imports.[6]

The most important Russian agro-food exports are sunflower seeds and oil, alcoholic beverages, and fish and marine products, which together accounted for over half of all agro-food exports in 1996 (with shares of 26, 17 and 12 per cent, respectively). The share of traditional Russian exports of fish and marine products has been constantly declining since 1992. Exports of sunflower seeds increased strongly in the 1990s to 1.8 million tonnes in 1996. More than 90 per cent was exported to OECD countries, mainly to the EU and Turkey.

Russian exports to the NIS include meat and meat products, processed food products (sugar, macaroni, canned foods), potatoes and grain. Moreover, Russia re-exports to other NIS such products as

◆ Graph II.5. ***Russia's agro-food exports by product, 1996***

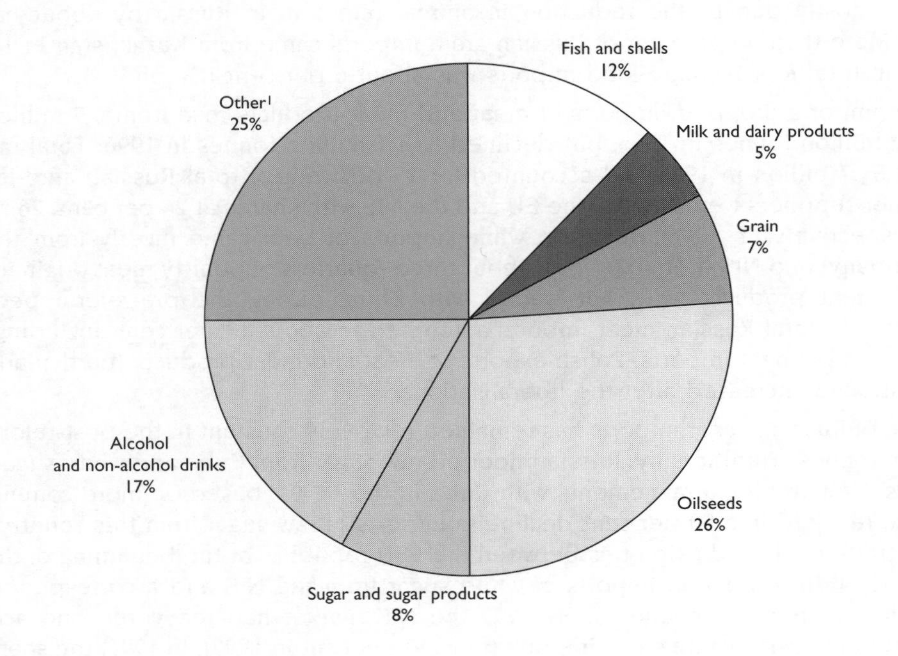

1. Groups of products with a share below 5 per cent each of the total.
Source: OECD.

tea, coffee, cocoa, and bananas. Russian exports to developed countries, including the EU, consist principally of such items as sunflower seeds, alcoholic beverages, sturgeon caviar, canned crab, fish, barley and buckwheat.

A sharp rise in grain output in Russia in 1997 (Part I) led to expectations of sizeable grain exports. Russia's export potential has mainly been for feed grains. However, there are some concerns about the quality of grains and an oversupply situation of feed grains on the world markets. Moreover, high transaction costs, barter deals, restrictions imposed by local authorities, weak law enforcement, high transportation costs and a lack of adequate market information can be considered as significant impediments to linking domestic producers with international purchasers and to realising export potential. Therefore, it is expected that in the 1997/1998 season with net imports of wheat of about 1.9 million tonnes and net exports of feed grains of about 1.4 million tonnes, Russia's net imports of all grains would be at about 0.5 million tonnes. The remaining grain surpluses will probably be used to rebuild domestic stocks, which had been depleted following poor harvests in 1995 and 1996. High domestic grain stocks and still declining animal production contributed to the decline in grain prices in 1997/1998 that may have negative impact on the 1998 grain harvests.

B. TRADE RELATIONS

1. Former trading arrangements

In the Soviet period the volume and value of trade were determined by the state and all trade was conducted through centralised, state-owned foreign trade organisations. Production and trade plans were defined nationally, but co-ordinated at the CMEA level within five-year frameworks. Production and trade were based on the 'division of labour' principle, though the allocation of product groups to specific countries was rather arbitrary, being based more on political considerations than on economic fundamentals, *i.e.* on international comparative advantage.

The Ministry for Foreign Trade was the major institution supervising the activities of 60 foreign trade organisations (*vnesnetorgovyie ob'edinenia*; VTO or VVO). The state fixed the product mix, value and volume of deals for each VTO which played the role of intermediary between domestic and foreign producers and purchasers. The most important VTOs in the food area were Exportkhleb (Bread Export), Soyuzplodimport (All-Union Fruit Import) and Skotimport (Animal Import).[7]

2. New trading arrangements[8]

The collapse of the CMEA, the disintegration of the Soviet Union and the subsequent declaration of independence by the former Republics, served to break the close economic and trading ties that had linked Russia with the other republics of the Soviet Union, Central and Eastern European countries and the other CMEA members for several decades. No longer bound by the old, often politically determined commitments, Russia needed not only to negotiate new trading arrangements with each of her traditional trading partners, but also actively to seek to reintegrate herself into the world economy and world trading arrangements.

i) Russia and the European Union

In June 1994, the Partnership and Co-operation Agreement (PCA) was signed between Russia and the EU. The PCA did not come into force until December 1997, but on 1 February 1996, an Interim Trade Agreement, the trade part of the PCA, came into force and placed bilateral relations in a new GATT/WTO-type environment. The EU and Russia accorded each other most favoured nation (MFN) status and goods traded between the two sides are to be free of quantitative restrictions. However, there are some exceptions in the steel, textile and nuclear sectors.

While EU import tariffs on Russian exports averaged about two per cent in 1997, the Russian weighted average tariff on imports from the EU was about 19 per cent. Moreover, while the EU's tariffs are bound in the WTO, Russia, as a non-WTO member, can raise its tariffs under certain conditions. Russia's current commitments, in particular under the PCA, oblige Russia to consult with the EU on changes in customs duties and/or the introduction of quantitative restrictions on certain imports from the EU, prior to their implementation. Under the agreement, Russia reserves the right to maintain existing relations with the NIS countries (more liberal than its relations with EU members) and to develop these within the framework of a customs union, without offering the same privileges to the EU.

Despite the PCA, the EU continued to insist that Russia maintain her "non-market economy status", which remained a contentious issue and might in some cases be harmful to Russian interests in trade disputes. The main consequences of Russia's non-market status may arise in antidumping investigations against Russian producers. For example, such cases concerned Russian fertiliser producers. Russian prices are then considered as non-market based, which means that to assess the dumping margin, another country's prices are considered (known as the "analogue country system"), without taking into account Russia's possible comparative advantage. In January 1998, during the first meeting of the EU/Russia Co-operation Council, a body created under the PCA, the European Commission submitted a proposal to remove Russia – as well as China – from the EU's list of non-market economies to reflect ongoing market reforms in both countries. The proposal was approved by EU trade ministers at the end of April 1998. However, under the new policy the two countries would not be designated as market economies, but the new rules will provide the means to take account of cases where market conditions exist. The EU suggested that it was important for Russia to join the WTO to resolve various bilateral trade issues and hold out the long-term possibility of a free trade zone between Russia and the EU.

ii) Agreements with the NIS

The disruption of previous inter-republican arrangements had a strong impact on the economic situation in a number of NIS countries and the normalisation of trade relations in the food sector remains a pressing problem. Under the Soviet system, Russia was not only the major USSR producer and supplier of agricultural inputs, but also provided the largest market for agricultural products produced elsewhere in the Soviet Union. Thus, with the dissolution of the USSR, each republic lost its own strictly regulated but stable market.

Attempts to re-establish and expand trade ties based on intergovernmental agreements rather than market exchanges have repeatedly been made among the successor states of the former Soviet Union, both within and outside the framework provided by the loose Commonwealth of Independent States (CIS), set up as a co-ordinating body of the NIS in December 1991 with its Executive Secretariat in Minsk, the Belarusian capital.[9] Nonetheless, since then intra-CIS trade has moved increasingly towards multilateral, market-based agreements, but until the establishment of a payments union can be agreed, the main mechanism of payment between CIS countries (which have a combined debt of about $9 billion to Russia), consists of barter deals which necessarily limit the promotion of free and open trade.

Russia continues to develop both bilateral and multilateral initiatives to strengthen its economic and political links with the rest of the CIS, but on the whole, these efforts have not been very successful. Russia has signed Free Trade Agreements (FTA) with all of the CIS countries and a number of agreements on regional economic co-operation, including one signed in 1993, by all the CIS, agreeing to form an Economic Union. Potentially the most important agreement, however, was that signed in January 1995 between Russia, Belarus and Kazakhstan to create a customs union (subsequently joined by Kyrgyzstan in 1996). While countries within the CIS have no tariffs on each other's goods the members of the customs union are expected to apply a common tariff against goods coming from outside and a joint procedure for regulating foreign economic activity (including the collection of foreign trade statistics). However, internal customs borders have only been abolished between Russia and Belarus and it has not proved possible to harmonise the legal basis of trade with third countries.

The relationship between the customs union and other agreements signed is not clear. For example, in 1994 Kazakhstan, Uzbekistan and Kyrgyzstan formed the Central Asian Union and in autumn 1997 Georgia, Ukraine, Azerbaijan and Moldova created so called GUAM. In both cases Russia was not invited to participate. In spring 1996, Russia and Belarus signed a Co-operation Treaty establishing a "common customs territory" and a matter of days later, Russia, Belarus, Kazakhstan and Kyrgyzstan signed a "common customs space" agreement. Moreover, in April 1997, the Presidents of Russia and Belarus signed a document which is to form the basis for a union between the two countries. Attempts to stimulate co-operation between all CIS countries undertaken during the last two summits in October 1997 in Chisinau, Moldovian capital, and in January 1998 in Minsk were not successful and the creation of new blocs inside the CIS seems to forcing it apart. On the whole, trade relations between CIS countries remain relatively non-transparent and contain some conflicting and discriminatory features, potentially prejudicial to compliance with WTO regulations.

iii) Formation of the common CIS agrarian market

In 1993 the creation of a CIS common agricultural market (CAM) was intended as a means to liberalise intra-CIS agrobusiness trade. A draft document called "The Agreement on a Common Agricultural Market" passed through several rounds of discussions between the CIS, but it was not signed by the heads of governments until the CIS summit in Moldova in October 1997. However, Azerbaijan and Uzbekistan decided not to be associated with the agreement. It is foreseen that the creation of the market should ensure the free movement of farm produce, foodstuffs, scientific and technical goods, technology, means of production and services for the agricultural sector on the basis of jointly developed rules and principles.

The agreement envisages a two-stage transition to the CAM. The first stage foresees the creation of a free trade zone followed by the second stage aiming at the formation of a common customs territory by those countries which would be ready for further integration. In the framework of a common customs territory all tariffs, licences and other obstacles to the movement of goods would be removed and the participating countries would operate a "co-ordinated price regulation system on the basis of market principles". The market intervention system foreseen in the document, based on minimum guaranteed prices and on intervention purchases and sales of agricultural products, resembles the EU's Common Agricultural Policy before its reforms of 1992 (see Butsykin, 1997). The already existing Commonwealth Council for Agro-Industrial Issues would coordinate activities to create and run the CAM.[10]

Taking into account the political and economic difficulties to co-ordinate overall trade policies between the CIS members, as presented in the previous section, it seems that the creation of the CAM has rather little chances to succeed. To prevent commodities from flowing to the country with the highest prices, there are growing pressures to introduce an internal system of tariffs that is in conflict with the CAM's principal aim of establishing an internal free trade area. Moreover, the limited nature of available funds calls into question the effectiveness of the planned measures. In addition, commitments made during initial WTO membership negotiations to reduce tariffs mean that there is little room to introduce a "co-ordinated price regulation system" with prices that would be substantially above world market prices.

iv) Agreements with the CEECs

Despite both sides' proclaiming efforts to create favourable conditions for restoring trade and economic relations, there has been little progress in developing trade agreements between Russia and the central and eastern European countries (CEECs). A number of the more advanced CEECs have created their own FTAs and trading agreements and rely rather on developing trade relations with the EU and on intra-regional trade, than on trading with Russia. CEFTA has been criticised by Russia as being excessively protectionist and creating a buffer zone between Russia and its major trading partners in Western Europe. In the context of a future EU enlargement, Russia is evoking the possibility of demanding compensation for trade losses incurred as a result of CEEC adhesion.

v) Agreements with the US

The two biggest concerns for Russia as regards current US legislation are the fact that Russia has yet to be granted unconditional most favoured nation (MFN) status and must seek annual renewal, and secondly that Russia continues to be classified as a non-market economy which, in anti-dumping and other trade disputes, may be prejudicial to Russia's interests (though the US has declared its willingness to grant Russia 'quasi-market' status, as and when independence from the former centralised planning system can be proven). The US and Russia have been able to develop a relatively good mechanism to resolve some trade controversies, such as on US chicken exports (see above), in the framework of the US-Russian Joint Commission on Economic and Technical Co-operation. In general it seems that attention to future trade relations between the two countries awaits the forthcoming negotiations on Russia's accession to the WTO, which both sides perceive as an opportunity to solve their respective problems.

vi) Agreements with other countries

In the framework of the General System of Preferences (GSP), preferential treatment was granted in 1994 by Russia to such countries as China, Cuba, Vietnam, Croatia, Mongolia, North Korea, Romania, and Slovenia, allowing a more favourable import duty to be applied to trade with these countries (75 per cent of the rate set for most favoured nations). In 1996, Russia concluded a three-year intergovernmental oil-for-sugar trade agreement with Cuba, under which it is to supply 10.5 million tonnes of crude oil in exchange for 4 million tonnes of sugar up to 1998.

vii) Russia and the WTO

Russia was not a member of the General Agreement on Tariffs and Trade (GATT) and did not participate in the Uruguay Round of negotiations. The decision to join the World Trade Organisation (WTO)[11] was taken by the Russian government in 1993 in order to widen Russia's access to international markets, and Russia formally applied for WTO membership in December 1994.

During the 1996 and 1997 meetings of a WTO working party, Russia's trade regime, economic policies, and laws were reviewed to determine their compliance with WTO rules and to develop the terms of accession. In its WTO negotiations Russia argues that the country's still unstable economic situation makes it difficult to accept definite commitments, including in the tariff area, and that the authorities need some room to manoeuvre in the future.

Selected problems related to Russian agricultural policy were discussed such as market access, internal support, and export subsidies, and more specific issues such as sanitary and phytosanitary (SPS) measures as well as technical barriers to trade (TBT). The Russian delegation declared that its intention was to secure a level of agricultural support and of tariff protection of Russian agricultural markets "comparable with the levels in other WTO-members". Moreover, it advocated that Russian agricultural producers should face "the same competitive conditions as producers of the same products in other WTO-members". In most cases, the EU's WTO commitments are taken as the reference point.

In 1996 the Russian delegation submitted to the WTO secretariat preliminary estimates of budgetary support to agriculture in the 1989-1991 and 1993-1995 periods. However, the Russian delegation insisted that the latter, "a crisis period of Russian agriculture", was not representative and maintained that the former represented "average, normal conditions" of budgetary support and should be taken into account as a base for further reductions of the support. Calculations by the Ministry of Foreign Economic Relations and the Ministry of Agriculture and Food put the level of state support for Russian agriculture in the 1989-1991 period at between US$80 billion and US$90 billion (EEM Moscow Bulletin No 12/7; 1997).

The market access issue was included in the 1997 memorandum agreed by Russia with the International Monetary Fund (IMF). Russia committed itself not to increase import tariffs above 30 per cent (with the exception of luxury foods, tobacco and alcohol) and to reduce its weighted average import duty by 20 per cent by 1998 and 30 per cent by 2000. Moreover, existing duties above 30 per cent are to be replaced by internal excise taxes. However, in the memorandum it was indicated that Russia's formal WTO obligations may differ from the commitments agreed with the IMF. Moreover, Russia declared that it was not prepared to comply with a tariff standstill for the period of WTO accession negotiations and wanted to bind its tariffs at higher levels than the tariffs currently applied. Moreover, as in its agreement with the EU, Russia wanted to preserve, "without any limitations", its existing trade arrangements with the NIS countries. In accordance with WTO rules, Russia will seek the acceptance of these arrangements by including them into derogations from the GATT-94 principles.

Despite significant progress made in liberalising its trade regime and market-oriented reforms Russia is still facing several outstanding issues related to the WTO membership negotiations:

- While Russia's **import regime** is relatively liberal with tariffs on agricultural products ranging from 10 to 30 per cent (albeit associated with minimum specific duties) (Part IV), its implementation creates trade barriers due to multiplicity of regulations and requirements, bureaucratic, time consuming and expensive certification, and complicated and subject to corruption valuation of imported products. Moreover, frequent changes to specific requirements and regulations, often introduced on an *ad hoc* basis, make trade policy not transparent for both domestic and foreign traders.
- Another area of concern are **trade relations with the NIS countries**; although based on free trade agreements, they often remain untransparent and involve fixed pricing and/or subsidies. Although diminishing, agricultural trade is still partly conducted through interstate agreements that specify trade volumes. Some NIS countries authorise a single company or agent to fulfil an inter-state agreement. While the use of a sole agent to trade on a non-commercial basis creates a

question of state trading, the inter-state barter trade agreement may be interpreted as trade discrimination (ERS/USDA, 1997).

- Specific problems arise from the **federal structure** of the country and the tendency for the regionalisation of some policy measures such as budgetary support, taxation and regional controls on agricultural flows, which may be in conflict with federal policies. Therefore, WTO members are concerned about whether regional policies will comply with trade concessions negotiated with the Federal Government.
- Since for acceding countries the base period for **domestic support**, as measured by the annual Aggregate Measure of Support (AMS), is "normally... the average of the most recent three year period" (WT/ACC/4) it is not expected that WTO-members would formally accept the 1989-1991 period as proposed by Russia. However, since the calculated level of support for 1989-1991 would constitute about 18-20 per cent of total Russian GDP in 1996, Russia would not be able to support agriculture at even one fourth of this level under current economic conditions.

WTO membership will assist the overall reform process in Russia by providing an institutional foundation ensuring the continuation and further consolidation of reforms, in particular the consistency, transparency and predictability of trade regulations and the adherence to multilateral rules and disciplines. These are crucial elements for domestic and foreign agents, including foreign investors, to establish their businesses in Russia on a firm basis. Moreover, membership will give Russia most-favoured-nation (MFN) trade status *vis-à-vis* all other WTO members, and equal treatment with other member countries in trade disputes in the framework of the WTO dispute mechanism. For example, access to the WTO dispute resolution process would be particularly useful in the case of charges of dumping made by various countries against Russia, which often result in import restrictions. Russia will also gain from the possibility of influencing future trade negotiations under the auspices of WTO and from better access to information on international trade systems and to the experience of other countries in dealing with trade issues.

NOTES

1. Including trade flows with other republics of the FSU. See *Narodnoie Khoziaistvo RSFSR*, 1989 and 1990 editions, Goskomstat, Moscow.
2. Some methodological problems related to internal trade statistics in the USSR are explained in World Bank (1993) and Tabata (1994).
3. In this and following sub-sections, trade data for the period 1994-1996 are based on the State Customs Committee information collected directly from customs declarations. As explained in note no.1, Table II.1. they differ from data published by the Goskomstat. The definition of food and agricultural products is based on the two digit Harmonised System categories applied in international trade statistics and includes the first 24 categories in this system. It should be noted that the data do not include so called "shuttle trade", *i.e.* purchases and sales made by professional and other traders travelling as tourists for the purpose of buying goods for resale on the home market.
4. From the end of the Soviet Union in December 1991 until mid-1993, Russia's trade relations with other NIS countries (except Georgia) operated through the Rouble zone. This was intended as a means of maintaining trade with the rest of the FSU. The Russian rouble was the common currency used for cash transactions, but the central bank in each NIS was able to produce its own non-cash roubles, the means of exchange for most inter-republican trade. The banks were also able to extend their credit lines in these roubles to pay for imports from Russia. Since NIS deficits accumulated quickly, Russia imposed credit limits on correspondent accounts of many NIS central banks. This resulted in defaults on payment by many enterprises. During this period, most enterprises stopped trading within the NIS or shifted to barter (OECD, 1995).
5. To the official trade flows between the NIS and Russia should be added the unofficial sales of alcohol and tobacco products as well as some other agro-food commodities produced in other NIS (especially Ukraine, Belarus and Moldova), and smuggled across the borders into Russia. The value of "shuttle" food trade has been estimated at as much as 30 per cent of Russia's importation of consumer goods and is particularly important in border regions and in large cities.
6. In 1996 disputes arose between Russia and the US on the quality of poultry exported to Russia. In February 1997 Russia threatened to ban US poultry imports on the grounds that American processing plants failed to meet Russian sanitary standards. In April 1997 however, the two countries agreed on new health standards, and the US authorities subsequently agreed to bring their poultry meat inspection procedures in line with Russian requirements.
7. For the total value of agro-food exports and imports, according to data provided by Goskomstat, see Table II.1.
8. However, the assortment they dealt with was broader than just food and agricultural products, and sometimes the name did not reflect the type of transactions they were involved in. *Exportkhleb*, for example, was the major VTO for grain imports which accounted for about 95 per cent of total turnover of this organisation.
9. This section draws partly on work carried out by the OECD Trade Directorate.
10. The New Independent States emerged in late 1991 with the collapse of the Soviet Union. They include Azerbaijan, Armenia, Belarus, Georgia, Kazakhstan, Kyrgyzstan, Moldova, Russian Federation, Tajikistan, Turkmenistan, Uzbekistan and Ukraine. Estonia, Latvia and Lithuania, which regained *de facto* independence, had *de jure* never ceased to exist as sovereign states, so they are not "new" independent states and are included in the group of central and eastern European countries (CEECs). Since the Commonwealth of Independent States (CIS) includes all NIS, there is no difference between the two and in the Russian language only one term is used (*Sotrudnitchestvo Nezavisimykh Gosudarstv* – SNG, which is CIS).
11. See: *Food and Agriculture Report*, Interfax, October 24-31, 1997; *Agra Europe. East Europe Agriculture and Food*, No. 182, November 1997.
12. Technically, it was still the GATT (General Agreement on Tariffs and Trade). The WTO was formed on 1 January 1995.

PRIVATISATION AND STRUCTURAL CHANGE IN THE AGRO-FOOD SECTOR

A. LAND OWNERSHIP IN RUSSIA – HISTORICAL OVERVIEW

Up to the middle of 18th century, the Tsar claimed ownership of all agricultural land in Russia although he granted its use to nobles or communities of "state peasant" households. The right to use the land by noble families was combined with the duty to serve the state, in particular the army. However, all real estate could be arbitrarily taken over by the state without compensation and land was not subject to sale or purchase transactions. In the European part of Russia private ownership of land started to develop in the second half of the 18th century when it became possible to use the land without the obligation to serve the state. Serfs providing labour for landlords were allowed to use a part of the land "owned" by the landlord. The land was used on a communal basis with three-field crop rotations, each peasant household using some strips of land in each field and crop rotations determined jointly by the whole village. The linkage between serfs and the land they used was partly recognised under the rule of Nicholas I (1825-1855) who prohibited the sale of serfs without "their" land. Until 1861, the serfs were under the jurisdiction of their landlords and the state usually did not intervene in the relations between them.

Under the terms of the 1861 emancipation act serfs were *de jure* freed from their landlord's autocracy and given some of the rights of citizens. Moreover, the land they used was allocated to them. However, each Russian village was forced to pay collectively for its newly acquired land through a mortgage arrangement with the state, which "paid the gentry for it in full and directly" (Figes, 1996). Only on full payment of the redemption dues, would the land pass into collective (village) ownership. Until these payments were made, no individual or family could leave the village without permission of the village meeting and the land could not be sold or used as collateral. Thus, peasants were still tied to the village commune, which enforced the old patriarchal order, were deprived of the right to own the land individually, and were legally inferior to most other groups in society.

Following the uprisings of 1905, the Russian government modified its agricultural policy to encourage the reorganisation of agricultural holdings. By a Law of 9 November 1906 peasants were given the right to convert their communal strips of land into family property on fully enclosed farms outside the village or consolidated holdings within it. The whole village could make this transformation by a two-thirds majority vote of the heads of households. The new holding could become family property and, under certain conditions, be sold.[1] State aid in this process of land consolidation was offered. Moreover, the reforms brought to an end the obligatory repayments for land acquired by peasant communes in 1861. In addition, vast areas of agricultural land, including part of the land belonging to the nobility, the Tsar's family and the state were opened up to farming by the peasants. The state also supported the move of four million people to previously unused land in Siberia. This process was generally known as the "Stolypin reforms", after the prime minister who initiated the policy.

By the time of the Bolshevik Revolution of November 1917, about 15 per cent of all the peasant households in European Russia had consolidated their land as private plots, either in groups or individually, bringing the share of peasant farms in hereditary tenure to between 27 and 33 per cent (Figes, 1996). Most individual farms were created in the west, the south and south-east of the country. However, the majority of the peasants in the central part of Russia were not affected by the reforms and continued their communal way of life with unaltered communal use of land based on three-field crop rotations.

The revolution put an end to the process of enclosure. One of the first decrees of the revolutionary regime announced that the land belonged to the whole nation and its use should be allocated on an equal basis to peasant families. The first Russian Federation land code in 1922 declared that land belonged to the state for the people as a whole and could not be a subject of private purchase or sale. This legal status, with further modifications introduced by the land code of 1928, remained until 1990.

Although land was state-owned, peasants claimed customary rights to use it and viewed the land they were farming as their "property". The New Economic Policy (NEP), announced in October 1921, prohibited rotational redistribution of land in the villages (*peredel*), and allowed the hiring of labour, the leasing of land for longer periods, and the selling of produce on the market. Before the start of collectivisation, about 97 per cent of all agricultural land was farmed individually.

Collectivisation of agriculture began on a mass scale in 1928-29 and imposed centralised economic and political direction on the villages, while eliminating or deporting many of the most skilled and prosperous (*kulak*) peasant families. Several million *kulaks* with their families were deported between 1929 and 1937. There was resistance to collectivisation, and government attempts to overcome that resistance led to the creation of rural institutions designed to be managed almost entirely on orders from above in which individual initiative was risky at best.

By 1937, collectivisation had been essentially completed with 83 per cent of peasant households consolidated into collective farms (*kolkhozes*). The average *kolkhoz* in the USSR included some 76 peasant households and had a total of 1 534 hectares of agricultural land (Wädekin, 1982). The *kolkhoz* amalgamation drive of 1950-51 consolidated some 254 000 farms into 97 000. Originally, most Soviet-era farms were collectives. However, estate land not distributed prior to collectivisation, newly-cultivated land (such as the Virgin Lands of East Russia opened up after 1953), experimental stations, breeding farms and other specialised units were generally organised into state farms (*sovkhozes*). Moreover, the transformation of many unprofitable collectives into state farms in the late 1950s and 1960s significantly increased the latter's importance.

B. FARM STRUCTURES BEFORE THE REFORM

1. Types of agricultural enterprises

As a result of the historical evolution described above, at the end of the 1980s there were three main organisational forms of agricultural enterprises:

- The collective farm (*kolkhoz*), theoretically a co-operative. It held land in "perpetual use" (*postoiannoe pol'zovanie*) from the state. Its production assets, buildings, etc., were the farm's collective and indivisible property (*nedelimye fondy*). In 1987, there were 12 124 *kolkhozes* (excluding fishing co-operatives) averaging about 6 300 hectares of agricultural land in the RSFSR.
- The state farm (*sovkhoz*), a state-owned enterprise involved in agricultural production. It operated on state land, and its production assets, buildings, etc., were state property. Its workers received wages directly from the state budget. In 1987, there were 12 810 *sovkhozes* averaging 15 600 hectares of agricultural land in the RSFSR.
- The interfarm enterprise (*mezhkhoz*), a joint holding of farms and other enterprises for some special purpose, such as feed lots or rural construction. The interfarm organisation was controlled by a board composed of representatives of the participating enterprises with voting rights based on their participation in the capitalisation of the enterprise.

In addition, three legal forms for individual plots and family gardens were recognised:

- The household plot (*priusadebnyi uchastok, lichnoe podsobnoe khoziaistvo*), held by households with family members employed on a collective or state farm. Limited to a small size (about 0.3 hectare), these plots not only provided farm families with much of their own food but also produced much of the agricultural produce requiring particular care or high labour input, such as potatoes, fruits and vegetables, as well as meat and milk.[2] This produce was either sold to the "union of co-operatives" (Centrosoyuz), a national organisation which also ran many rural food processing enterprises, stores in towns, and stores and consumer services in the rural areas, or directly by individuals on so called "*kolkhoz* markets" organised by the state in the cities. Part of plot production could also be contracted to a collective or state farm in exchange for inputs, money and services or given to relatives living in towns for free or in exchange for other goods and services.

- The garden plot (*sadovyi uchastok*) and vegetable plot (*ogorodnyi uchastok*) (usually between 0.04 and 0.08 ha) held by urban families not employed on the large farms. These plots were usually assigned and physically laid out in groups. "Gardening societies" were organised by city factories or other enterprises to distribute land and secure basic infrastructure.
- The dacha plot, a plot of land allotted to families for the construction of a summer cottage. Land around these houses was also used to produce food.

2. Organisation and management of collective and state farms

The collective farm was theoretically the voluntary creation of its members who pooled their fields, work stock, and tools for common use. In practice, however, membership was compulsory, and once property had become part of the farm's indivisible fund it could not be redeemed. Generally, the peasant family's house in the village and any outbuildings remained private property, although the building lot was owned by the state. An individual member or family could not quit the collective without a domestic passport, which in general was not available to rural families until the Khrushtchev era. In either case, a person who left the farm had no right to any share of the collective's land or productive assets. *Sovkhoz* workers were hired like any other workers in state enterprises, with no rights at all to any of the farm's land or assets. The introduction of state-guaranteed wages and pensions for collective-farm members in the late 1960s made *kolkhozes* effectively indistinguishable from *sovkhozes*. In the 1970s many *kolkhozes* changed their legal status into *sovkhozes*.

The state closely regulated the structure and operations not only of the state farms, but of the nominally independent collective farms as well. The Kolkhoz Model Charter (*primernyi ustav*) codified collective farm organisation. Government, or party and government, resolutions confirmed each Model Charter. After 1956 some deviations from the model charters were allowed to reflect local circumstances. New model charters were adopted by USSR-wide congresses of collective farmers in 1969, 1988 and (for Russia) in 1992.

The chief management body of a *kolkhoz* was its elected management board (*pravlenie*), headed by an elected farm chairman. *Sovkhoz* managers, including the director, were state employees, like the workers. Although most state farms were supervised by the Ministry of Agriculture, substantial numbers of farms belonged to the Ministry of Defence, Ministry of Railroads, and other institutions. Until the Communist Party was banned in 1991, local Communist Party authorities approved the election of *kolkhoz* chairmen just as they appointed state farm directors. This party power to appoint and dismiss farm managers was its key lever of central control over farm managers.

For each collective and state farm annual and five-yearly plans were drawn up by the authorities. The plans usually "came down" from above, although managers of good farms had some room to bargain. Plan targets were supposed to have the force of law.

Moreover, each collective and state farm had to fulfil major social and economic functions in the village. The so called "social sphere" on a Soviet-type farm included a wide variety of services, facilities, and payments. Some of these goods and services were available to all community members, but some depended on a discretionary decision by the director of the enterprise.

C. THE PROCESS OF LAND AND AGRARIAN REFORM

1. Purposes and goals

The Russian agrarian reform began at the end of 1980s with the explicit purpose of increasing productivity within the existing system. At the beginning of the 1990s the scope of goals was broadened to include: creation of private agricultural producers; creation of conditions for entrepreneurial activity in rural areas; privatisation of land; promotion of various forms of land ownership; equal opportunities for development of various forms of management; the "independence of agricultural producers"; increase of equity for the rural population by "ensuring them a stake in what they, and their ancestors, had produced"; and, promotion of democracy in rural areas. In more practical terms, the reform was intended to make it possible to run privately owned farms for those who wished to do so, to transfer the land and non-land assets to the people who lived and worked on the farms at the time of the reform, and to transform *kolkhozes* and *sovkhozes* into more market oriented legal entities. No restitution of land to heirs of peasants collectivised in the 1930s and especially no restitution to heirs of pre-1917 landowners was provided for. The continuing evolution of the reform mechanisms, reflects these sometimes contradictory or poorly articulated purposes. Moreover, efforts to change fundamentally land tenure, large-scale farm organisation and management as well as legal ownership clashed with attempts to preserve the previous system.

2. Institutional framework

Land privatisation, and privatisation of agricultural enterprises in general, has to some extent been carried out separately from overall privatisation, in part because non-agricultural privatisation began after the agrarian reform programme was adopted. The State Land Committee (Goskomzem) has been responsible for all land privatisation; the Ministry of Agriculture and Food (MAF) has been responsible for privatisation of non-land assets in agriculture; and the State Committee on Management of State Property (GKI) has been responsible for general privatisation, including preparation of legal documents, and privatisation of local and regional upstream and downstream industries. This division of responsibility has resulted in some lack of co-ordination between the institutions involved. In addition, the transformation of agriculture has been increasingly regionally diversified. According to the 1993 constitution, issues related to the ownership, use and disposal of land fall within the joint jurisdiction of the Russian Federation and regions. However, regional land laws have, in practice, taken precedence over national legislation (Annex III).

3. Legal framework

Laws enacted in the Russian Federation since the late 1980s have formed the legal basis for land reform and restructuring of large agricultural enterprise. The laws address three major issues: distribution of land between state, municipalities and private owners; transfer of land and non-land assets from large farm collectives to individuals; and creation of conditions for private enterprise activity in agriculture. Legal ambiguities and contradictions regarding each of these three components reflect both the complex process of legislation in Russia and the transitional status of land and legal issues more generally. The legal framework consists of the constitution, laws, decrees, resolutions, and other documents issued at governmental levels from the federal to the local (Brooks, 1994). The structure and precedence of various forms of legislation in Russia are presented in Part I.

The legally specified procedures for collective and state farm reorganisation, have evolved through **three major stages** (Box I.1). During the first stage (1989-1990) the Soviet era legislation allowed the creation of a non-state enterprise as a co-operative, denationalised land and non-land assets by transferring them legally from the state to *kolkhozes* and *sovkhozes*, and established through the November 1990 Law on the Peasant Farm the legal basis for individual (family) farming. During the second stage (1991-1993) the legislation concentrated on privatisation of *sovkhozes* and *kolkhozes*, establishing procedures for the determination of land and non-land entitlements and securing their holders wide-ranging rights. As the process of entitlement determination had been largely completed in most areas of the country, during the third stage (1994-1996) legislative attention shifted to creating a more precise legal framework for deep restructuring of large agricultural enterprises and to ensure that holders of entitlements know their rights and are actually able to dispose of their entitlements as they wish. Each iteration of developing legal procedures has more precisely defined the framework as legal difficulties and practical problems have been discovered and addressed. However, some fundamental legal problems and contradictions in this respect had not been resolved by February 1998, as for example the Civil Code adopted in 1993 allows sales of agricultural land, while the Land Code adopted by Parliament but vetoed twice by the President does not (see below).

Box III.1. **Reorganisation of large-scale agricultural enterprises: the evolution of the legislation framework**

Soviet era legislation (1989-1990): limited privatisation and structural change in the agro-food sector of the USSR began. There were several attempts to change the internal structure of *kolkhozes* and *sovkhozes* to create new incentives for workers and farms as a whole. Adopted in 1988, the All-Union Law on Co-operation in the USSR, for the first time since the NEP in the 1920s, created the legal possibility to set up a non-state agricultural enterprise as a co-operative. The Principles of Land Legislation of the USSR, adopted in 1990, gave some guarantees for individual farming, provided procedures of withdrawal from the state and collective farms and introduced basic legislation for a land lease system. In the same year, on the basis of an amendment to the USSR Constitution, the land was legally transferred from the state to *kolkhozes* and *sovkhozes* for collective use (*pol'zovanie*) and non-land assets for collective ownership (*sobstvennost'*). The RSFSR approved its Land Code, laws on the Peasant Farm and on Agrarian Reform, based on the All-Union legislation, at the end of 1990. The November 1990 "Law on the Peasant Farm" detailed the procedure of exit from the collective and state farms with land and non-land assets and established the legal basis for individual farming.

The 1991-1993 legislation: Following the formal dissolution of the Soviet Union in December 1991, the newly independent Russian Federation developed its own agrarian reform programme. Presidential Decree No. 323 of 27 December 1991 ("On Urgent Measures for Implementing Land Reform in the Russian Federation"); Government Resolution No. 86 of 29 December 1991 ("On the Procedure for the Reorganisation of *Kolkhozes* and *Sovkhozes*"); Government Resolution No. 708 of 4 September 1992 ("On the Procedure for Privatisation and Reorganisation of Enterprises and Organisation of the Agro-Industrial Complex"): required all large-scale farms to undergo legal re-registration up to the end of 1992 and established the entitlement determination and exit procedures; these decrees were followed by Presidential Decree No. 1767 of 27 October 1993 ("On the Regulation of Land Relations and the Development of Agrarian Reform in Russia") which defined real estate to include land and all that is attached to it, guaranteed land entitlement holders the right to sell, lease, give away, mortgage and bequeath their land entitlements. Moreover, it required the issuance of titles (*svidetel'stvo*) to land entitlement holders, and guaranteed ownership rights.

The 1994-1996 legislation: while the December 1993 Constitution of the Russian Federation (articles 9, 36, 72) legalised private ownership of land,[3] the Civil Code of the Russian Federation of November 1994 defined legal forms of enterprises, forms of ownership of land and procedures for exercising ownership. Government Resolution No. 324 of 15 April 1994 ("On the Experience of Agrarian Reform in the Nizhny Novgorod *Oblast*"); Government Resolution No. 874 of 27 July 1994 ("On Reorganisation of Agricultural Enterprises Based on the Experience of Nizhny Novgorod *Oblast*) and the Government Resolution No. 96 of 1 February 1995 ("On the Method of Exercising Rights to Land and Property by Owners"): incorporated the work of the Nizhny Novgorod project ("Program of Land Privatisation and Reorganisation of Agricultural Enterprises in Nizhny Novgorod Oblast") into the Russian federal legislation and aimed to provide a mechanism for transferring ownership to individuals and to establish the basis for a land market. In March 1996, the Presidential Decree No. 337 ("On Guarantees of Constitutional Rights of Citizens to Land") emphasised that land entitlements could be freely traded and required that all users of agricultural land conclude formal purchasing or leasing contracts with every individual land entitlement holder and that local authorities give land certificates to land owners by the end of 1996.

4. The Law on the Peasant Farm

In December 1990, the RSFSR Congress of People's Deputies adopted a package of reform legislation. The most important item in that legislation, the Law on the Peasant Farm, developed the legal basis for individual farming (*krest'ianskoe (fermerskoe) khoziaistvo*).[4]

The law defined an individual farm as an independent legal entity, representing individual citizens, families or a group of other persons, whose activity was to carry out production, processing and sale of agricultural production, using non-land property and land plots for which they have the user rights (*nakhodiaishchikhsia v ikh pol'zovanii*), including lease, lifetime heritable tenure or as private property.

The Law on the Peasant Farm provided the basis for individuals and families to leave (the right to exit) the collective and state farms with a share of the large farm's land and property, although the shares were determined only theoretically (on paper) in terms of amount of land and value of property, without being designated in physical terms. They were determined physically (specifying plot of land and piece of property) only if somebody decided to leave a large-scale farm to establish an individual farm. The law also created a *raion* land redistribution fund based on unused or underused land of the collective and state farms in order to provide land to individuals wishing to begin farming but who had not previously been collective farm members or state farm workers. People receiving land for individual farms from this fund were required to use it for agricultural purposes and to preserve soil fertility and prevent erosion. The final stage of setting up an individual farm was its registration, the farm could open its own bank account, and got its own seal to record officially its documents. Only after this registration did a farm become a legal entity.

The reformers expected that the Law on the Peasant Farm would create private owners in rural areas and stimulate the restructuring of *kolkhozes* and *sovkhozes*. However, the number of peasants wishing to establish their own individual farm was relatively small and, consequently, pressure to determine individual entitlements was weak. Changes inside large agricultural enterprises, if any, were to a large extent spontaneous and aimed at giving more economic independence to the working teams (brigades and smaller units) within large enterprises. The development of private activity in agriculture was hampered by a lack of initial capital to establish a farm, unfavourable changes in farm input-output prices and a high level of uncertainty as to the direction and durability of reform.[5] A very important reason was the scepticism and outright hostility with which local, regional and ministerial officials received the reforms, including the creation of individual farms. Another factor was the lack of a tradition of operating a family farm. According to surveys conducted by Moscow's Agrarian Institute in 1990-1992 only about 10 per cent of workers employed in large agricultural enterprises were interested in farming independently on their own land and leaving the Soviet-type farms.

5. The large farm reorganisation process

To speed up the reform process, the December 1991 presidential decree and accompanying regulations required all large-scale farms to undergo legal reorganisation and to determine individual entitlements by the end of 1992. While the reorganisation concerned a change in legal structure and in formal ownership, it also provided for the determination of land and property entitlements (called also conditional land and asset shares.[6]) The decree also required the Ministry of Agriculture and State Committee on the Management of State Property (GKI) to draw up a list of state farms exempt from the process.

i) Initiating reorganisation

According to the 1991-1993 legislation, the reorganisation process was to be controlled by *oblast* and *raion* commissions established for the purpose. The head of the *raion* administration chaired the *raion* commission, which included representatives from the *raion* council, GKI, agricultural administration, State Land Committee (Goskomzem), the agricultural bank (Agroprombank in most areas) and the *raion's* collective and state farms.

Formally the reorganisation started with a decision by the general assembly of the large farm members on the legal form the new enterprise should take. Each large farm was to establish its own "Farm Commission" to lead the process and to do most of the work of re-registration. The farm commission was to be established by a general meeting of the farm workers, but its membership and its key actions were to be confirmed by the *raion* commission. The composition of the farm reorganisation commission was mandated to include the farm's director, who chaired the farm commission, at least four other members of the farm, representatives of the *raion* land committee (Raikomzem), any major creditors, and relevant local agricultural administrators. The mandated positions for (at least) four other members of the farm often went to senior staff members (agronomist, livestock specialists, economists, bookkeepers, engineers and veterinarians). This meant that almost all decisions related to the farm's reorganisation were made *de facto* by its leadership (Butterfield, 1995). Approval was required by the general meeting of large farm members, but anecdotal evidence indicates that this was rather a formality. Once a decision on the corporate form was made, the next step was the complex process of calculating land and asset shares for each farm member.

ii) Calculating of land and property entitlements

To calculate the land and non-land property entitlements the farm commission was charged with drawing up lists of individuals eligible to receive land and property entitlements. In this respect the legislation was changing over time. As for land entitlements, the December 1991 Government Resolution provided that permanent workers and pensioners who continued to live in the area were eligible to receive the entitlements. Such entitlements could also be distributed to people who were absent for good reasons (military service, study), social workers, "and others" as the general meeting decided. The September 1992 resolution slightly changed the eligibility list to include farm workers, pensioners living on the farm territory, workers of the social facilities located on the territory of the farm, people temporarily absent for good reason, and individuals laid off from work on the farm after 1 January 1992.

As for non-land property entitlements, the December 1991 Government Resolution provided that people then working on the farm, workers temporarily absent for good reason, and the farm's pensioners (with no reference to where they may now be living) were all eligible for these entitlements. In addition, the general meeting could decide to include social workers, as well as people who had worked on the farm in previous years on the list of those eligible. The September 1992 resolution added that the farm's general meeting could also decide to include farm workers who had been laid off since 1 January 1992. Once the lists of eligible land and non-land property shareholders were established, the commission was required to hear appeals or complaints.

The next step was an inventory of land and non-land property. The Raikomzem was responsible for carrying out the land inventory and deciding on the amounts of land to be transferred to the municipality and to be left as state land. The land transferred to the municipality included plots to be used as common pasture, for the enlargement of household plots, and for building construction, parks, sport stadia, cemeteries, etc. Land left under state ownership included forested area, ponds, plots used for testing seeds, etc. The remaining land was divided by the number of eligible people. The amount of land resulting from this calculation constituted the land entitlement size as long as it did not exceed a *raion* norm which was the upper limit that could be obtained as a land entitlement. The mechanism for setting the upper *raion* norm was not clarified until March 1992. At that time the President's Decree specified that the amount of land a citizen could receive free of charge was limited by a *raion* norm determined by dividing the total amount of agricultural land within the boundaries of the *raion's* agricultural enterprises by the total number of farm workers, farm pensioners, and workers in on-farm social assets.[7] If the size of land entitlement exceeded the upper *raion* limit, the size of the entitlement had to be reduced to the *raion* limit and the excess had to be transferred to the state redistribution fund at the *raion* level (Figure III.1).[8]

Land entitlements were calculated in hectares and so-called "point-hectares" (*balo-gektary*), reflecting the quality of land. All arable land in Russia is scored according to quality, and these quality scores are appended to the size of land entitlements so as to equalise imbalances in land quality. The result is

◆ Figure III.1. ***Land division during the reorganisation process of an agricultural enterprise,*[1] *1991-1993***

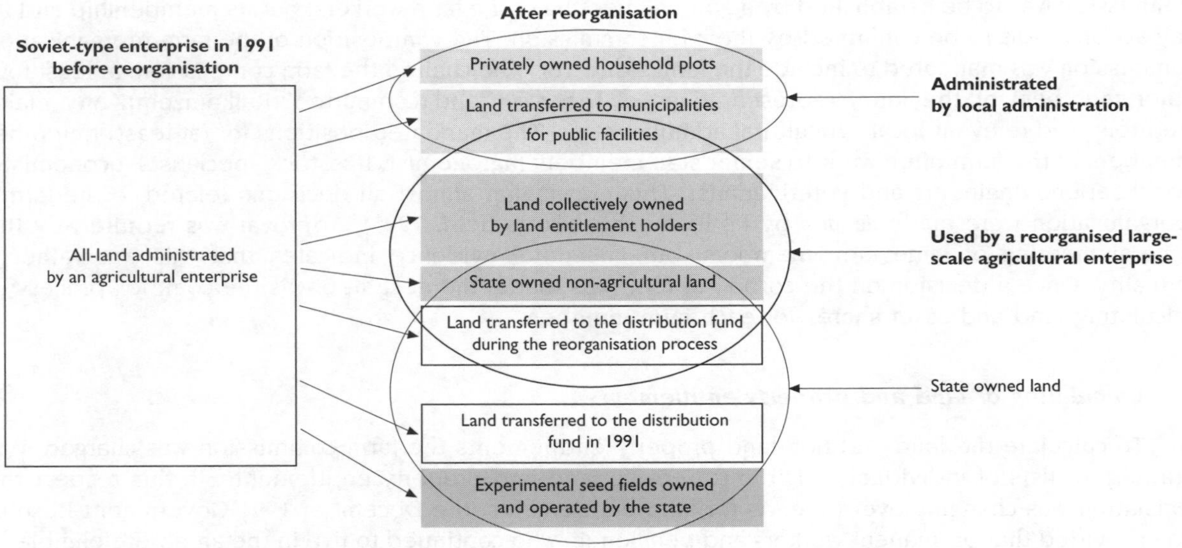

1. The figure is illustrative and does not represent the percentage distribution of land after the reorganisation.
Source: OECD Secretariat.

that the farmer deciding to convert his/her land entitlements into physically determined pieces of land receives a somewhat smaller amount of land if the quality of land is good, while one who receives poorer quality land receives somewhat more.

Determination of property (non-land asset) entitlements was more complicated. First, the commission had to determine the overall value of the farm's assets. This involved an inventory of all assets.[9] The value of any assets connected with social services ("social sphere") and the physical infrastructure which were to be turned over to local government were to be deducted from the overall value of assets.[10] Moreover, while in *sovkhozes* the value of housing was excluded from the overall value of assets, in *kolkhozes* it could be included and covered by property entitlements (see below).

Once the commission had approved the inventory, the property entitlement fund (the value of property to be divided up into entitlements) was calculated according to the following formula:

$$P = A + B + C - D - E$$

where,

P = Total value of property to be divided;
A = Total residual book value of fixed assets;
B = Total value of current assets;
C = Total value of cash and securities
D = Total value of all social sphere and public utilities objects to be transferred to the local administration free of charge;
E = Total value of debts.

The total value of fixed and current assets was to be calculated based on the latest accounting balances and results of the inventory. Individual property entitlements were not equal in value and were differentiated depending on the individual's contribution to creating the property to be divided.

Therefore, each individual's work contribution had to be calculated separately before calculating his/her individual property entitlements. The methodology for calculating these contributions was chosen by the farm commission and had to be approved by the general meeting. There were several methods recommended but in principle a formula was used that considered both the total number of years worked by an individual and the level of salary.[11] This calculation generated the value of a member's contribution to the development of the farm. A coefficient was determined that indicated each member's contribution relative to the entire collective's work contribution. The value of the asset entitlement fund was multiplied by each person 's coefficient to determine the value of that person's property entitlement.

Once all the calculations were completed by the farm commission, the farm had to conduct a general meeting to discuss and approve by majority vote the calculation of non-land property entitlements and the final lists of those eligible for land and non-land entitlements.[12] The general meeting had also to agree to dissolve the old legal entity and acted as the founding meeting of the new farm entity that was to emerge.

At this stage, if the farm had not chosen to break up, the general meeting adopted a new charter and other founding documents. The founding documents, including the list of eligible property shareholders of the new enterprise, had to be sent to the registration department of the *raion* administration. Once the *raion* administration had approved the results the previous enterprise was erased from the register and a new one created. The new enterprise was registered in accordance with the Law on the Enterprise and Entrepreneurial Activity (till the end of 1994) and later in accordance with the Civil Code (since 1 January 1995). All the founders of the new enterprise were to receive a document indicating the

Box III.2. **Rights of land and non-land property entitlement holder***

Each **land entitlement** holder may (without the consent of other entitlement holders):

- apportion a plot of land in kind for the creation of an individual farm;
- mortgage the land entitlement;
- lease the land entitlement to other users for the production of agricultural produce;
- use the land entitlement to extend a household plot;
- exchange a land entitlement for a non-land property entitlement;
- bequeath the land entitlement;
- sell the land entitlement;
- give the land entitlement;
- pass over the land entitlement to another holder in exchange for rent in money or in kind;
- invest the land entitlement or the right to use it into the charter capital of an enterprise.

Each holder of a **non-land property** entitlement may:

- receive the non-land property in kind to create an individual farm, agricultural enterprise, service enterprise, or to undertake another individual entrepreneurial activity;
- contribute a non-land property entitlement to an agricultural enterprise to be founded;
- sell a non-land property entitlement, give to an individual farm or any other land and property entitlement holders;
- exchange a non-land property entitlement for a land entitlement;
- bequeath a non-land property entitlement;
- exercise other rights specified by the legislation of the Russian Federation.

* As of February 1998. The new Land Code, if approved, may reduce the rights of land entitlement holders.

Source: Government Resolution, No. 96, 1 February 1995; President Decree, No. 337, 7 March 1996.

value of their non-land stake in the new enterprise, to sign for their entitlements in the farm's register book, and to be given a certificate of land contributed to a new enterprise. The rights of land and non-land property entitlement holders are summarised in Box III.2.

At the moment of reorganisation members had again a possibility, as earlier when the Peasant Law started to be applied, to exit which meant resigning membership in the collective and withdrawing land and property entitlements by converting them from "conditional" to "actual" (physically determined) entitlements. The exiting member might then establish or join a peasant farm or sell and/or lease his/her entitlements to a peasant farmer. However, for the same reasons as indicated above, the vast majority of farm workers did not choose to exit, which meant that they made a formal decision to contribute their land and property entitlements (conditional shares) to the enterprise to be created. Independently from the legal status of the new farm, the land entitlement holders retained the right to exit with the plot of land, provided that the land had not been invested into the charter capital of the enterprise. In the case of non-land property, the 1995 Civil Code provided that those who want to exit should be compensated in money terms for their property entitlements invested in the enterprise. The right to exit with land has been a major issue at stake in the discussion on the new draft Land Code since early 1994.

It is important to note that the reorganisation process meant legal privatisation of land and assets in the farms that underwent the process. It provided the means by which the majority of land and assets in the Russian farming sector were converted legally into the property of shareholders.

iii) Restructuring

By the end of 1993 about 95 per cent of large-scale farms had re-registered in a new legal form, but few actually broke up into separate successor enterprises and relatively few farm members exited to form individual farms. The total number of agricultural enterprises increased from 25.5 thousand at the end of 1991 to 26.9 thousand at the end of 1993, that is by 5 per cent only (Goskomstat, 1996).

Where farm members decided to implement a more fundamental internal restructuring, including a break up of the large farm, it turned out to be very difficult to match actual land and physical assets with entitlements and to regroup or trade entitlements to create new farms. A methodology for solving this problem, based on existing federal legislation, was designed and implemented on a small number of pilot farms in Nizhny Novgorod *Oblast* between 1993 and 1994. The project was initiated by the Nizhny authorities and prepared in co-operation with the International Finance Corporation (IFC) and financial support from the British Know How Fund. The key elements of the project were fourfold: transparent legal procedures; information on the process available to all interested parties, in particular members of collectives; an auction as the mechanism for land and asset distribution among successor enterprises; formal contracts between entitlement holders and the newly created enterprise. The successive stages of the farm restructuring according to the Nizhny procedures are shown in Box III.3. While the preliminary work, including calculation of entitlements, and creation of new enterprises does not go beyond the reorganisation process undergone in 1992 and 1993, the obligation to distribute certificates (Stage 1) and to conclude contracts (partly Stage 2) as well as procedures included in Stages 3 and 4 are new in the model and describe clearly further steps allowing for deep restructuring of former enterprises.

The Nizhny procedures include competitive bidding. However, only those items for which no prior agreement between newly emerging business units could be reached are being auctioned. These are closed auctions within the former farm collective and open sales of farm land and assets are not foreseen. The procedures require a formal contract for all transactions made by entitlement holders, *i.e.* transactions have to be recorded, transparent, and enforceable by all parties. Moreover, the procedures stress the importance of an information campaign to let the entitlement holders know which option they have with their entitlements (Box III.2).

The Nizhny model was extended to several other *oblasts* between 1994 and 1997 and received formal acknowledgement and recognition in government resolutions. These resolutions suggested that the principles used for farms in Nizhny Novgorod *oblast* be examined and, if found appropriate,

Box III.3. **The Nizhegorodskaya Farm Restructuring Procedure**

Preparatory Stage

- Draft a new structure for the farm
- Vote to reorganise
- Inventory land and property
- Transfer the social sphere to local authorities
- Clarify land and property entitlement lists
- Calculate land and property entitlements
- Approve preparatory work

Stage 1: Distribution of Certificates

- Distribute land and property certificates to qualifying individuals

Stage 2: Creation of New Enterprises and Concluding Contracts

- Prepare foundation agreements and register new enterprises
- Conclude contracts

Stage 3: Auctions

- Form land and property lots
- Submit applications for land and property
- Distribute land and property through a closed auction

Stage 4: Transfer of Land and property to New Owners

- Transfer the land and property to new enterprises
- Issue new land certificates to land owners
- Employ workers
- Liquidate the old farm
- Adjust new balance sheets

Source: *Land Privatisation and Farm Reorganisation in Russia*, IFC and ODA, Washington, D.C. 1995.

extended to other farms in Russia. Information on the procedures applied in Nizhny was sent to all farms in Russia accompanied by a recommendation letter from the Ministry of Agriculture. In parallel, some other ways of restructuring are being attempted in various parts of Russia based on local experiences.

6. Results

i) Agricultural land ownership pattern

In November 1997 of the 221 million hectares of agricultural land in Russia, 137 million hectares (62 per cent) were considered privately owned, while the remaining 84 million hectares (38 per cent) were still owned by the State or local municipalities. The majority of "privately" owned land, representing above half the total, was in the form of collective shared ownership (*obshchaia dolevaia sobstvennost'*). These are re-registered large-scale farms where the land and non-land assets are owned by the enterprises, and the enterprises in turn owned by shareholders, who are employees, pensioners and social workers entitled to participate in the distribution of land and asset shares. The rest of privately owned land was owned by individual farms and household plots. Of the 38 per cent of non-privatised land 9 per cent belonged to municipalities and 15 per cent to various types of agricultural and non-

◆ Graph III.1. ***Agricultural land ownership pattern in Russia, November 1997***

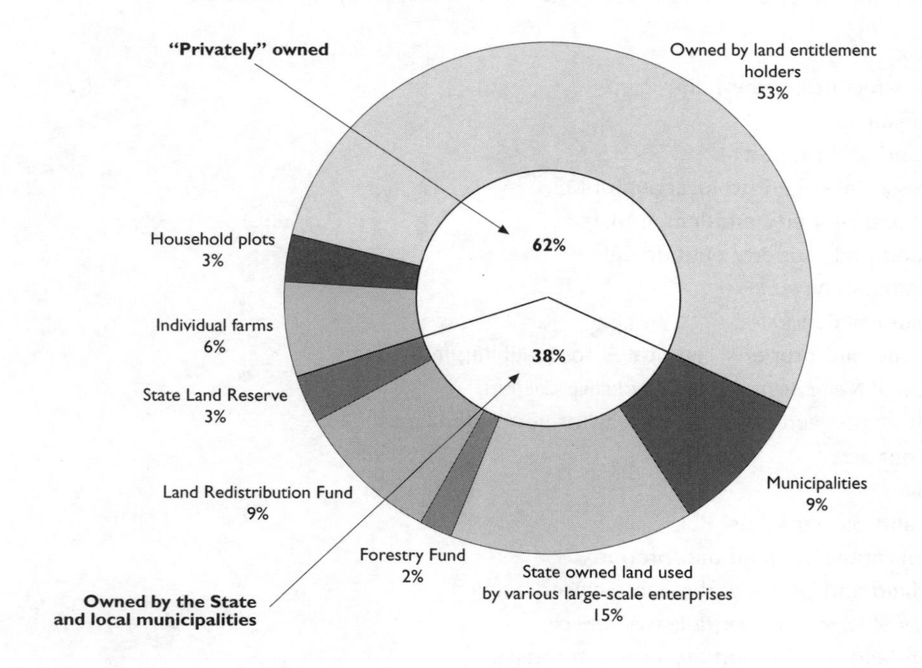

Source: OECD.

Table III.1. **Land privatisation and reorganisation of agricultural enterprises, 1 January 1997**

Type of reorganisation	Number of enterprises	Area	
		Million hectares	Per cent
All types of large-scale enterprises dealing with agricultural production [1]	41 322	151.9	100
a. Reorganised with privatisation of land, including:	23 479	116	76
– enterprises with no limits for land entitlement holders' rights	17 994	90.2	59
– enterprises with limits for land entitlement holders' rights [2]	1 356	6.6	4
– enterprises which retained their previous status [3]	4 129	19.2	13
b. *Kolkhozes* and *sovkhozes* not reorganised for various reasons	3 601	17.7	12
c. Reorganised without privatisation of land [4]	2 122	8	5
d. Enterprises not covered by land privatisation and reorganisation [5]	12 120	10.2	7

1. Including about 27 000 agricultural enterprises (with at least 70 per cent of total revenues originating from agricultural production) and more than 14 000 other enterprises partly engaged in agricultural production.
2. Enterprises producing poultry, tea, grape, hops, animal fur, and pharmaceutical plants as well as enterprises multiplying new crop and animal breeds for further distribution to other farms. In these enterprises the exiting land entitlement holder may be compensated for his land in money terms but he cannot exit with land in physical terms; the limitation may concern the whole or a part of his/her land entitlement.
3. *Kolkhozes* and *sovkhozes* which were reorganised but retained their previous status. Most of them could be allocated to the group of enterprises with no limits on land entitlement holders' rights.
4. Mostly northern territories and Islamic republics of the Russian Federation where local authorities did not allow land privatisation in order to preserve common land ownership. In the North, these areas include: Nenetsia, Komi, Yamalia, Taymyria, Evenkia, Yakutia; and in the South: Mari El, Kalmykia, Dagestan, Kabardino-Balkaria, North Ossetia, Ingushetia, Chechenia, Bashkorstan, Tuva. Moreover, land was not privatised in agricultural enterprises located within the boundaries of large cities.
5. All types of agricultural enterprises belonging to non-agricultural enterprises and other institutions, such as research, experimental and breeding stations, education institutions, railways and army.

Source: State Land Committee (Goskomzem), 1997.

agricultural enterprises and institutions in which land was not privatised for various reasons (Graph III.1).

ii) Large-scale enterprises

In January 1997 there were about 27 000 large-scale agricultural enterprises operating on 134 million hectares of agricultural land, with an average area of about 4 950 hectares.[13] More than 50 per cent of them had adopted the legal form of a joint stock company, while the others were registered as agricultural co-operatives (12 per cent) and associations of private farms (4 per cent). About 30 per cent of farms retained their previous legal status as *kolkhozes* or *sovkhozes*. In addition, there were over 14 000 other enterprises and institutions dealing with agricultural production such as subsidiary agricultural enterprises, mostly run by urban industrial enterprises for supplying food to their workers, agricultural research and scientific institutions and other enterprises run by army, railways and education institutions, operating on about 18 million hectares of land, averaging about 1 300 hectares. These enterprises were either not covered by land privatisation and reorganisation or were reorganised but without privatisation of land (Table III.1).

Following the March 1996 presidential decree requiring local authorities to give land certificates to land owners and all users of agricultural land to conclude formal purchasing or leasing contracts with every individual land owner within a one-year period, 10.8 million people (93 per cent of the total number of land entitlement holders) received official certificates for their entitlements by October 1997, of which 4.3 million contracted their shares to the users of land, mostly under leasing arrangements, with contracts being registered officially. Some entitlement owners decided to contribute their entitlements to the enterprise's charter capital, but very few sold entitlements. The 6.5 million remaining land entitlement owners did not sign contracts, mostly due to the reluctance of the large-scale farm managers to pay rents for the use of land. Some managers expect that the new and long-discussed Land Code would give land users (*i.e.* large-scale enterprises) ownership rights to the non-contracted land.

The vast majority of entitlements have been contracted to their original, but now re-registered farms. However, between 30 and 50 per cent of land entitlements are owned by pensioners and non-farm workers who have more flexibility for allocating their shares to other enterprises since they are not employed by the farm enterprise using their entitlements. As a result, there is some evidence to suggest that entitlement holders prefer to lease their plots to neighbouring large enterprises or individual farms in expectation of higher rents per land entitlement. This creates some room for the reallocation of land use to those enterprises and/or individual farms which are more efficient and are able to pay higher rents. However, the scope for such reallocation is limited because pensioners and non-farm workers are still strongly dependent on supplies of various inputs and services from the original farm. Another obstacle is location. The would be lessors have difficulty getting land entitlements designated in a plot relatively close to their existing farms. Many agricultural enterprises have designed a particular field from which land can be allocated to any land entitlement owner who wishes to withdraw land in kind, either to establish his own farm or for other purposes. In most cases, it is the least accessible land that limits the possibility to lease the land to another user. Rents are usually paid in kind and/or in the form of services provided by land users to land entitlement holders.

By the end of 1997, in the vast majority of large-scale enterprises (about 90 per cent of the total number), "reorganisation" had not proceeded beyond re-registration of the original farm under a new form and has changed very little in terms of the institutional structure of the farm, management practices, and agricultural techniques. Moreover, the differences among various new legal forms undertaken by re -registered *kolkhozes* and *sovkhozes* are not perceived by participants and appears to be arbitrary. In practice, regardless of the variety of legal forms and new names, the majority of farms have been converted into production co-operatives with fixed assets belonging to the collective under the terms of share-based ownership; management selected on the basis of co-operative principles ("one member one vote"); and profits distributed mostly on the basis of work rather than share contribution (Serova, 1996).

By the end of 1997, a minority of farms (about 10 per cent of the total number) underwent substantial restructuring, most often in one of the following ways:

- Break-up: the most radical process, resulting in complete liquidation of the old farm and its division into smaller, but still large-scale, technologically integrated production units such as owner-operated family farms, agricultural co-operatives and joint stock companies, private or co-operative service enterprises. As a rule, these new units jointly use and maintain the production infrastructure of the former farm and are often based on previously-existing territorial or production subdivisions. By mid-1997, in about 4 per cent of large-scale farms, members decided to implement this form of reorganisation. On those farms where solid preparations were undertaken before the break-up, the process went relatively smoothly. However, there were frequent cases in which division was carried out spontaneously, without any preparation, creating strong tensions between various players and groups of the local population. The final stage of this radical path is legalisation of the division and the signing of new contracts between land and property entitlement holders and new enterprises.
- Concentration of large farm property and land entitlements in the hands of small groups of owners through the purchase, exchange and leasing of shares. These farms have retained skilled and motivated personnel, capable of and willing to make profits in agriculture. In the future these enterprises may become large commercial farms controlled by a small group or even one effective operator, leasing land from local residents, employing a rather small number of permanent workers and hiring a large number of seasonal workers.
- Partition: agricultural enterprises preserve their previous status (regardless of their actual name, *kolkhoz* or joint-stock company), but in fact peasants split off and rely on their own household plots, most often expanded somewhat by land taken over formally and informally from the large farm. The collective part of the assets is used only to serve the household plots for such purposes as input supply, maintenance and use of specialised machinery, and marketing. This type of mutual ties corresponds to the "Association of Individual Farms and Small Enterprises" provided for in the national legislation and has become characteristic for some provinces most affected by the economic crisis in the agrarian sector (such as Pskov *oblast*). In some cases partition of former large-scale farms has meant that bankrupted farms have completely broken up and all members have taken their land and property shares. Some of the resulting individual farms have regrouped into new, smaller associations.

Most probably, the remaining majority of large-scale enterprises will in the future be forced by further market orientation to opt for one of the above forms of restructuring. As indicated in Part I, between 60 per cent and 80 per cent of large-scale farms are considered loss-making in recent years. Most of them are unable to pay current debts, and many are unable to pay wages to their workers for several months. In evident cases of insolvency the enforcement of the bankruptcy law could lead to the break-up of nonviable enterprises into smaller productions units. It is expected that the law on insolvency, effective from 1 March 1998, will provide the impetus for the next round of enterprise restructuring. When other ailing large-scale enterprises come to the conclusion that their mode of operation is not sustainable, spontaneous reorganisations may become more common easing the flow of resources to more efficient farm structures.

iii) Household plots

In 1997, about 5.7 million hectares, *i.e.* 2.6 per cent of total agricultural land was divided among 16 million household plot owners, resulting in an average of just 0.4 hectares.[14] As indicated in Part I, their share in total agricultural production has increased significantly since the transition started, to about 50 per cent of GAO in 1997. So far, production on household plots is mostly for family use, but a growing part of it is being commercialised. The repeal of mandatory deliveries of agricultural products to the state before selling them to any other buyer left more produce for large scale farms to pay in kind for labour delivered by workers or to sell to workers at discount prices. In turn, households may resell

Table III.2. **Services supplied by farm enterprises to rural residents**[1]

Services	Per cent of enterprises supplying the services to each recipient category		
	Employees	Other villagers	Private farmers
Farm machinery for contract work	98.7	85.9	49.1
Transport	99.1	85.9	41.9
Pasture and hay	90.2	72.2	26.9
Consulting	76.5	66.2	51.3
Credit	18.4	2.6	0.4
Veterinary services	92.7	80.3	53
Product marketing	66.7	45.3	12.4
Fuel for farm uses	33.8	15.3	12.4
Construction materials	85.9	49.1	17.5
Heating fuel	46.6	26.5	7.3

1. The survey was conducted in 1994 and covered 234 large scale agricultural enterprises.
Source: Brooks (1996).

the products received from the large farms to whatever buyer they may find. Owners of household plots pay a land tax, very low in Russia, but do not pay income tax on income earned from the plot.

Following the Presidential Decree of March 1996, the household plots are free to use their land entitlements to expand their scale of operation to the upper limit fixed by the local administration (usually between 6 and 12 hectares), but few of them use this opportunity and even fewer show any interest in developing into more independent, family type farms, except for such cases as the partition of large farms described in the previous section.

The activity of household plots is still strongly determined by formal and informal ties with large-scale agricultural enterprises. In most cases, the links between village inhabitants and the large enterprise may be delineated into five major types:

- the large enterprise uses land legally belonging to village inhabitants constituting the majority of land entitlement holders;
- the large enterprise uses non-land property legally invested in the enterprise by village inhabitants constituting the majority of non-land entitlement holders;
- the large enterprise employs village inhabitants and pays them wages;
- the large enterprise provides social and technical infrastructure for households in the village;
- the large enterprise provides an ample part of services and inputs for the economic activity of the household plots owned by village inhabitants (Table III.2).

iv) Family farming

The family farm sector, operating on about 6 per cent of agricultural land in Russia in 1997, remained of rather minor importance. Due to lack of initial capital, legislative uncertainty, difficult macroeconomic conditions, a rather unfavourable political climate in particular at the local level (except for some regions), difficult access to information, credits and markets, and a lack of tradition and experience with individual farming in Russia, only a small proportion of farm workers decided to establish private farms. Moreover, potential individual farmers are afraid of losing access to production infrastructure (storage, repair service, grain drying facilities etc.) located on large farms and to social infrastructure provided through employment contracts with large farms.

The number of individual farms increased rather quickly to 270 000 in 1994, stabilised at about 280 000 between 1995 and 1997 and fell to 274 000 at the beginning of 1998 (Table III.3). It seems that those most eager and best prepared managers and workers to establish individual farms had done so by 1994.[15] The initial rise in the number of individual farms was stimulated by several factors which overrode the major problem of a lack of capital to establish a farm. At the beginning of the 1990s,

Table III.3. **Development of individual farms in Russia, 1991-1998**

	1991	1992	1993	1994	1995	1996	1997	1998
Number of farms (thousands; 1 January)	4.4	49	182.8	270	279.2	280.1	278.6	274
Average size (ha; 1 January)	41	42	43	42	43	43	44	48
Yearly increase in number (thousands)	44.6	133.8	87.2	9.2	0.9	–0.9	–4.6	n.a.

n.a.: not available.
Source: Goskomstat.

individuals who wanted to leave large farms could relatively easily exercise their right to withdraw property in kind or receive the cash value of their property entitlements. The machinery on large farms was still in rather good shape and the financial situation of large farms was still relatively good. Moreover, by 1993 individual farmers also had the possibility of receiving preferential credits supported by the state budget.

In more recent years the situation changed. Since the majority of large agricultural enterprises have a shortage of working machinery due *inter alia*, to a lack of funds to make repairs, they are reluctant to give property entitlement owners any functioning machinery or other capital assets. Moreover, while the situation may differ across various legal forms of agricultural enterprises, property entitlements on the majority of large farms have not been adjusted for inflation for several years making the real value of the entitlement shares extremely low, sometimes undervalued by a factor of several thousand (Prosterman, 1997). In addition, due to evidence of fraud and violation of financial rules, the preferential credit system for individual farmers was discontinued in 1994 (Part IV).

Therefore, due to the deteriorating financial conditions to establish individual farms and despite the Presidential Decree of March 1996 greatly facilitating the ability of land entitlement holders to exercise their land ownership rights, in 1997 the number of newly established individual farms was lower by 4 600 than the number of those liquidated. However, the total amount of land operated by individual farms, including under leasing contracts, increased by 7 per cent to about 13.1 million hectares. As a result, the average size of individual farms rose from 44 hectares in January 1997 to 48 hectares at the beginning of 1998 (Table III.3). About 55 per cent of individual farms have less than 20 hectares, while more than one-third have 21-50 hectares and below 10 per cent over 100 hectares. The largest individual farms are located in the regions of Lower Volga, and Western Siberia, while the smallest are in Central Russia and Northern Caucasus, where land is used more intensively and access to land is more difficult.

The increase in the total amount of land used by individual farms was mostly due to the above mentioned Presidential Decree, which gave a farmer the possibility to conclude leasing contracts with land entitlement holders to enlarge his farm.[16] According to some field research, individual farmers with an average payment of 0.23 tonnes of grain per hectare of leased land were paying about four times as much as large agricultural enterprises. Assuming an average entitlement of 7 hectares, the payment constituted the equivalent of an additional six months of pension payments in autumn 1996 (Prosterman, 1997). However, as indicated above, both problems with the physical allocation of land available for leasing, and still existing links between land entitlement holders and the large-scale enterprise on which they continue to work or used to work for decades limit to a large extent the possibility to reallocate the land use to individual farms. To facilitate such reallocation, both intensive programmes to inform land entitlement holders (including pensioners) about their rights and options for using land entitlements and, in particular, the transfer of social and technical infrastructure to local authorities (see below) would be needed.

7. Legal impediments for further restructuring

During the privatisation process in Russia, agricultural land and non-land assets have been denationalised and transferred from the state to the collectives of workers, pensioners and social workers.

The legislation also provided the possibility for land and non-land assets to be allocated to farm members who decided to establish their own individual farms. However, the second major step, *i.e.* distribution of land and non-land assets from collectives to individuals, combined with the distribution of land certificates and signing formal contracts with certificate holders, has not been finalised by a large part of enterprises. Particularly in these enterprises, the relations and respective rights in the triangle: collectives (collective land and property owners), individuals (individual holders of land and property) and agricultural enterprises as legal entities (enterprises which are corporate users of legal land and property) are not sufficiently clear cut. The main problems are the following:

- Although the 1993 Constitution and the as-yet-unimplemented Article 17 of the Civil Code explicitly permit the purchase and sale of agricultural land, there is as yet no implementing legislation specifying a sale mechanism.[17] This mechanism is to be specified in the new Land Code. However, an earlier version of the Land Code, restricting land market transactions and conflicting with the new Russian Federation Civil Code, has passed the Duma twice only to be vetoed by the President. The last veto was overturned by the State Duma, but failed to be overturned by the more moderate Federation Council. In December 1997, a trilateral conciliation commission was set up involving representatives of the two chambers of the Parliament and the government. As a result, on 22 April 1998, the State Duma approved the new version of the Land Code, which, however, was more restrictive than the President's proposals. In fact, the version approved by the Duma allows farmers to sell agricultural land only to the state and limits their rights to pass it onto their heirs. Therefore, it was unlikely that the President would sign this version into law, and by mid-May 1998, the *impasse* had not been resolved.[18] Therefore while sale of land is legal, the conditions are not sufficiently specified. This is an impediment to land transactions and to the development of a land market.
- Although land ownership by individuals and workers' collectives is allowed, it is unclear whether current legislation provides for land ownership by private agricultural enterprises as legal entities. This problem becomes apparent when an agricultural enterprise opts to buy, sell, or mortgage land.
- Although the re-registration process has been completed, the reorganisation of farms is continuing. The rights of land and property share holders to withdraw shares from newly created private enterprises are not sufficiently defined, giving rise to disputes between re-registered enterprises and nominal land owners. Lawyers, administrators and farm managers are confused about how to reorganise successor organisations (*i.e.* joint stock companies, associations, production co-operatives etc.). For example, in some cases the land held in land entitlements is said to be individually owned; in some cases it is said to be owned by the enterprise. In addition, much room for manoeuvre was given to farms when they developed their own charters to fix conditions of exit for individuals deciding to establish their independent farm. Some new problems arose with the 1995 Civil Code, which imposed limits on the legal forms of enterprises into which re-registered enterprises could be transformed during further restructuring processes, for example a joint stock company can be transformed into a limited liability company or production co-operative only.[19] Moreover, all enterprises should change their legal status and internal organisation to conform with the new Civil Code and other laws regulating the internal organisation of an enterprise.

D. PRIVATISATION IN THE UPSTREAM AND DOWNSTREAM SECTORS

1. Upstream and downstream structure before reform

In the Soviet period agricultural upstream and downstream enterprises were organised into state monopolies, structured pyramidally from Moscow to local agencies, divided along functional lines with each enterprise directly subordinated to a branch ministry or another central institution. Although these various monopolies were periodically combined into larger agencies or split into smaller ones, the basic hierarchical arrangement remained the same until 1991. The whole structure was strongly politicised with enterprise managers involved in bargaining processes with bureaucrats and many of the economic functions normally performed by enterprises in a market economy were in fact handled by branch ministers.

Farm machinery production was concentrated in several specialised production units, quite often combined with military equipment production, supplying the whole of the USSR with tractors, grain combines and other machinery. An extensive planning system was supposed to match amounts produced with quantities of machinery and other inputs requested by agricultural producers, but in practice "demand" was always larger than "supply"; prices often did not cover production costs; and production was directly and/or indirectly subsidised. As a result permanent shortages of machinery, chemicals and other inputs prevailed. This system was inefficient, administratively unmanageable, nonresponsive to farmer needs, and supplied inputs of low quality (World Bank, 1992).

The food processing industry was consistently undercapitalised and was among the weakest links in the food distribution system. As elsewhere in the Soviet economy, centralised and remote administration hindered the efficient direction and utilisation of resources.

Some attempts to reform the system started during the *perestroika* period. In 1988 and 1989 selected enterprises changed their legal status under the co-operative legislation. Then the law on enterprises opened the door for "spontaneous" or "nomenklatura" privatisation through which enterprises were rented to management and employees and next converted into joint stock companies with exclusive insiders' ownership rights. Many agricultural downstream enterprises were affected by this type of privatisation.

2. Objectives of privatisation

The Russian privatisation programme started in mid-1992 had both political and economic objectives. Political purposes were even more important than economic and in the first instance included "depoliticisation" of Russian enterprises by depriving branch ministries of their ownership of and control over enterprises. The government also wanted to take control over the rampant process of "spontaneous" or "nomenklatura" privatisation. Another political purpose was to create a new class of property owners in such a way as to make the privatisation irreversible. To gain the political support of major social groups the programme granted substantial concessions to each of the major participants. As a result, workers and managers, and, in the case of some agricultural upstream and downstream enterprises, agricultural producers, were given pre-emptive rights on substantial parts of the assets of the privatised enterprises. They had the possibility of acquiring shares under preferential terms, including blocks of shares distributed free or at a discount and the possibility of paying in instalments which, given high inflation, represented a further price cut. Moreover, each Russian citizen received the right to acquire assets using vouchers distributed for a symbolic low fee.

Economic objectives, although less pronounced than political aspects, included improved efficiency of enterprises, a demonopolised and competitive environment for enterprises, increased budget revenues, and a larger inflow of foreign investments. However, given the concessions made to employees, the budget revenues from privatisation were not significant. Similarly, at the beginning of the process, foreign investors could not be attracted by the minority shareholding left after satisfying the rights of management, employees and other groups of Russian society. The programme paid little attention to the role of corporate governance and, in most cases, the initial distribution of shares did not give any participant enough of a shareholding to exert sole control over the enterprise (OECD, 1995).

3. Legal framework and methods

i) General privatisation programme[20]

General rules governing privatisation in Russia were fixed in the "Law on Privatisation" adopted by the Russian Supreme Soviet on 3 July 1991 and specified in the "Government Programme on Privatisation of State and Municipal Enterprises in the Russian Federation in 1992" adopted by the Russian

Supreme Soviet on 11 June 1992. The latter was largely repeated in the privatisation programme of 24 December 1993 with some modifications reflecting a more advanced stage of the process.

The Russian privatisation programme divided all enterprises into three groups: enterprises subject to mandatory privatisation; enterprises whose privatisation was discretionary, depending on the decision of the Federal Government, local government or the State Committee on Management of State Property (GKI); and enterprises which were exempt from privatisation. Enterprises in the first two groups were divided into three categories subject to different types of privatisation. Small enterprises, *i.e.* enterprises with up to 200 employees and book value of assets less than Rb1 million as of 1 January 1992, were to be sold in their entirety through competitive bidding or lease buy-out. Large-scale enterprises, *i.e.* enterprises employing more than 1 000 workers and with a book value of assets above 50 million as of 1 January 1992, were to be transformed into joint-stock companies (JSCs) and their shares sold or distributed in accordance with the provisions of the mass privatisation programme. Medium size enterprises could be privatised through either the direct sale or the corporatisation method. Most assets were to be transferred into private hands within a period of 18 months. It was considered that a decentralised approach, based on the initiative of each enterprise to prepare and carry out its own privatisation plan, would be the best way to meet this ambitious deadline.

Mass privatisation was mandatory for 5 000 large enterprises and optional for 16 000 to 20 000 medium-sized enterprises. Enterprises had to initiate privatisation by submitting a privatisation plan, their asset valuation and joint stock charter by October 1992 at the latest. Among other elements, the privatisation plan had to specify which option for employees was to be followed (Box III.4). Meanwhile, 150 million vouchers were distributed for a small fee of 25 roubles each to all Russian citizens. Each voucher had a nominal value of 10 000 roubles, compared to the average monthly wage at the time of 6 000 roubles. Vouchers were bearer documents, fully tradable for cash. Their validity was initially set to expire by 31 December 1993 and then extended to 30 June 1994. Vouchers could be used to buy small-scale businesses and enterprise shares at auctions, and to pay for housing.

After corporatisation, the sale of enterprise assets was carried out in three steps: first, a closed subscription through which employees could buy shares of their enterprise within the limits specified in the option they had chosen; second, the sale of given amounts of shares at public voucher auctions; and, third, the sale of the remaining shares at cash auctions.

Mass privatisation was under the responsibility of two administrations. First, the GKI was responsible for preparing and implementing the programme. Second, the Federal Property Fund (FPP) was the title owner of the state assets and their legal sellers. The FPP was initially subordinate to the Parliament but, as from December 1993, both the GKI and the FPP became directly accountable to the government. Both administrations had *oblast* and municipal branches.

ii) Specificity of privatisation in agricultural upstream and downstream enterprises

The general principles of the Russian privatisation programme were applied to the privatisation of agricultural upstream and downstream enterprises. Most of these enterprises were included in the group of enterprises covered by mandatory privatisation. Some downstream enterprises, in the cooling industry and elevators responsible for storing government stocks, were included in the group of enterprises whose privatisation depended on the discretionary decision of the Federal Government. At a later stage the list of these enterprises was shortened and privatisation was allowed provided that state ownership in total share stock was secured at the level of 51 per cent. Since 1997, the government's share may be even lower provided that the government receives a "Golden Share" allowing it to make crucial decisions concerning the enterprise.

However, more than one fourth of upstream and downstream enterprises, those which were involved in direct supplies (input traders and service suppliers) to and direct purchases from agricultural producers, were covered by special provisions giving preferences to agricultural producers in privatisation of these enterprises. Input producers and second stage food processors, for example bread producers buying grain from elevators, were not covered by these provisions. In general terms these special provisions were included in the 1992 Privatisation Programme and next specified in the

Box III.4. **The three options of employee preferences in the Russian privatisation programme**

In the process of corporatisation of large- and medium-size enterprises the employees were given pre-emptive rights on purchases of parts of the assets on preferential terms according to one of the following options:

Option 1: employees (workers and managers) were given 25 per cent of non-voting shares free of any charge. In addition, workers were entitled to buy for cash or vouchers another 10 per cent of voting shares at closed subscription. These shares were to be sold to them at a 30 per cent discount of the January 1992 book value and could be paid in instalments over three years.[21] Moreover, up to 5 per cent of voting shares could be sold to management at nominal prices.

Option 2: workers and managers could buy 51 per cent of voting shares at preferential prices fixed at 1.7 times the book value of the enterprise as at 1 July 1992. These sales could be made for vouchers and cash.

Option 3: applied only to medium-sized enterprises. A managing group (made up of workers, managers or any physical or legal persons) took the responsibility of meeting the privatisation plan's objectives while ensuring the solvency and the retention of the employees of the enterprise for at least one year, subject to the right to buy 30 per cent of the voting capital. Another 20 per cent was to be sold to employees and managers (including those belonging to the group) on preferential terms (30 per cent discount and instalment payments over a period of three years[22]). If the managing group could not meet the objectives of the privatisation plan, its shares were to be sold at cash auctions after the completion of the voucher privatisation.

Moreover, from 5 to 10 per cent (depending on the option) of voting shares were held for subscriptions financed by the Share Fund for Employees of the Enterprise (FAPR) and enterprise Privatisation Accounts. Only employees of the enterprise were eligible to receive or to buy shares from the FAPR. For enterprises initiating privatisation after 1 February 1994, the creation of FAPR was not envisaged. In total, the closed subscription gave enterprise managers and employees pre -emptive rights over 40 to 60 per cent of the capital of their enterprises. The decision on which option to choose had to be made by the general meeting of the work collective. If the second or third option was to be selected, it had to be approved by at least two thirds of the work collective. In the absence of such an approval, the first option had to be followed.

Source: OECD *Economic Surveys*: *The Russian Federation* 1995, Paris, OECD, 1995.

regulations "On Privatisation of Enterprises Involved in Primary Processing of Agricultural Produce, Fish and Sea Products and Enterprises Involved in Production and Technical Servicing, and the Material and Technical Supply of the Agro-Industrial Complex" approved by the Resolution of the Government of the Russian Federation No. 708 of 4 September 1992. The regulations stated that the enterprises should be privatised in accordance with one of the three options of preferences given to management and employees and all remaining shares should be sold at nominal prices (valued at the 1 January 1992 book value of assets) at closed auctions to agricultural and fish producers not later than three months after the decision on privatisation had been made. Payments for shares could be made in instalments over three years. The maximum number of shares which could be bought by a producer could not exceed his share in the total amount of services provided by the supplier or in total purchases by the downstream buyer being privatised. No less than 10 per cent of the shares were to be reserved for individual farmers and other newly emerging agricultural enterprises. In case of a lack of purchasers at the closed auction the shares could be sold through other means specified in the regulations.

These provisions were slightly modified in 1993 and next in 1994. The President's Decree No. 1767 (27 October 1993) and the Government Privatisation Programme for 1993-94 (24 December 1993), stated that the compulsory corporatisation of upstream and downstream enterprises (as defined above) should be made in accordance with the first or the third options of employee preferences (Box III.4), thus excluding the second option giving a clear majority of share ownership to the employees, and that rural populations living in "the raw material zone of a privatised enterprise" should be covered by the preferential distribution of shares on the same terms as agricultural producers. Moreover, the programme specified that shares not distributed during the closed auctions should be sold on open auctions.

Later, the President's Decree no. 2205 (20 December 1994) declared that in order "to protect agricultural producers against monopoly power of processing industries and of service suppliers" the upstream and downstream enterprises (again, as defined above, but including additional ones in accordance with the lists approved by the GKI) should be privatised through corporatisation according to the first option of preferences given to employees and all remaining shares should be sold at closed auctions to agricultural producers (the rural population was not mentioned). The decree also recommended that upstream and downstream enterprises transferred into joint stock companies (JSCs) at an earlier stage and not complying with the decree, should organise the second emission of shares and distribute them free of charge among agricultural producers.

4. Privatisation of supply and service enterprises

The vast majority of agricultural input producers were privatised in accordance with general privatisation laws. By the end of 1996, out of 173 agricultural machine producing enterprises 134 (77 per cent of the total) were completely privatised, 13 (about 8 per cent) were still state owned and 26 (15 per cent) were partly owned by the state. In this last group were such enterprises as the tractor enterprise in Lipetsk, Rostselmash and the combine enterprise in Tula. However, Rosagrosnab, currently a joint stock company with shares partly owned by the State, enjoyed monopoly power over the distribution of farm machinery in the framework of the government leasing programme until 1996 (Part IV).

Input distribution and service enterprises were privatised according to the specific rules giving agricultural producers privileged access to shares. On the basis of information available from 71 regions, about 80 per cent of such enterprises were privatised by the end of 1996 (Table III.4). About 75 per cent of those privatised were transformed into JSCs, most often according to the second option of employee preferences (Box III.4), thus giving the employees the majority in total shareholding. Agricultural producers bought all shares for which they were eligible in just above one half of the total number of enterprises transformed to JSCs, and in 12 per cent of the JSCs, they acquired the majority shareholding. In just 74 cases, about 1 per cent of the total number of privatised enterprises, the enterprises issued additional shares to meet the President's recommendation of December 1994 (see above).

5. Privatisation of processing enterprises

The food industry had the largest number of industrial enterprises in the former RSFSR (about 5.7 thousand, *i.e.* more than 20 per cent of the total) and the smallest number of employees per enterprise (273 compared to the average of 780 in 1990), which was an advantage during the privatisation process. Most food processing enterprises were included in the general privatisation programme adopted in 1992.

By 1996, 96 per cent of food processing enterprises, producing 93 per cent of food products and employing 91 per cent of labour employed in the food industry, were fully or partly privatised. While the percentage share for the number of privatised enterprises was at the average for the whole of Russian industry, the other two shares were significantly above the averages. This indicates that privatisation in food processing industry has been proceeding more rapidly than in the industry as a whole. However, about 12 per cent of enterprises, mostly large ones, producing 40 per cent of food products, were mixed enterprises with state participation in their capital (Table III.5).

Primary food processing enterprises, involved in direct purchases of agricultural produce from agricultural producers, were to be privatised in accordance with special provisions giving agricultural producers privileged access to shares. On the basis of information available from 71 regions, more than 90 per cent of such enterprises were privatised by the end of 1996. Almost 90 per cent of them were

Table III.4. **Privatisation of upstream and downstream enterprises covered by the special rules of privatisation, number of enterprises as of 1 January 1997**[1]

Type of activity	Total	Privatised		Transformed into Joint Stock Company						Organised second emission of shares
					Option of preferences for employees			Enterprises in which agricultural enterprises fully used their rights to buy shares on preferential terms		
		Total	Partly state owned	Total	I	II	III	Total	With agricultural enterprises' majority in total shareholding	
UPSTREAM										
Repair	1 710	1 568	256	1 355	382	947	26	789	156	28
Chemicals supply and services	1 398	1 105	217	968	346	618	4	580	145	11
Construction of electric lines	726	468	20	89	47	37	5	766	43	–
Transport	632	575	78	515	99	414	2	234	48	18
Building construction	2 481	1 637	163	930	286	621	23	259	91	–
Animal medicaments supply	153	49	10	49	16	33	–	21	7	–
Vegetable seeds supply	93	28	3	23	5	18	–	18	4	–
Machinery, spare parts, fuels and other material supply	1 366	1 199	179		207	799	38	640	109	17
TOTAL	8 559	6 629	926	4 968	1 388	3 487	93	2 729	603	74
DOWNSTREAM										
Meat industry	529	487	80	449	65	367	17	282	45	24
Milk industry	1 522	1 399	257	1 245	268	955	22	939	227	46
Bakery	1 333	1 197	411	1 053	483	514	56	604	173	18
Linen production	23	19	6	17	13	4	–	11	2	–
TOTAL	3 407	3 102	754	2 764	829	1 840	95	1 836	447	88

1. Based on information provided by 71 regions.

Source: Ministry of Agriculture and Food, 1997.

Table III.5. **Ownership structure of food processing enterprises, 1996**

Type of ownership	Share in total number of enterprises	Share in total food production	Share in total employment in food industry
State	3.1	5.3	7.4
Municipal	1.2	1.4	1.5
Co-operative	0.1	0.1	0.1
Private	81.2	44.8	43.7
Mixed without foreign capital	11.5	40.0	44.3
Mixed with foreign capital	2.9	8.4	3.0
Total	100	100	100

Source: *Russia in Figures 1997,* Goskomstat, Moscow.

transformed into JSCs, most often according to the second option of employee preferences, thus giving the employees the majority in total shareholding. Most of them had already been rented by management and employees in the *perestroika* period with the ultimate right to purchase. There have been widespread reports of agricultural producers not taking ownership of their shares in the processors. By the end of 1996 they bought all shares for which they were eligible in about 60 per cent of enterprises transformed into JSCs and in 16 per cent of them they acquired the majority shareholding. In 88 cases, about 3 per cent of the total, the enterprises issued additional shares to meet the President's Recommendation of December 1994 (Table III.4). Consequently, in 1997, the MAF prepared a proposal of so called "second stage privatisation" which goes along the President's Recommendation lines and proposes that small-scale dairy plants, grain storage elevators and some small scale processing enterprises located in rural areas should issue additional shares to give majority shareholding to farmers.

As a result of the creation of the new enterprises and, to some extent, due to the privatisation and reorganisation of old ones, the number of enterprises in food processing industries increased from 5.7 thousand in 1990 to 13.9 thousand in 1995 and the average number of employees declined from 273 in 1990 to 108 in 1995 indicating somewhat better adaptation to more fragmented farming structures and more differentiated consumers' needs. However, the food processing industry is still in very poor shape. While between 1990 and 1996 food production almost halved, the total number of employed remained roughly at the 1990 level meaning significant overemployment. Moreover, the decline in production meant very low capacity use resulting in high fixed costs. These factors, combined with a lack of proper management skills and a lack of capital, resulted in low efficiency of the industry, undermined its international competitiveness and contributed to significant implicit taxation of agricultural producers (Part V).

As explained in Part I, the inflow of Foreign Direct Investment (FDI) into Russia has been small. By 1996, only about 3 per cent of food processing enterprises were privatised with the participation of foreign capital (Table III.5), reflecting both the shaky economic and political situation in Russia and rather unfriendly privatisation legislation giving strong preferences to insiders and agricultural producers. However, in recent years the inflow of FDI to the agro-food sector (including food retailing and catering) increased from US$682 million in 1995 to US$831 million in 1996, and US$980 million in 1997. The stock of FDI in the agro-food sector at US$3.1 billion in 1997 accounted for 24 per cent of the total FDI stock in Russia. Most FDI in the agro-food sector was concentrated in food processing (58 per cent of the total) and in food retailing and catering (40 per cent). Agricultural production attracted less than 2 per cent of the total. US and UK investors are the most active on the Russian market. Foreign investors are becoming more and more involved in the production of high value products such as confectionery, tobacco products, baby food and ice cream. It may be expected that with the continuing macroeconomic stabilisation, the amount of foreign direct investment will significantly increase in Russia.

6. Development of wholesale trade

In the Soviet period about 70 per cent of food products was channelled through state wholesale and retail outlets, about 25 per cent through co-operative outlets and about 5 per cent through so called *kolkhoz* markets. The first two channels were strictly regulated and vertically organised into four basic chains of institutions subordinated to the Ministry of Trade, to the Ministry of Procurements (for grain and grain products), to other ministries (*e.g.* Ministry of Railways) supplying their workers with food and other products in so called remote areas, and to Centrosoyuz (for co-operative organisations). The *kolkhoz* markets were located in towns and were supplied with products from household plots and, to a limited extent, from surplus production originating from *kolkhozes* and *sovkhozes*. Prices and margins were fixed by the state at all levels of the food chain with the exception of products sold on the *kolkhoz* markets which were less controlled. Both the state and co-operative channels had special divisions at the local, *oblast* and federal levels, set up on a territorial and commodity basis, organising the work of retail shops, financing their activities and providing them with goods. To carry out the latter function they also operated a network of storage facilities.

At the end of 1991 many of these vertical structures were reorganised, but they remained state-owned and the reorganisation brought only a few changes into the relations between producers and purchasers. More profound changes started with the 1 July 1992 President's decree "On organisational procedures of transforming state enterprises and their associations into joint stock societies" and the government regulation, dated 3 July 1991, "On Privatisation of the State and Municipal enterprises in the RSFSR". In particular, the President's decree requested that previous vertical structures regulating purchases of agricultural products from farmers, as well as food production and food distribution, should be transformed into joint stock companies within a period of three months. In practical terms, this meant that previous branch ministries and committees and their regional sub-units were transformed into open joint stock societies with shares owned by processing enterprises and other enterprises and institutions. Their functions changed from direct regulation to activities servicing shareholders, such as supplying them with inputs, searching for markets, and conducting research and development. Some departments of the ministry of trade (*e.g.* meat and dairy department) operated as dealers: they contacted buyers and sellers for a charge. As reforms progressed, the regulating functions were discontinued at the federal level, but field research indicates that in some towns the local divisions still exist and implement the wholesale functions. However, since a large number of new private agents appeared and the trade activities linking processors and retailers became highly competitive, the remaining local administrative units are not able to exercise their monopolistic position any more.

With the dissolution of the vertical structures and their forced corporatisation, combined with the liberalisation of prices, the role of the state in purchases of agricultural products from farmers has declined significantly in recent years and new sale channels have appeared. In 1994 a federal programme for the development of private agro-food wholesale markets was prepared, but due to a lack of budgetary support the programme was implemented on a very limited scale (Part IV). In general, the link between agricultural producers and processors remains one of the weakest in the whole food chain and producers quite often face a semi-monopsonic position of food processing enterprises.

7. Restructuring of retail trade

Privatising and restructuring of retail food trade underwent two parallel processes: establishment of new retail outlets and privatisation of state and co-operative shops, canteens and restaurants. Particularly just after the liberalisation, a large number of private retail outlets emerged, at first small kiosks and shops which then, in the process of restructuring, were often transformed into larger units.

In accordance with the 1992-1993 privatisation programme, the vast majority of state enterprises in retail trade, food service and consumer service were subject to mandatory small-scale privatisation. However, since these enterprises were subordinated to local authorities, the real privatisation process was strongly influenced by various regional approaches. In many regions these enterprises were already privatised in 1992, but in many other regions they were transferred to municipalities and their privatisa-

Box III.5. **Privatisation of Centrosoyuz**

The privatisation of Centrosoyuz, in the past a dominant operator of retail stores in rural areas with numerous outlets in some urban areas, was a specific case. Although the organisation was formally a consumer co-operative, and individuals in rural areas bought shares in order to use the services of Centrosoyuz, its activities were strongly regulated by the state, and shareholders did not participate in its management. Centrosoyuz bought directly from farms and individuals, processed and then resold either to rural or urban dwellers. The organisation was privatised in accordance with the June 1992 Russian Federation Law "On Consumer Co-operation" with similar procedures to those applied to the privatisation of *kolkhozes, i.e.* property shares had to be calculated and the right to exit with shares was formally assured. By that law the organisation was privatised as a single unit while the restructuring of local co-operative shop networks was decided by the local authorities.

In general, retail trade in the countryside is still neglected. Centrosoyuz has not been able to maintain its existing network and, despite relatively good equipment being at its disposal, has had difficulties in competing with small scale traders selling their products just outside the Centrosoyuz shop. It has therefore requested subsidies and privileges from local authorities. For example in the Orel *oblast*, during the bread price regulation period between 1992 and 1994, Centrosoyuz shops received special privileges for the exclusive selling of bread at controlled prices. By contrast, in the Nizhny Novgorod *oblast* and in one of the Moscow region districts, the local authorities gave the former share-holders of Centrosoyuz the right to privatise small facilities in the villages, such as shops and canteens, which stimulated the development of retail trading in rural areas in these regions.

tion continued even up to 1995. It was for municipalities to sell these enterprises for cash and vouchers through auctions and tenders. The scope of small-scale privatisation was restricted in various ways. First, commercial tenders gave the municipalities the possibility of imposing specific restrictions, for example, on layoffs or production profiles; second, municipalities had the discretion not to allow foreign participation; third, incumbent management and workers were given a 30 per cent discount in cases where their bid was successful and could pay over three years. Moreover, the municipalities tended to keep ownership rights to land and buildings, renting these assets to the owners of businesses this being not only a means of extracting revenues for local budgets, but also of regulating the range of goods and even prices.

In the majority of cases former state retail outlets were taken over (privatised) by the employees and no effective owners emerged. However, loss-making shops in particular, are being sold to individuals and emerging retail shop chains. The chains are being established by both retailers and processors and are amalgamating new and previously existing individual shops and restaurants. Catering systems and food retail trade are becoming more and more attractive for foreign investment. Many fast food chains have established their networks in Russia, including McDonalds, Kentucky Fried Chicken, and into Pizza-Hut. In 1995, the inflow of FDI into retail trade and food service was US$438 million, *i.e.* 23 per cent of the total FDI in Russia that year. In 1996, the amount declined to US$255 million, but in 1997 it increased again to US$455 million.

8. Changes in foreign-trade enterprises

In the Soviet period, most important foreign trade organisations in the food area were Exportkhleb (exports of grains, but in practice occupied mostly with grain imports, see Part II), Soyuzplodoimport (imports of fruits) and Skotimport (imports of animal products). All these were state owned, administered by the central institutions and managed in the overall framework of the state monopoly on foreign trade transactions (for more details see Part II). After the liberalisation of 1992 the status of so-called special exporters was preserved, especially for grain and oilseed exports. This status was given to selected foreign trade enterprises and other organisations registered in the Ministry of Foreign Economic Relations. They enjoyed wide-ranging customs preferences, officially to allow the organisations to earn foreign currency for investment. Up to 1994, Exportkhleb and Roskhleboproduct were given the status of government agents for centralised imports of grain. Moreover, state import guarantees, available to selected importers only, were given to other traders, for example sugar importers. Partly due to these special arrangements, large-scale organisations, both privatised former Soviet foreign trade organisations and new ones emerging during the reform period, tend to dominate in foreign trade transactions in Russia.

On the sugar market the major operators are such organisations as MENATEP-ImpEx, Alfa-Eco, Trade House "Russia's Sugar", Soyuzcontract, and Sucden. These companies, private and established during the reform period, control around 75-80 per cent of the white sugar market in Russia. Over several years the Ministry of Finance provided them with government guarantees for sugar imports. MENATEP and Alfa were also the agents on government authorised oil for sugar deals with Cuba. Periodically, "short-lived" companies appear on the sugar market for short-term operations. Usually they do not have their own processing or transporting facilities and disappear when the transaction is done.

On the grain market there are also several large traders with the largest remaining Exportkhleb, which is now privatised with a small state share in its equity. Another important grain trader is OGO, which emerged in the reform period. Both MENATEP and Alfa are also engaged in the grain trade business. In major grain production areas, state-owned regional food corporations were quite often directly involved in grain exports.

Soyuzcontract and Tyson from the USA are the main importers of poultry meat into Russia. The biggest meat plants, such as Cherkizov meat plant in Moscow, are directly importing raw meat for further processing.

At the beginning of the reform period, Russia was exporting large quantities of skimmed milk powder and casein. A large number of rather small companies was engaged in these operations. For example, in Moscow alone there were between six and eight companies collecting these products all over the country and exporting mostly to EU countries.

Roscontract is the major trader dealing with the NIS, fulfilling inter-governmental clearing agreements between these countries. It has received duty cuts and preferential credits. In 1995 the state limited its share in the equity from 51 per cent to 25 per cent. Roscontract is the major importer of meat from the NIS with about a two-thirds share in meat imports from these countries in 1994.

E. PRIVATISATION AND REORGANISATION OF THE SOCIAL AND TECHNICAL INFRASTRUCTURE

In the Soviet period, *kolkhozes* and *sovkhozes* were charged with the supply of a wide range of services in rural areas and so called "social sphere" facilities were provided, maintained and to a large extent financed by agricultural enterprises. These facilities included housing, clubhouses, medical centres, libraries, kindergartens, schools, playgrounds, recreation equipment and others. Moreover, *kolkhozes* and *sovkhozes* were also providing various public utilities such as electricity, gas, water and heat supply systems, fire stations, sewerage systems, roads between villages and the telephone network.

According to the 1991/1992 legislation (in particular Resolutions No. 86 and No. 708), state and collective farms, as well as their successors, may transfer (*mogut peredavat'*) their social sphere and utility assets to the relevant provincial and local authorities. Upon transfer, the cost of operating and maintaining the social assets and utilities becomes the responsibility of the local administration or the appropriate government agency. However, there is no legal obligation on the part of the local authorities to assume the assets (Brooks, 1994).

During the reorganisation process of agricultural enterprises, the farm commission was charged with preparing a list of property that was to be transferred to the local authorities and, thus, excluded from the property entitlement fund (see above). The farm commission was obliged to make preliminary recommendations concerning the public agency to which the social assets and utilities should be transferred. The final decision on such proposals was to be made by the local administration which was

Table III.6. **Transfer of utilities and social assets to local authorities by 1 January 1997**

	Percent transferred
Housing	19
Schools	38
Kindergartens	36
Clubs	21
Water supply network	20
Gas supply network	6
Electric power network	6
Roads	24

Source: Ministry of Agriculture and Food, 1997.

also to make a decision on the allocation of funds for maintenance of these assets from the local budget and/or request financing from the regional budget.

Privatisation of housing was a specific case. In the event that tenants were unable to buy their housing, the ownership of houses (or apartments) had to be transferred to the local administration as part of the social sphere under the payment terms agreed between the farm and local administration. However, there was a distinct difference between *sovkhozes* and *kolkhozes*. In *sovkhozes* as well as in enterprises emerging from them during reorganisation, houses and/or apartments were not the property of workers' collective but were owned by the state. Therefore, housing privatisation in state farms had to be carried out in accordance with federal and regional legislation on the privatisation of state-owned housing. In *kolkhozes* and enterprises emerging from them, housing was owned by the collective. Therefore, the general meeting of the enterprise's collective could choose the method of housing privatisation. For example, the value of housing owned by a collective enterprise could be included both in the entitlement fund and in the calculation of property entitlements. In this case, any tenant had the first rights to use his/her property entitlements to obtain the housing of his/her residence (IFC, 1995).

While situations may differ across regions, the transfer of the social assets and utilities to local authorities has been very slow. According to MAF data at the beginning of 1997, agricultural enterprises continued to be responsible for about 70 per cent of their pre-transition stock of social assets and utilities. Only 6 per cent of gas and electric supply systems had been transferred by that time (Table III.6). The most important reason is a lack of finance at local level to put such transfers into practice. A very small federal budget allocation for this purpose is distributed in regions between rural and urban areas with the preference given to the latter. Other reasons quoted include a general lack of institutional framework at the local level to take the assets over, inertia, high prices charged by some of the new owners of already transferred utilities that make farm managers reluctant to transfer the utilities, and reported cases of cuts in gas and electricity supplies by the new owners to the isolated villages. In some regions, such as Nizhny Novgorod, the local administration creates communal services to take over and maintain the utilities. Some assets, previously on the balances of *kolkhozes* and *sovkhozes*, such as bread bakeries, shops, canteens, transport and repair equipment have been privatised, most often in exchange for asset entitlements distributed in the process of the reorganisation of large-scale enterprises. The small-scale enterprises emerging in this way service both village dwellers and agricultural enterprises and add to the few non-agricultural employment opportunities in rural areas.

The sooner the process of releasing agricultural enterprises from non-production related activities is finalised, the better for the economy. The current situation in which social assets and public utilities are still on the balances of agricultural enterprises diverts large farms' financial and management resources from the commercial functions of the enterprise; hampers the restructuring process of agricultural enterprises; and makes village inhabitants continue to depend on services provided by the enterprise making them less interested in reallocating their land and non-land entitlements to other, possibly more efficient, enterprises and/or family farms.

NOTES

1. In 1910, the State Duma adopted a law which regulated land transactions more precisely: peasant land was not considered as personal private property, but as household property. This meant that any land transaction could be made only by the decision of all adult members of the family concerned. The land could be sold only to a neighbour or member of the same village community. The maximum size of land property could not exceed five times the average land plot in the community. All financial operations connected with the peasant land market, including the use of land as collateral, had to be conducted through the state peasant bank. Private banks could not operate in this area.
2. Although these plots have commonly been referred to as "private plots," this land was no more privately owned than that of the large farm which allocated it. However, houses and other structures on the "private plot" were privately owned.
3. The Constitution provides for private, state and municipal ownership of land and guarantees that land owners may freely possess, utilise and dispose of land within limitations set out by law. For example land must be farmed.
4. "Peasant farm" tends to imply, in Russian as in English, a subsistence farm which is not market-oriented. These new farms were intended to be producing for the market, and so their advocates revived a Russianised English term, *fermerskoe khoziaistvo*, literally a "farmer's farm," to describe them. Since the most important distinction is between the large farms run like factories on the land and the "peasant" farm which is owner-operated, these new farms are called "individual" farms here.
5. 1991 was a particularly volatile year: it started with price liberalisation; in June Yeltsin was elected president of the Russian Federation; in August there was a coup attempt; in November the Communist Party was prohibited; at the end of the year the Soviet Union collapsed.
6. Conditional land and asset shares are shares that exist on paper and can be converted to actual land plots and assets (or monetary equivalents) if the shareholder decides to exit the collective.
7. This decree had the additional effect of making all social workers unconditionally eligible for land entitlements.
8. The redistribution fund is under the supervision of the *raion* land reform committee, which allocates the land from the fund in several ways. One way is to lease some of it to individual farmers who have already exited to supplement their (usually small) land shares, thereby creating a more rational economic unit. Another is to allocate it to migrants wishing to establish individual farms. However, when collectives transfer their excess land to the *raion* land committee, they choose the least accessible land with no infrastructure. Land that is not distributed to create or supplement individual farms is leased back to the collective to which it originally belonged.
9. These were to include fixed and unfixed assets. Buildings, livestock, machinery and equipment were part of the former; crops in the field or in storage, seed, chemicals and cash were part of the latter.
10. The 1995 Law on Agricultural Co-operatives provided for the creation of an indivisible fund which could be used for the creation of service co-operatives to service agricultural production co-operatives and other agricultural enterprises and/or private farms emerging from the reorganisation of the previous large *kolkhoz* or *sovkhoz*. Depending on the decision made by the general meeting, the indivisible fund could consist of storage capacity, repair service, drying machinery, mixed-feed production capacity, etc.
11. There were actually three variations of the formula but all used the same variables. Salary calculations were to be adjusted for inflation in the period immediately preceding the reorganisation.
12. It was recommended that the non-land property inventory list and the list of non-land property entitlement holders be approved by the *Raion* Reorganisation Commission. The land inventory and land entitlement holders' lists had to be approved by the *Raion* Land Committee (*Raikomzem*). Up to October 1993, the *Raikomzem* had to prepare a State Act on common land ownership for the entire farm. Lists of land entitlement holders had to be

attached. After October 1993 each land entitlement holder has to receive an individual certificate for land ownership.

13. The total number of large-scale farms and their average size have been calculated on the basis of data provided by the Goskomzem (see in particular Table III.1). According to Goskomstat, in 1996 the total number of large farms was 26 874 with the average size of 5 909 hectares. However, it is not clear to what extent other enterprises, partly engaged in agricultural production, and the land cultivated by them are included into the average.

14. In addition, in 1997 there were about 22 million garden plots *(sadovyi uchastok)* and vegetable plots (*ogorodnyi uchastok*) held by urban families. In total they cultivated about 2.5 million hectares.

15. By that time a relatively large number of individual farms was also established by Russian immigrants coming back from the "near abroad" countries after the collapse of the Soviet Union.

16. Earlier such a possibility existed, but was rather theoretical; now the contracts are signed and private farmers may participate in the process of reallocation of land certificates.

17. Current legislation allows for sales and purchases of land entitlements which can be followed by transfers of land in kind.

18. Article 72 of the Constitution assigns "issues relating to the ownership, use and disposal of land" to the sphere of joint federal-regional jurisdiction, leaving the possibility for the regions to act in the absence of federal legislation. Profiting from such a possibility, a law providing for the purchase and sale of land, including agricultural land, was passed by the regional Duma in Saratov and signed by Saratov's governor in November 1997. The first land auctions took place in the Saratov region at the beginning of March 1998. Saratov's example seems to be attractive for other regions. In April 1998, the parliament of the internal Russian Republic of Tatarstan adopted a Land Code which allows the free sale and purchase of land, including to foreign individuals and companies.

19. This situation restricts the possibility for a reorganising agricultural enterprise to take the legal form that would be the most suitable for it under new circumstances. This is due to the fact that during the 1992-1994 reorganisation, quite often formal and enforced by legislation, many agricultural enterprises did not care about the legal status under which they were reregistered. Joint stock companies became most popular, but now according to the Civil Code joint stock companies cannot be reorganised into other than limited liability companies or production co-operatives.

20. The description of the Russian privatisation process draws on: *OECD Economic Surveys: The Russian Federation 1995*, Paris, OECD, 1995.

21. For enterprises privatised according to this option in 1993 and later the period was shortened to three months.

22. In 1993 and later, the period was shortened to three months.

AGRICULTURAL AND FOOD POLICY OBJECTIVES AND MEASURES

A. AGRICULTURAL POLICY FRAMEWORK

As the Russian economy has been progressively transformed from a command system into a market oriented one, the main goals of agro-food policy have been altered. Although policy instruments applied in the transition were often modified and some of them remain non-transparent, the systemic transformation from a centrally planned economy to a market-oriented one underlies the entire discussion of specific policy measures that follows in this part of the report.

1. Agro-food policy in the pre-reform period

The overall purpose of Soviet-era farm policy was to ensure a guaranteed supply of foodstuffs to the population, especially in the cities, at relatively low and stable prices. For a long period, retail food prices were subsidised to keep them stable in spite of the rise in nominal per capita incomes and low growth rates in gross agricultural output. The retail food prices set in the 1950s were raised only in 1962 and again in the late 1980s in attempts to reform the planned economy. In the meantime, some increases in prices were introduced in a "hidden" way, through changes in the range of food products sold or an increase in the share of higher priced products sold through Centrosoyuz.

The policy framework was determined by the planned economy. As for the other sectors of the economy, the state authorities were heavily involved in the decision making process affecting agricultural and food production and distribution as well as the allocation of inputs and investment within the agro-food chain. Instead of the market playing the key role in determining prices and allocating resources, state planning authorities at central and regional levels made these decisions.

Most agricultural production had to be delivered to state procurement agencies at fixed prices. The only markets where prices were freely negotiated were *kolkhoz* markets. However, they served mainly as town markets where the rural population sold production from their household plots. Prices of products delivered to Centrosoyuz were contracted but some were negociated. The state also set prices for agricultural inputs. These prices represented nothing more than accounting devices, as most of the inputs were distributed based on decisions of the planning authorities.

A policy of "equal profitability" was applied towards agricultural producers. Prices across regions were differentiated on a cost-plus formula basis to ensure theoretically the same level of farm profitability nation-wide. Moreover, wages were regulated by the state on the *sovkhozes* and to some extent on the *kolkhozes*. Finally, the profit allocation to various farm funds – reserve fund, social fund, cultural fund – was prescribed by administrative rules. Such a rigid administrative system failed to provide the necessary incentives to trigger the increase in agricultural production that had been the main goal of Soviet agro-food policy. To stimulate agricultural production, direct product subsidies, as well as input and credit subsidies were increasingly channelled into agriculture, without consideration of real production costs or efficiency.

Attempts to reform agricultural policies were made in the 1980s under the centrally planned Soviet economy. This was due mainly to increasing budgetary constraints to financing expensive agricultural policies (payments to agriculture and substantial support to food consumers).[1] However, the need for a real, radical agrarian reform became manifest in Russia in the beginning of the 1990s, in line with the radical reform of the whole economy, which started the process of transition from a centrally planned to a market economy.

2. New agricultural policy objectives in the reform period

The economic reform process started in Russia in 1992. The key elements of the reform were abolition of consumer subsidies, price and trade liberalisation (of both domestic and foreign trade) and privatisation (Part I). However, reforms in the agro-food sector were implemented only gradually. Retail prices for the main foodstuffs remained under state or regional control and the system of state procurement for the main agricultural products was maintained (although narrowing in scope over time).

One of the first objectives of agricultural reform was to transform the former collective and state farms into independent, market oriented production units, including the possibility of creating private individual farms. For this purpose a special concept of farm restructuring was developed, including the privatisation of up and downstream enterprises and transfers of social infrastructure assets from farms to local government (municipalities) (Part III).

As far as the other policy objectives are concerned there was no clear concept for coherent agricultural policy during the first years of the transition. Agricultural policy was limited to *ad hoc* measures addressing the most immediate problems of the agricultural sector. In reaction to worsened output-input price ratios for agricultural producers, the government introduced livestock and input subsidies. The lack of financial resources in agriculture was addressed by credit subsidies. However, most of the underlying problems were consequences of macroeconomic instability and inadequate institutional reforms, so that the agricultural policy measures were addressing only the symptoms but not the cause of the problems.

A "Programme for Stabilisation and Development of Agricultural Production in the Russian Federation for 1996-2000" was approved by a Presidential decree in June 1996. The programme sets the main objectives for agricultural policies over a five year period. Although the objectives in the Programme were very general, the policy priority was to stop the decline of agricultural production (defined as a crisis in the agricultural sector) and to create conditions for its stabilisation during 1996-1997 and development in 1998-2000. The programme's clear goal is to increase Russian food self-sufficiency and reduce dependence on food imports. The main tools envisaged to achieve these objectives were various types of input subsidies and market intervention.

In reality many of the specifics of the programme have not been applied due to the lack of finance. Although the Presidential decree approving the programme provided for the Russian cabinet to "appropriate funds to carry out the programme" in drawing up the annual budget, it did not make providing such funds a priority or "protected".[2] Therefore, although the funds to finance the programme were earmarked in the budget, most of the programme measures were not implemented in 1996 and 1997, due to persisting tax collection difficulties.

3. Basic policy instruments

During the reform the set of policy instruments changed from administrative ones to those more appropriate to a market economy. The main instruments of agrarian policy applied at the federal level are the following:

- input subsidies (direct payment to farms partially compensating input costs; a system of installment payments in so-called "leasing"; reduced prices for some inputs and services for farms);
- direct product subsidies (mainly for livestock products);
- soft credits for agricultural producers (subsidised interest rates, favourable credit terms, commodity credit granted by the government, debt restructuring and write-offs);
- import tariffs;
- direct state investments;

– general services such as support to create a national market information system, education and extension programmes, etc.

B. PRICE AND INCOME SUPPORT MEASURES

1. Soviet period

In the final years of Soviet central planning the following prices existed within the agro-food market:

– procurement prices for agricultural products with a range of supplementary payments;

– negotiated prices for agricultural products;

– market prices for agricultural products;

– wholesale prices;

– retail prices, applicable to goods sold to final consumers through state and co-operative retail networks.

All prices, apart from negotiated and market prices, were state imposed either by the USSR State Price Committee and by the Council of Ministers in the case of certain basic products or by the State Price Committee of the Soviet Republics (for less important products).

i) Procurement prices and supplementary payments

Basic procurement prices

Up to 1992, almost 100 per cent of production marketed by the *kolkhozes* and *sovkhozes* (except for potatoes and vegetables for which the share was lower) was sold to the state at procurement prices. Procurement prices for specific products were fixed by the government and were applied to all mandatory deliveries to state organisations (state procurement agencies, state food industry enterprises, *etc.*). The procurement price consisted of a base price and bonuses related to quality and, for some products, also to timely delivery. State procurement prices were fixed centrally but differentiated by production zones, in order to ensure the same level of "product profitability" (cost-plus concept) in all regions. From the introduction of this differentiation in 1964, until 1990 the number of price zones and price variations within them grew. For 1991 the number of zones for specific products was substantially reduced (Table IV.1). Since 1992 the centrally applied price zoning system has been abolished and the procurement system changed radically (Section IV.B.2).

Table IV.1. **Number of price zones for main agricultural products in Russia**

	1990	1991
Wheat	20	3
Barley	20	4
Oats	20	3
Rye	20	4
Sunflower seed	5	2
Sugar beet	6	4
Cattle	12	8
Pigs	16	4
Sheep	8	2
Milk	15	11

Source: L.M. Semiletov, "Purchasing prices for agricultural products in Russian Federation", *Planning and accounting in agricultural enterprises*, 7/1991, page 10.

Supplementary payments

Until 1989, the administered prices were accompanied by an untransparent and complicated system of supplementary payments over and above the procurement price:

- premiums intended to stimulate production;[3]
- additional payments provided to farms that showed poor financial results in order to ensure them an adequate level of return ("equal profitability concept").

In 1989, the diversified system of various supplementary payments to the administered price was somewhat simplified and replaced by a common system of so-called differentiated supplementary payments. These differentiated supplementary payments compensated for the two above mentioned supplementary payments, as well as for some other discontinued payments from the previous period such as:

- investments for development of rural areas;
- compensation of the social insurance payments for low-profit and loss-making *sovkhozes*;
- payments compensating expenses for certain inputs and investments (soil improvement, some types of machinery);
- compensation for administrative increases in wholesale input prices.

The value of transfers under these programmes was reallocated into the new differentiated supplementary payments system. In addition to the reform of prices and supplementary payments, farm debts

Table IV.2. **Total value of supplementary payments to administered prices for main agricultural products**

	1986	1987	1988	1989	1990	1991
Millions of roubles in current prices						
Total	11 783	12 375	19 399	23 825	21 206	5 651
Crop products	2 517	2 235	2 480	3 419	992	n.a.
of which:						
Grains	1 672	1 291	1 324	1 811	0	478
Sunflower seed	47	189	213	350	6	29
Sugar beet	260	316	429	492	381	71
Potatoes	285	183	133	289	46	191
Livestock products	9 267	10 140	16 918	20 406	20 214	n.a.
of which:						
Cattle	3 303	3 796	5 949	7 353	5 396	1 621
Pigs	931	1 042	1 648	2 082	1 556	436
Sheeps and goats	122	166	344	430	282	80
Poultry	147	187	96	115	135	180
Milk	4 325	4 497	7 816	9 321	11 439	1 945
Wool	387	395	842	841	991	209
Shares by product in %						
Total	100.0	100.0	100.0	100.0	100.0	100.0
Crop products	21.4	18.1	12.8	14.4	4.7	
of which:						
Grains	14.2	10.4	6.8	7.6	0.0	8.5
Sunflower seed	0.4	1.5	1.1	1.5	0.0	0.5
Sugar beet	2.2	2.6	2.2	2.1	1.8	1.3
Potatoes	2.4	1.5	0.7	1.2	0.2	3.4
Livestock products	78.6	81.9	87.2	85.6	95.3	
of which:						
Cattle	28.0	30.7	30.7	30.9	25.4	28.7
Pigs	7.9	8.4	8.5	8.7	7.3	7.7
Sheeps and goats	1.0	1.3	1.8	1.8	1.3	1.4
Poultry	1.2	1.5	0.5	0.5	0.6	3.2
Milk	36.7	36.3	40.3	39.1	53.9	34.4
Wool	3.3	3.2	4.3	3.5	4.7	3.7

n.a.: not available.
Source: Summary Annual Reports of Farms, various years.

were restructured and a special loan of Rb one billion was allocated to provide farms with financial assistance.

In 1991, the system was further simplified, as all supplementary payments were abolished and incorporated directly into newly set basic procurement prices. Simultaneously there was a reform of the entire wholesale price system, and the estimated growth of agricultural production costs (estimated at Rb 31 billion more than in 1990, due to the rise of wholesale input prices) was reflected in increased procurement prices. The total value of various supplementary payments linked to procurement prices in the period from 1986 to 1991 is shown in Table IV.2.

ii) Negotiated prices for agricultural products

Negotiated prices were used primarily for state purchases of quantities in excess of state orders and for direct inter-enterprise trade. The official consumer co-operative, Centrosoyuz, was traditionally the major single buyer at negotiated prices. Centrosoyuz purchased from farms and individuals, processed, and sold through its own retail network to consumers at prices higher than state retail prices. Negotiated prices were determined in bilateral discussions between the buyer (Centrosoyuz) and the sellers (agricultural producers) when signing delivery contracts. However, the state administration retained control over negotiated prices as well, mainly using price ceilings. No supplementary payments were paid for production sold at negotiated prices. In general the prices negotiated with the agricultural producers were higher than the state procurement prices (including all supplementary payments).

Quantities of most commodities sold at negotiated prices were small. For potatoes and vegetables, however, after state orders were abolished in 1990, all production was sold at negotiated prices either to the state or traded between private enterprises.

iii) Market prices for agricultural products

Products sold directly by individual producers to consumers at the *kolkhoz* markets were exchanged at market clearing prices. Local councils had the authority to impose price ceilings, but did so only in exceptional cases, since it was widely expected that prices on the free markets would be higher than in the state outlets. Producers had to market their output directly and no paid intermediation was allowed (although existed in practice). Individuals who did not wish to market their output directly could sell through the local state or collective farm, or to Centrosoyuz.

Table IV.3. **Share of town sales in total retail trade in Russia**

Per cent

	1985	1988	1989	1990
Meat	3.7	3.8	3.6	3.0
Milk	1.2	1.2	1.0	0.5
Potatoes	26.5	24.4	21.4	21.2
Vegetables	9.6	11.6	11.1	11.5

Source: *Statistical Yearbook of the Russian Federation*, various years.

Free market transactions accounted for only limited quantities of products, although for some (potatoes, fruits, vegetables, eggs) the share of free market sales was considerable (Table IV.3).

iv) Wholesale prices

Under the Soviet central planning system, wholesale prices for agro-food products were not fixed directly but represented a residual price calculated as the difference between fixed retail prices and fixed trade margins plus turnover taxes. The trade margins were set by the state for all types of goods to cover the estimated selling costs and to ensure a minimum profit for the distributor (cost and profit normative). However the calculated production cost was not based on the procurement price actually paid to farmers but rather on country-wide average (so-called calculated prices – *raschetnyie tseny*).

Since the retail prices of food products did not change, corresponding wholesale prices also stayed constant. Hence, increases in production costs, related to increased procurement prices paid to agricultural producers, had to be offset by direct payments made from the state budget to the food processing enterprises. The amount of compensation was fixed as the difference between the "calculated price"[4] and the fixed wholesale price. This system of compensation payments from the state budget had the objective of maintaining wholesale and retail food prices at fixed levels (regardless of increases in production costs) and effectively constituted consumer subsidies.

v) Retail prices

Food products were sold to consumers through three channels: state retail outlets, state controlled co-operative shops, and *kolkhoz* markets. Retail prices were fixed in the state stores, negotiated but administratively controlled in the co-operative shops (Centrosoyuz), and free in the town markets. According to some estimates about 70 per cent of food purchases were made at state retail outlets at the end of the 1980s.

Retail prices for food products sold through the state retail network were set centrally and differentiated according to three broad geographic zones of the country. These prices changed little from the early 1960s, with the exception of substantial increases for a few specific items (alcohol, tropical fruits, coffee, etc.) and the introduction of new products at higher prices. State prices showed little seasonal variation and poorly captured differentiated product qualities (*e.g.* beef of different cuts).

The stability of retail food prices combined with an increase in nominal wages (Table IV.4.) and a lack of spending opportunities for other goods and services led to an increased demand for food and shortages in the state retail stores. Higher prices on the parallel market and the growing shortages in state stores created pressures for an increase in official retail prices. Also the increasing gap between

Table IV.4. **Nominal wage and food retail price indices in the former USSR**

Per cent (1970 = 100%)

	1980	1985	1986	1987	1988	1989	1990
Average monthly wages	174	212	220	229	249	281	332
Foods retail prices	103	105	102	101	101	102	105

Source: Sinelnikov S., *Budget crisis in Russia: 1985-1995*, Eurasia, Moscow, 1995, pp. 20-24.

Table IV.5. **Indices of agricultural procurement prices (including supplementary payments) and of food retail prices in the former USSR**

Per cent (1986 = 100%)

	Procurement prices	Retail prices
1986	100	100
1987	103	102
1988	115	103
1989	123	104
1990	138	108

Source: Goskomstat.

rising procurement prices and nearly constant state retail prices during the 1980s (Table IV.5) resulted in unsustainable levels of consumer subsidies.[5]

The possibility of selling an increased share of agricultural production without state procurement at negotiated prices was only a partial remedy to the above mentioned problems of food shortages and budgetary pressures. The price reform in April 1991, which almost doubled the food retail prices in state outlets, was not sufficient to prevent the collapse of the state food distribution system in autumn 1991.

2. Price and income support in the reform period

i) Agricultural producer prices

The 1992 reforms abolished the previous system of centrally administered food prices and compulsory deliveries to the state. However, the state retained some control over prices for products delivered to government reserves. After the abolition of compulsory deliveries the federal state procurement system was used to ensure food supply to the large cities (Moscow and St. Petersburg), northern regions of Russia, the army, prisoners, etc. State procurement at the federal level served also to maintain strategic reserves. Since 1995 the Federal Foodstuff Corporation (FFC) has handled the procurement for federal food stocks and in each region the FFC has its branch or representative company to handle procurement purchases.

Between 1992 and 1995 the federal government did not apply any price regulation to products purchased to replenish federal stocks, with the exception of grains. For grains the government continued to fix purchase prices for deliveries to federal stocks in 1992 and for a short period in 1993. However, the government lacked the financial resources necessary to honour its price obligations and therefore in 1994 ceased to fix prices for grains as well. In 1995, following enactment of the federal law

◆ Graph IV.1. ***Share of state purchases in total marketed production of selected products, 1995, 1996 and 1997***

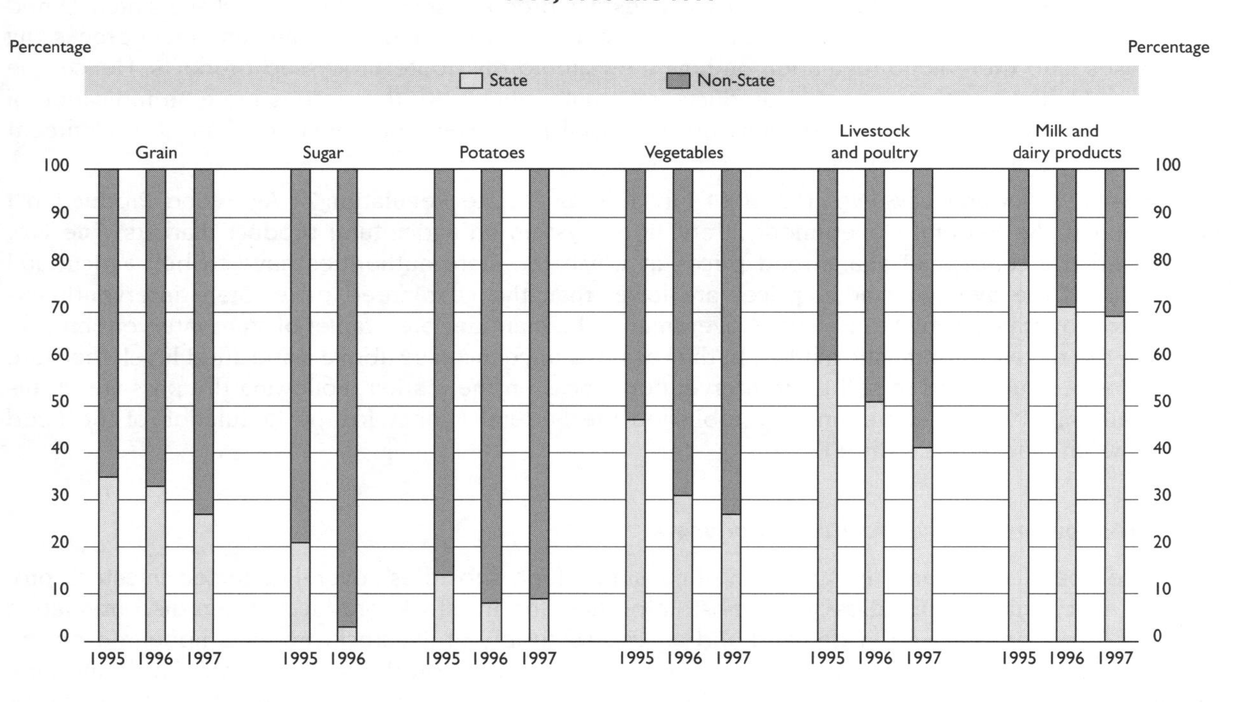

Source: Goskomstat (1997*a*) and Goskomstat (1998).

Table IV.6. **Share of state purchases in total sales of main agricultural products**

Per cent

	1992	1993	1994	1995	1996	1997
Grains	56	56	29	35	33	27
Sunflower seeds	65	32	n.a.	19	3	n.a.
Sugar beet	81	75	32	21	3	n.a.
Potatoes	57	47	33	14	8	9
Vegetables	70	67	54	47	31	27
Cattle, pigs, sheep, poultry (liveweight)	90	88	79	60	51	41
Milk and dairy products	95	96	93	80	71	69
Eggs	86	91	85	87	79	70

n.a.: not available.
Source: Goskomstat.

on "Purchases and Deliveries of Agricultural Products, Raw Materials and Foodstuffs for Government Needs", the Federal Government introduced guaranteed minimum prices[6] for grains, oilseeds, milk and meat products purchased for federal reserves. According to the law these prices were set quarterly and were in general lower (with the exception of milk) than market prices, thus reducing incentives for producers to sell to federal reserves.

With the introduction of the system of federal and regional stocks producer prices were increasingly influenced by interventions at the regional level. There were wide differences in price regulations among regions with no co-ordination from the federal level. The minimum guaranteed prices applied to purchases for federal stocks were only "recommended" for deliveries to regional stocks. As a result prices offered by local governments were considerably differentiated across regions. The overall level of state intervention, expressed as a share of state purchases (federal and regional) in total purchases, has been declining (Graph IV.1 and Table IV.6).

Moreover, deliveries of animal products to the state reported by the official statistics are rather misleading. According to available statistics all livestock products sold to official purchasers are defined as deliveries to state reserves. Official purchasers are usually privatised local dairy and meat processing enterprises, and there is no regulation on how they should distribute processed products. Hence, the fact that a high proportion of animal products is formally bought for the state is more an indication of the large number of enterprises designated as official purchasers for the state, than of government involvement in market regulation.

In 1997, a Federal Law (No. 100 of 14 July 1997 on "State Regulation of Agro-food Production") strengthened the Federal Government intervention system on agricultural product markets. The Law introduced the concept of guaranteed prices at which the state authorities have to buy agricultural production if the average market prices are lower than the guaranteed price. State intervention is limited to a quota fixed by the Federal Government. The main declared target of state intervention is to stabilise the prices on domestic market. In the case of a price increase above a specified level, the state authorities are supposed to sell their intervention stocks on the market. Following the passage of this law, in autumn 1997 the government established the Federal Agency for the Regulation of the Food Market which is to replace the FFC.

ii) Direct payments to agricultural producers

Price liberalisation and the abolition of consumer food subsidies severely affected livestock producers. As consumer subsidies were most important for livestock products, their discontinuation resulted in a sharp decline in demand and an excess supply of livestock products appeared on the domestic market. This situation depressed prices for livestock products. To counter declining farm revenues, the government introduced subsidies paid to livestock producers in March 1992. The subsidies were calculated on the basis of production delivered (milk, beef, pork, poultry, eggs) to state procurement (designated official purchaser of livestock production – see above) and were paid directly to producers. These subsidies were paid as a rate per tonne for products delivered to the federal and regional food procurement agencies. As there was no reliable way of distinguishing livestock products destined for state procurement from other purchases, all livestock production delivered to authorised purchasers was eligible for subsidies. Only products sold on *kolkhoz* markets or to a plant situated in another region (even if designated as an authorised purchaser for that region) were ineligible for subsidies. To a lesser extent, direct payments were also provided to subsidise wool, flax and soya production.

In January 1993, livestock subsidies were transferred to the regional level. However, regional governments were not strictly compelled to use the relevant transfers from the federal budget for livestock subsidies. In some cases the regional authorities reallocated the transfers to other purposes (*e.g.* for consumer subsidies in Nizhny Novgorod, for targeted investments in agriculture in Tyumen region, etc.). Together with the different schemes of market intervention across the regions, the differentiated treatment of livestock subsidies reinforced the incentives of local governments to limit inter-regional flows of food products and thus contributed to an effective break-down of the Russian market. Livestock subsidies were one of the most important programmes supporting agriculture. In 1992, livestock subsidies accounted for around 26 per cent of total budgetary expenditures to agriculture and the share declined slightly to 17 per cent in 1996. In 1997 the transfers from the federal budget to regions to finance livestock subsidies were abolished, but the subsidies as such were continued to be paid from the regional budgets.

iii) Wholesale prices

In January 1992, agro-food wholesale prices were liberalised and all the subsidies provided to food producers to compensate them for low prices were abolished. However, in many regions there were attempts to control the margins at the processing, wholesale and retail levels. Moreover, processing plants with local monopolies were obliged to report the changes in their production costs monthly to the local authorities. All these administrative measures (relics of central planning) did not prevent the rise of wholesale prices and processors' margins. The real limitation on increased margins was imposed by increasing competition between food processors for agricultural raw materials.

iv) Retail prices

At the beginning of the reform process retail prices for basic foodstuffs were not completely liberalised. In January and February 1992 the Federal Government controlled retail prices basically by setting "mark-up limits"[7] and limits on price increases for wholesalers and retailers. These restrictions were mostly related to products like bread, milk and other dairy products, and meat. Local authorities were empowered to modify the federal list of controlled products and also to impose stricter limits on retail price increases. The outlays for implementation had to be financed from their own budget.

Since March 1992, food price control has been delegated completely to the regional authorities. The Federal Government retained control only for bread retail prices through the end of 1993. The regional governments have been empowered to:

- establish the list of specific foodstuffs, whose prices are to be controlled, from federal regulatory list;
- extend this list at their discretion; and
- select the mechanism used to control retail prices.

In practice, the regional governments were given the power to regulate retail prices without any federal co-ordination of such a policy. The factor limiting retail price regulation at the regional level was budgetary resources. By the end of 1992, due to budgetary pressures, most of the regions limited retail price control to bread, milk and sugar. However, differences among the regions persisted with respect to both the level of price control and the products covered. Regional governments also adopted a wide range of tools to control retail prices: setting wholesale and retail mark-up limits, introducing maximum price-increase coefficients, setting retail price ceilings, but, their impact was mixed across the regions. Recent surveys show that two regions Orel and Ryazan (both in the Central region of Russia) had similar retail prices for a set of 19 main foodstuffs, although the Orel local administration regulated eleven of them and the Ryazan administration only two of them. On the contrary, regional authorities in Ulyanovsk kept food prices artificially low and applied a food rationing system up to mid 1996.

3. Purchase and retail-price regulation for specific agricultural and food products

i) Grains

Grains are the most important crop in Russian agriculture. Although the grain area declined by almost 10 million hectares between 1990 and 1996 its share in total arable land remained rather stable around 54 per cent. The grain output varies strongly from year to year due to the variation in yields (Table IV.7).

Until 1991 grains had to be delivered to state procurement agencies at administered prices. The share of supplementary payments in the overall procurement price was relatively low for grains. All marketed grain was delivered to the state agencies immediately after the harvest. In the period 1986-1990, on average, one third of total grain production was sold to state procurement and the rest was used on farm for feed and human consumption. The grain procurement prices were increased several times in the 1980s. However, the wholesale and retail prices remained unchanged. Consequently payments from the budget compensating for the difference between the procurement price and the calculated price (derived from the wholesale price) increased as well. Table IV.8 illustrates the difference between these prices in 1988. It is notable that even wholesale prices (and also the retail prices of grain based products) were far below the prices paid to farms. The larger gap for rye, barley and oats than for wheat may partly explain why wheat production contracted less during the reform period in response to subsidy cuts.

With the liberalisation of the economy, obligatory deliveries were reduced in 1992 and completely abolished as of 1993. Deliveries to state procurement agencies were continued on a voluntary basis. In

Table IV.7. **Grain area and production in Russia**

Thousand hectares, thousand tonnes

	1990	1991	1992	1993	1994	1995	1996	1997
Total grain area	63 068	61 783	61 939	60 939	56 280	54 705	53 388	53 611
of which:								
Wheat	24 244	23 152	24 284	24 666	22 190	23 909	25 707	26 026
Barley	13 723	15 282	14 564	15 478	16 604	14 710	11 793	12 493
Rye	8 008	6 480	7 592	6 000	3 903	3 247	4 147	4 005
Oats	9 100	9 032	8 540	8 402	8 333	7 928	6 904	6 436
Maize	869	733	810	800	524	643	622	918
Total grain production	116 676	89 094	106 855	99 094	81 297	63 406	69 341	88 505
of which:								
Wheat	49 596	38 899	46 166	43 547	32 129	30 119	34 917	44 188
Barley	27 235	22 174	26 988	26 843	27 054	15 768	15 933	20 771
Rye	16 431	10 639	13 887	9 166	5 989	4 098	5 934	7 480
Oats	12 326	10 372	11 241	11 556	10 757	8 562	8 346	9 381
Maize	2 451	1 969	2 155	2 441	892	1 738	1 088	2 671
Total grain yields (t/ha)	1.85	1.44	1.73	1.63	1.44	1.16	1.29	1.65
of which:								
Wheat	2.05	1.68	1.90	1.77	1.45	1.26	1.35	1.70
Barley	1.98	1.45	1.85	1.73	1.63	1.07	1.35	1.66
Rye	2.05	1.64	1.83	1.53	1.53	1.26	1.43	1.87
Oats	1.35	1.15	1.32	1.38	1.29	1.08	1.21	1.46
Maize	2.82	2.69	2.66	3.05	1.70	2.70	1.75	2.91

Source: Goskomstat.

Table IV.8. **Procurement, calculated and wholesale prices for grains in 1988**

Rouble per tonne

	Procurement price	Calculated price	Wholesale price	Procurement/calculated, %
Wheat	164	111	130	147.7
Rye	234	118	133	198.3
Feed barley	146	76	91	192.1
Oats	169	86	101	196.5

Source: *Pricing in Agro-industrial Complex*, Agropromizdat, Moscow, 1989.

Table IV.9. **Grain production and volumes of state procurement**

Million tonnes

	1986-1990 average	1991	1992	1993	1994	1995	1996	1997
Total grain production	104.3	89.1	106.9	99.1	81.3	63.5	69.3	88.5
State procurement	34.3	23.6	26.1	28.2	12.4	9.5	8.8	n.a.
Share of state procurement in total production, per cent	32.9	26.5	24.4	28.5	15.3	15.0	12.7	n.a.

n.a.: not available.
Source: Goskomstat.

1993 the state fixed so-called recommended prices for buying into federal reserves. These prices, when announced, were set at higher levels than market prices; however, due to a lack of budget resources, the state delayed payments for the grains delivered. The delayed payments combined with the high level of inflation caused serious financial losses for farms. As a result, from 1994 farms tended to keep more grain for their internal use such as barter trade with input suppliers, feed for animals, payments in kind to farm workers, etc. The share of grain production sold to state procurement remained relatively high in 1992 and 1993 and was substantially reduced as from 1994 (Table IV.9).

In the early years of the reform Russia applied high export duties for grains. These export duties were originally set at 70 per cent. On 1 October 1993, they were lowered to 10 to 25 per cent and abolished in September 1995. In March 1994, Russia introduced a 1 per cent tariff for grain imports and since 15 May 1996 a five per cent import tariff is applied. A higher import tariff set at 10 per cent has been applied for grain products (bread, flour, etc.) since 1995.

ii) Oilseeds

The major oil crop in Russia is sunflower, which accounts for 90 per cent of total oilseed output. The production of sunflower is concentrated in only a few regions (93 per cent of production originates from three regions including 57 per cent from the North-Caucasian region). In contrast to other crops the sunflower area has been steadily increasing from 1991 to 1995. The record sunflower production in 1995 (4.2 million tonnes) was mainly due to sharp increases in the area sown (by 32 per cent over 1994) and yields (by 20 per cent over 1994) and was rather exceptional as confirmed by the evolution in 1996 and 1997 (Table IV.10).

From 1986 to 1989 price support was provided through procurement prices and related supplementary payments. The share of supplementary payments in the overall procurement price increased between 1986 and 1989 and was larger than for grains. Most of the supplementary payments were related to over-plan deliveries, while the payments to low-profit or loss making farms were rather low. The procurement prices were fixed for five zones and the difference between the zones with the highest and lowest price was 45 per cent. In 1990, the system of price support was modified. Procurement prices were increased to compensate for the suppression of supplementary payments and the number of

Table IV.10. **Sunflower area, production and procurement**

Thousand hectares, thousand tonnes

	1986-1990 average	1991	1992	1993	1994	1995	1996	1997
Area	2 530	2 576	2 889	2 923	3 133	4 127	3 874	3 585
Total production	3 120	2 896	3 110	2 765	2 553	4 200	2 765	2 824
State procurement	2 379	1 801	1 042	569	150	357	n.a.	n.a.
Share of state procurement in production %	76.3	62.2	33.5	20.6	5.9	8.5	n.a.	n.a.
Yields, tonne per hectare	1.2	1.1	1.1	0.9	0.8	1.0	0.7	0.8

n.a.: not available.
Source: Goskomstat.

price zones was reduced from five to two. Also, the gap between the zone prices was reduced to 16 per cent. In addition to the procurement price, supplementary payments were paid for three years as an incentive to producers who started to cultivate sunflowers.

In 1992, obligatory deliveries to state procurement agencies were abolished and the related price support ceased. Oilseeds were traded at prices agreed between producers and buyers (oilseed crushers or procurement organisations). Market liberalisation led to a sharp decrease of state procurement (Table IV.10). In 1993, the Ministry of Agriculture announced so-called recommended prices for oilseeds to be delivered to federal funds. However, as with the case of grains, the government failed to meet its commitments. Also the minimum guaranteed prices announced in 1994 and 1995 were applied only to a very limited extent due to the lack of public finance.

Under such circumstances export remained the most attractive outlet for sunflower producers. However, up to 1993 a 15 per cent export duty was applied on oilseed exports. The duty was removed in 1994 and 40 per cent of oilseed production was exported. Simultaneously Russia imported increasing volumes of vegetable oil and domestic crushers lacked raw materials. This situation resulted from the low efficiency of the Russian crushers and oil producers and their inability to pay for deliveries of oilseeds on time. Since 1995 price support has also been provided through border protection with a 15 per cent import tariff for oils, 10 per cent for sunflower seed and 5 per cent for other oilseeds.

iii) Sugar – sugar beet

Historically Russia was a net importer of sugar, with imports covering some 40-50 per cent of consumption. Domestic production of refined sugar was to a large extent based on imports of raw cane sugar from Cuba. At the beginning of the 1990s, production of sugar, from domestically produced sugar beet, was around 3 million tonnes while consumption was around 7 million tonnes. The sugar beet production area was around 1.5 million hectares at the end of 1980s and declined to less than one million hectares in 1997. Sugar beet production was stable at around 25 million tonnes in 1991-1993 but dropped sharply in 1994-1997 (Table IV.11).

From 1986 to 1989 price support for sugar beet was based on fixed procurement prices differentiated for 9 price zones (there was a 60 per cent difference between the highest and the lowest prices). In addition to the supplementary payments for over-plan deliveries, low income farms, etc., sugar beet growers were granted *premia* for early deliveries (30 per cent of the basic procurement price for deliveries before 1 September and 20 per cent before 15 September). Farms introducing sugar beet cultivation received additional supplementary payments of 30 per cent. In 1990 procurement prices were increased and price zones were reduced from nine to four with lower price variations across the zones (37 per cent).

In January 1992, fixed state procurement prices, were abolished and the share of state procurement dropped (Table IV.11). Since 1993, obligatory deliveries to the state reserves have been abolished and deliveries are regulated by contracts signed between producers and sugar enterprises. The recom-

Table IV.11. **Sugar beet area, production and procurement**

Thousand hectares, thousand tonnes

	1990	1991	1992	1993	1994	1995	1996	1997
Area	1 460	1 399	1 439	1 333	1 104	1 085	1 060	935
Total production	32 327	24 280	25 548	25 468	13 949	19 072	16 166	13 841
State procurement	25 118	18 647	10 730	7 055	879	629	n.a.	n.a.
Share of state procurement in total production, %	77.7	76.8	42	27.7	6.3	3.3	n.a.	n.a.
Yields, tonne per hectare	22.1	17.4	17.8	19.1	12.6	17.6	15.3	14.8

n.a.: not available.
Source: Goskomstat.

mended prices set in 1993 and "minimum guaranteed prices" in 1994 and 1995 have only a limited influence on market prices as the state lacks the finance to intervene in the market.

In the early years of the reform Russia applied high export duties for sugar. These export duties were originally set at 70 ECU per tonne, from 1 October 1993 they were slightly lowered to 60 ECU per tonne and abolished since September 1995. Since 1994 domestic prices have been supported through border measures. In 1994, a 20 per cent import tariff was introduced for refined sugar, increased to 25 per cent in 1995 (but no less than 0.07 ECU per kg). The support goes mainly to sugar plants as the import tariff for raw (cane) sugar was set at only 1 per cent.

iv) Potatoes

Traditionally potatoes have been one of the basic food items in Russia. In contrast with some other crops, the total area planted and production of potatoes have steadily increased during the reforms. The major part of potato production comes from household plots and is consumed directly within the households. In 1986-1990 the share of household plot production was around 60 per cent of the total. During the reform period this share increased to reach 90 per cent in 1995. While total production in 1995 was 11 per cent higher that the 1986-1990 average, household plots show a 61 per cent increase and agricultural enterprises fell by 75 per cent over the same period. The share of state procurement also dropped sharply (Table IV.12).

Although fixed procurement prices existed until 1989 (11 price zones plus various supplementary payments), negotiated prices had been widely used for potato sales since 1986. Compared with other crops, the share of supplementary payments in the procurement price was relatively low between

Table IV.12. **Potato area, production and procurement**

Thousand hectares, thousand tonnes

	1986-1990	1991	1992	1993	1994	1995	1996	1997
Total area	3 331	3 187	3 404	3 548	3 337	3 409	3 404	3 350
Total production	35 881	34 330	38 330	37 650	33 828	39 739	38 652	37 015
of which: on household plots	21 384	24 869	30 196	31 431	30 114	36 084	n.a.	n.a.
agricultural enterprises	14 497	9 461	8 134	6 219	3 714	3 655	n.a.	n.a.
State procurement	7 742	4 758	2 995	1 670	740	695	526	n.a.
Share of state procurement, %								
a) in total production	21.6	13.9	7.8	4.4	2.2	1.7	1.4	n.a.
b) in production of agricultural enterprises	53.4	50.3	36.8	26.9	19.9	19.0	n.a.	n.a.
Yields, tonne per hectare	10.8	10.8	11.3	10.6	10.1	11.7	11.4	n.a.

n.a.: not available.
Source: Goskomstat.

1986 and 1989. As of 1990 the fixed procurement prices were abolished and negotiated prices introduced for all types of deliveries. Together with vegetables, potatoes were the first product for which state control and price support were abolished.

Traditionally Russia imported only limited amounts of potatoes (around 3 per cent of consumption) and during the reform period potato imports were further reduced. Moreover, to support domestic prices an import duty of 25 per cent was introduced in 1995.

v) Fruits and vegetables

For fruits and vegetables the production trends, state procurement system and marketing are similar to those for potatoes. From 1991 to 1994 vegetable production declined slightly but then increased in 1995 to reach the average of 1986-1990. There has been a significant change in the structure of production with the household plots' share in total production increasing from 28 per cent in 1986 – 1990 to 74 per cent in 1995. For fruits, vegetables, and potatoes alternative marketing channels (*kolkhoz* markets, fairs) to the state procurement system already existed in the 1980s.

In the Soviet period fixed procurement prices were applied to deliveries from large enterprises (*kolkhozes* and *sovkhozes*) to state procurement agencies. In 1986-1990 state procurement represented 92 per cent of the marketed production of large enterprises and 67 per cent of total marketed production of vegetables. Also, supplementary payments were applied to deliveries to the state procurement system. As in the case of potatoes, state control and price support for fruits and vegetables were abolished at the federal level as early as 1990. However, local authorities retained the power to fix upper limits to contracted prices, in order to limit price increases in a situation of high demand and relatively low supply of fruits and vegetables. This was clearly against the logic of the market price function. Free market prices were established only for production sold through the producers' own retail outlets and on *kolkhoz* markets.

Since 1994 domestic prices for fruit and vegetables are supported by border protection. For vegetables the import tariff is fixed at 15 per cent. For fruits, a 1 per cent import tariff was introduced in March 1994, and increased to 5 to 10 per cent as of 1 July 1995. Imports of apples are subject to a tariff of 0.2 ECU per kg.

vi) Milk

With 41 per cent of the total value of livestock output (in 1995), milk has been the most important livestock product in Russian farming. Milk production is spread through all regions of Russia, with some important production zones, such as the non-black soil areas in the central part of Russia. The policy of regional self-sufficiency, applied under the former regime, even allocated milk production to regions with a traditional deficit of animal feed. During the reform period the drop in demand for dairy products led to a reduction in livestock and milk production. From 1990 to 1997 the number of cows declined by 19 per cent, milk yields by 25 per cent with the resulting fall in milk production of almost 40 per cent (Table IV.13). So far statistics do not indicate any significant improvement in resource allocation in the dairy sector as the regions are still pursuing a policy of local self-sufficiency.

In the Soviet economy, milk and dairy products together with beef meat, were under strict governmental control and were highly subsidised both at the producer and consumer levels. High consumer subsidies stimulated consumption to economically unjustified levels. Between 1986 and 1989 milk and beef accounted for around 70 per cent of total supplementary payments and in 1990 their share increased to almost 80 per cent.

In the Soviet period, price support to milk producers was granted through procurement prices and supplementary payments. Up to 1990, procurement prices were set for 15 zones. The number of zones was reduced in 1991. However, for milk the number of zones remained relatively high as there were eleven price zones fixed with a substantial variation between the zone prices (the highest price was 3.4 times higher than the lowest one). The share of state procurement in total milk produced was around 70 per cent in 1986-1990.

Table IV.13. **Cattle numbers, milk production and state procurement**

Thousand heads, thousand tonnes

	1985	1990	1991	1992	1993	1994	1995	1996	1997
Cattle inventories (1 Jan.)	60 044	58 841	57 043	54 677	52 226	48 914	43 297	39 696	35 103
of which: cows	22 000	20 760	20 557	20 564	20 243	19 831	18 398	17 436	16 874
Milk production	50 169	55 715	51 886	47 236	46 524	42 174	39 241	35 819	34 066
State procurement	35 000	40 100	34 200	26 100	24 600	18 900	15 900	12 553	n.a.
Share of state procurement in production, %	69.8	72.0	65.9	55.3	52.9	44.8	40.5	35.0	n.a.
Milk yields (kg/cow/year)	2 280	2 684	2 524	2 297	2 298	2 127	2 133	2 054	2 019

n.a.: not available.
Source: Goskomstat.

As of 1992, milk prices were liberalised, and all payments linked to procurement prices and consumer subsidies were abolished. In 1992 and 1993 attempts were made in the regions to control the margins of milk processors and to set a price paid to milk producers acceptable to both processors and producers. In many regions commissions of milk producers and processors were established under the patronage of local governments in order to set the prices paid for milk. However, since the reduction of producer prices was mainly a result of the drop in real incomes of the population and cuts in subsidies, the impact of these commissions on price levels was rather limited and the commissions were dissolved after 1993. Efforts undertaken since 1995 by the Ministry of Agriculture to disseminate market information should improve market transparency and, thus improve price formation in the sector.

At the beginning of the reforms a 15 per cent export duty was applied on milk and dairy products (up to October 1993). In 1993 a 10 to 15 per cent import tariff was applied on milk and dairy products. Due to increasing surpluses of butter on the Russian market, tariffs on butter were increased from 15 to 20 per cent in 1995 (but not less than 0.3 ECU per kg).

vii) Beef and veal

Meat output represented almost 60 per cent of total livestock production in 1986-1990. Traditionally beef had the highest share of total meat output – around 43 per cent in 1986-1990. The share increased to 50 per cent in 1996 due to a strong decline in total meat production which fell 48 per cent between 1990 and 1996, compared to a 40 per cent decline for beef and veal over the same period. Cattle inventories declined at a somewhat lower rate (33 per cent) (Table IV.14). In 1997 cattle inventories and beef meat production continued to decline at the same rate as in 1996.

As for milk, beef and veal production and consumption were highly subsidised in the pre-reform period. There was no specialised beef meat production in Russia as dairy and dual-purpose cattle were

Table IV.14. **Cattle inventories, beef production and procurement**

Thousand heads, thousand tonnes carcass weight

	1985	1990	1991	1992	1993	1994	1995	1996	1997
Cattle inventories (1 Jan.)	60 044	58 841	57 043	54 677	52 226	48 914	43 297	39 696	35 103
Beef meat production	3 572	4 329	3 989	3 632	3 359	3 240	2 733	2 630	2 338
State procurement	2 840	3 131	2 684	2 178	1 807	1 322	896	n.a.	n.a.
Share of state procurement in total production, %	79.5	72.3	67.3	60.0	53.8	40.8	32.8	n.a.	n.a.

n.a.: not available.
Source: Goskomstat.

widely used. Therefore, beef meat production was closely linked to milk production. Until 1991 beef and veal production was supported through the system of state procurement similar to that for other products. Basic procurement prices were fixed for 12 price zones and the highest zone price was 2.9 times higher than the lowest one. The share of various supplementary payments in the procurement price had been increasing up to 1989 and was slightly reduced in 1990. In 1991 the number of price zones for beef was reduced to eight and all previous supplementary payments were consolidated into the basic procurement price. The share of state procurement in total production was around 75 per cent between 1986 and 1989.

Following price liberalisation in 1992, obligatory deliveries to state procurement were abolished but subsidies for deliveries of beef and veal to state procurement agencies were introduced. The share of beef sold to state procurement agencies (central and regional) has been declining since 1990 and in 1995 it represented around one third of total production.

As for milk and dairy products all meat and meat products were subject to a 70 per cent export duty before October 1993. Since March 1994 domestic beef prices have also been supported by border measures. On 15 March 1994 meat imports (beef, pigmeat) were subject to an 8 per cent import tariff. Since July 1995, the import tariff has been increased to 15 per cent and since 15 May 1995, the tariff remains at 15 per cent but not less than 0.15 ECU per kilogram.

viii) Pigmeat

The fall in inventories and output in the pig sector was more pronounced than for beef. Between 1990 and 1997 the numbers of pigs declined by 52 per cent and the pigmeat production by 55 per cent.

Between 1986 and 1990, pigmeat production was subject to state procurement with obligatory deliveries to the state agencies. Basic procurement prices were set for 16 regions and the price difference between the regions was lower than for beef (the highest zone price was 1.9 times higher than the lowest one). Also, the share of various supplementary payments in the procurement price was lower than for beef. However, up to 1990, pigmeat production was heavily subsidised through deliveries of cheap grains. In 1991 the number of price zones for pigmeat was reduced to four.

Following price liberalisation in 1992, obligatory deliveries to state procurement agencies were abolished and subsidies linked to deliveries to state procurement agencies were introduced. However, the level of subsidies granted to pigmeat was lower than for milk and beef. The share of pigmeat sold to state procurement was around 55 to 60 per cent between 1986 and 1990. Since 1991 the share of state procurement (central and regional) has been declining and in 1995 it represented 21 per cent of total pigmeat production.

Since March 1994 domestic pigmeat prices have been supported by border measures. The levels of border protection and instruments applied are the same as for beef and veal.

Table IV.15. **Pig inventories, pigmeat production and procurement**

Thousand heads, thousand tonnes carcass weight

	1985	1990	1991	1992	1993	1994	1995	1996	1997
Pig inventories (1 Jan.)	38 732	39 982	38 314	35 384	31 520	28 557	24 859	22 631	19 115
Pigmeat production	2 960	3 480	3 190	2 784	2 432	2 103	1 865	1 705	1 565
State procurement	1 770	1 861	1 501	1 035	754	536	392	n.a.	n.a.
Share of state procurement in total production, %	59.8	53.5	47.1	37.2	31.0	25.5	21.0	n.a.	n.a.

n.a.: not available.
Source: Goskomstat.

Table IV.16. **Poultry flock, poultry and egg production and procurement**

Million heads, thousand tonnes carcass weight

	1985	1990	1991	1992	1993	1994	1995	1996	1997
Poultry flock (1 Jan.)	616.6	653.6	659.8	652.2	568.3	566.8	491.0	422.6	371.9
Poultry meat production	1 527	1 801	1 751	1 428	1 277	1 068	859	690	632
State procurement	1 066	1 187	1 045	824	794	670	489	n.a.	n.a.
Share of state procurement in total production, %	69.8	65.9	59.7	57.7	62.2	62.7	56.9	n.a.	n.a.
Egg production (million pieces)	44 277	47 470	46 875	42 902	40 297	34 473	33 830	31 902	31 884
Egg production (thousand tonnes)	2 459.8	2 637.2	2 604.2	2 383.4	2 238.7	1 915.2	1 879.4	1 772.3	1 771.3
State procurement (million pieces)	30 758	30 763	30 506	24 329	24 256	21 761	18 360	16 799	n.a.
State procurement (thousand tonnes)	1 708.8	1 709.1	1 694.8	1 351.6	1 347.6	1 208.9	1 020.0	933.3	n.a.
Share of state procurement in total production, %	69.5	64.8	65.1	56.7	60.2	63.1	54.3	52.7	n.a.

n.a.: not available.
Source: Goskomstat.

ix) Poultry and eggs

The poultry flock numbers dropped by 43 per cent between 1990 and 1997. However, the drop in poultry meat production was much larger at around 65 per cent over the same period, while egg production dropped by one third (Table IV.16).

Between 1986 and 1990 poultry and egg production was subject to state procurement with obligatory deliveries to the state. The share of state procurement in total production in that period was around 70 per cent both for poultry and eggs. For poultry and eggs there were only two price zones (for the Northern territories and the rest of Russia). The difference between the basic procurement prices in the two zones was 30 per cent. Supplementary payments to the basic procurement price were low. As in the case of pigmeat, significant support to poultry and egg production was provided through deliveries of cheap grain and protein feed.

Following price liberalisation in 1992, obligatory deliveries to state procurement were abolished and subsidies linked with the amounts of poultry and eggs delivered to state procurement agencies were introduced. However, the level of subsidies granted to poultry and eggs was less important than for milk and beef. The share of state procurement in total production has been declining since 1991 but remains higher than for beef and pigmeat. In 1995, it represented 57 per cent of total poultry meat production and 54 per cent of egg production.

Since March 1994, poultry sector prices have been increasingly supported by border measures. In March 1994 a 20 per cent import tariff was set for poultry. The import tariff was increased to 25 per cent in July 1995 and to 30 per cent (but not less than 0.30 ECU per kg) in May 1996. The level of border protection provided for the poultry sector is substantially higher than for beef and pigmeat.

C. FOREIGN TRADE MEASURES

1. Pre-reform period

Under the Soviet system, all foreign trade was centrally controlled and import and export decisions were not made at the enterprise level but at the level of the central planning authorities and other institutions of the state administration. Hard currency export earnings were collected centrally and used for financing centrally planned imports. All export and import operations and related finance operations were handled by state monopoly agencies (specialised export and import companies, banks, etc.). The system completely isolated domestic producers and markets from developments on world markets. The isolation of the enterprise sphere from foreign market competition led to low productivity growth and increasing lack of competitiveness.

In the second half of the 1980s, the state administration made some attempts to de-centralise and partly liberalise foreign trade activities, under the centrally planned economy. The main goal was to stimulate exports. In the first phase, foreign trade activities were delegated to the level of 21 branch ministries (including the Ministry of Agriculture) and to all Soviet republics. The trade in strategic resources (fuels, raw materials) remained under the control of the Ministry of Foreign Trade of the USSR. There was an increasing demand from enterprises to operate directly on foreign markets. From 1989 onwards producers were given the right to participate in foreign trade activities. The list of products to be exported directly by the enterprises was extended and by the end of 1990 some 26 000 enterprises were registered as having a licence for foreign trade activities. However, the share of enterprises' foreign trade in total foreign trade was marginal (around 1 per cent in 1988-1990).

2. During the reform

i) General measures

The economic reforms of 1992 focused on price and domestic and foreign trade liberalisation. The reforms created a new environment for the implementation of trade policies and trade measures, and the role of the state foreign trade administration was gradually modified. In the beginning of 1992, the Russian state continued to intervene heavily in foreign trade. Export goods (primarily oil and gas, metals and other raw materials) were purchased by state trading organisations at domestic prices and sold on the world market at international prices with high profits. Through the state budget these profits were used to supply the domestic market with imported goods at low domestic prices. This mechanism was phased out during 1992 and 1993, when domestic prices approached world market price levels, contributing to the hyper-inflation rates of that period.

In 1992 and partly in 1993 foreign trade was still largely in the hands of state trading organisations, and not subject to competition and market mechanisms. Thus until 1993, trade policy was conducted through direct centralised controls. From 1993 to 1996 direct state control was gradually replaced by price based measures such as tariffs and taxes. Administrative export controls were removed, although certain administrative procedures still apply to exports, motivated by tax and capital flight concerns.

ii) Import measures

Up to 1993 imports of agro-food products were controlled through centralised state controls. Moreover, in order to increase availability of imported food and agricultural inputs, subsidies were applied which covered the difference between high import and low domestic prices. In June 1993 the subsidies applied to the main imported foodstuffs were abolished.

Already in 1993 a 15 per cent duty was applied on imports of milk and dairy products. From July 1994 import tariffs ranging from 1 per cent (grains) to 20 per cent (poultry, sugar) were also applied to some key products.[8] In July 1995, new import tariffs were applied. The tariffs for foodstuffs were raised from 13.2 per cent to 15.5 per cent on average. Import tariffs were set for all imported food products and a minimum import tariff was set at 5 per cent (Table IV. 17).

New specific import duties were introduced in May 1996 both in percentage terms and in ECU per unit (as a minimum import duty). The main purpose of introducing a minimum import duty in ECU terms was to avoid underpricing of declared goods by importers and to increase budget revenues. In fact, the introduction of specific duties meant an increase in border protection. Moreover, in July 1996 all privileges granted to food importers, such as three month delays in payments of duties and VAT, were abolished. In 1997, the percentage tariffs and specific import duties remained unchanged, but as of February 1998 the scope of products covered by a minimum import duty increased (Table IV.17).

A food labelling law was introduced in May 1997, requiring all imported foodstuffs into Russia to bear labels in the Russian language and Cyrillic script. The labels are also required to provide product information, the date of manufacturing and the country of origin. So far no import quotas[9] or selective import licences have been applied to limit imports.

Table IV.17. **Import tariffs for basic agro-food products in Russia**

	From 15 March 1994 per cent	From 1 July 1995 per cent	From 15 May 1996		From 1 Feb. 1998	
			Per cent	Minimum ECU/kg	Per cent	Minimum ECU/kg
Beef meat	8	15	15	0.15	15	[0.15-0.2] [2]
Pigmeat	8	15	15	0.15	15	[0.15-0.25] [2]
Poultry meat	20	25	30	0.3	30	0.3
Milk and dairy products	10-15	10-15	10-15		15	
Yogurts	10	10	10		10	0.12
Cheese	15	15	15		15	0.3
Butter	15	20	20	0.3	20	0.3
Grains	1	1	5		5	
Sunflower	1	10	10		10	
Other oilseeds	–	5	5		5	
Sunflower oil	–	15	15	0.09	15	0.09
Soya oil	–	15	15		15	
Rapeseed oil	–	15	15		15	
Olive oil	–	10	10		10	
Other oils	5	5	5		5	
Potatoes [1]	25	25	25		25	
Fruits	1	5-10	5-10			
Apples	0.2 per 1 kg	0.2 per 1 kg		0.2		[0.1-0.2] [2]
Juices	5	10	15	0.07	15	0.07
Raw cane sugar	1	1	1		1	
White sugar	20	25	25	0.07	25	0.07

1. Excepting seed potatoes with zero tariff.
2. Depending on season.
Source: Russian Federation Customs Committee, 1998.

iii) Export measures

Export duties, quotas and licences for agro-food products were in place in Russia until 1995 (some export duties even beyond 1995). However, the list of products subject to export restriction measures was shrinking during the years of reform. Already in October 1993 export duties for all livestock products were abolished. Export duties applied to grains, sugar and some other foodstuffs were removed in 1994. Nonetheless, at the end of 1995, due to the low grain harvest, export duties for some grains were reintroduced (Table IV.18). Due to the lack of co-ordination in border protection measures within the

Table IV.18. **Export duties applied for agro-food products**
Per cent

	Before 1 Oct. 1993	From 1 Oct. 1993	From 1 Sept. 1995 until 31 March 1996
Meat and meat products	70	0	0
Milk and dairy products	15	0	0
Grains	70	10-25	0
Durum wheat	0	0	17
Other wheat	0	0	7
Maize	0	0	10
Oilseeds	15	0	0
Flour and oilseed products	0	0	7
Sugar beet	0	15	0
Butter	15	0	0
Vegetable oil	70	0	0
Sugar, ECU/tonne	70	60	0

Source: Russian Federation Customs Committee, 1997.

customs union with Belarus and Kazakhstan, these export duties were largely ineffective. The government abolished all export duties (except for oil and gas) from 1 April 1996.

During the first years of the reform (1992-1995) the so called "special exporters" system (companies having a state licence to export specific products) was maintained, especially for grains and oilseeds (besides agro-food products the "special exporter" scheme was applied mainly to oil and gas exports). Companies with export licences were given a specific export quota. In March 1995 the system of "special exporters" was abolished, which represented an important step away from state trading towards trade liberalisation.

Up to 1997 no export subsidies were applied under the Russian trade policies. Subsidies contemplated to compensate part of the transport costs of exported grain were not applied due to a lack of funding.

D. REDUCTION OF INPUT COSTS

1. Credit policies

i) Privatisation in the banking sector

During the Soviet period, Gosbank, a state banking monopoly, with large regionally-based branch structures, played a crucial role in the functioning of the economy. It combined the functions of a deposit, credit, and monetary centre. Gosbank effected all payments between its clients, controlled the use of wage and salary funds in enterprises and organisations, and was the sole money emitter and the cashier for the state budget. All enterprises and organisations were allowed to have only one payment account in the bank and special accounts for budget allocations that could be used strictly for purchases of inputs according to the yearly plan. Transfers of these allocations between accounts of enterprises directly followed deliveries of goods and the fulfilment of the plan. All payments were made on accounts, and cash could be withdrawn by enterprises from the bank in amounts equal to wage and bonus funds after the bank automatically made various tax and social security deductions. At the end of each year Gosbank credited ministries for settling mutual debts between their subordinate enterprises. A special state savings bank (Sberbank) held the savings of the population. There were also some other banks, dealing with investment and foreign trade, but their activities were subordinated to the Gosbank monopoly.

In 1987, a decentralisation of the system started with four banks being separated from the Gosbank to fulfil some of its former functions and to deal with investment and construction (Stroibank), social expenditures (Zhilsotsbank), foreign trade payments (Vneshekonombank) and agriculture (Agro-Industrial Bank – Agroprombank). In 1988 a two-tier banking system was established with Gosbank becoming Russia's Central Bank and former branches of Gosbank transformed into state-owned commercial banks. In 1990 two laws – "On Banks and Banking Activity" and "On the Central Bank of the Russian Federation" – were passed that gave a start to the reform of the banking system in the country. Specialised (former state-owned) banks were transformed into joint-stock companies. A network of private banks began to form and by 1996 their number reached 2 500.

In 1988, the USSR Agroprombank took over all Gosbank activities in agriculture, including deposit-taking, lending and acting as fiscal agent for the government (World Bank, 1992). In 1991, the bank was formally transformed into a joint-stock commercial bank and renamed Rosselkhozbank (Russian Agricultural Bank). The same year the Rosselkhozbank reorganised itself again, substantially modifying its charter and financial base (bad loans were written-off by the state), and returned to its original name Agroprombank. With its extended subsidiary network in the countryside (the second after the Savings Bank – Sberbank) Agroprombank became one of the largest banks in Russia. In 1996, the bank's system included more than 1 200 affiliates and 62 regional branches – former *oblast* and *rayon* divisions.

Due to weak credit evaluation and low credit repayments, Agroprombank accumulated bad debts and by mid-1996 it had a balance deficit of Rb1 trillion (about US$200 million). In the fall of 1996 in an effort to ensure the survival of Agroprombank, the Federal Government allowed the bank to issue additional stocks, which were sold on open tender. The buyer was *Stolichny bank sberezheny* (SBS –

Capital's Savings Bank) which afterwards was renamed SBS-Agro. The sale was made on the condition that Agroprombank continue to devote 60 per cent of its lending portfolio to farmers and the understanding that SBS would clean up Agroprombank's balance sheet. SBS also agreed temporarily to transfer a 24.5 per cent stake back to the government (in addition to the government's retained 1.5 per cent stake) with the intention that the government would share the burden of losses associated with pre-existing bad loans. Being wholly-owned by SBS-Agro, Agroprombank continues its activities under its proper name.

ii) Credit policies in the pre-reform period

Under the Soviet regime, short-term credits to finance the running cost of farms were made available at low interest rates. Investment credits, also at low interest rates, were given to finance investments decided by central planners. Under that kind of system, credit had a limited economic role as no real financial market existed. Interest rates did not express the price (scarcity) of financial resources as these were mostly centrally allocated. Interest rates were set to cover the operational costs of the state banking sector. Within the centrally planned system all credits channelled to agriculture were handled Agroprombank.

Up to 1991 Agroprombank issued short and long-term credits to *kolkhozes* and *sovkhozes* and financed fixed capital investments of the state farms from the budget funds (*kolkhozes*, not being state enterprises, were supposed to repay investment loans). Budget allocations were the primary source of long-term credits issued by the Agroprombank, while short-term credits were financed using funds kept by enterprises on the bank's accounts. However, the crediting was not of a commercial nature: in many cases loans were not repaid and repeated write-offs meant that "credits" were simply a transfer of resources.

Long-term credits extended by Agroprombank to finance capital investments in agriculture were not financed from its own capital resources but from funds channelled to the bank from the state budget. The interests on long-term credits were fixed at 0.75 per cent for planned investments, and at 2 per cent for other investments. Long-term credits to finance large-scale buildings to keep livestock were granted for up to 20 years, and loans to finance purchases of agricultural machinery were repayable within eight years. Long-term credits for capital investment in agriculture financed from the state budget were granted only to *sovkhozes* (as state companies) while *kolkhozes* financed their capital investments from their own resources.

Restructuring and debt write-offs were a common practice under the centrally planned economy. By 1988 total debt of Rb97 billion had been accumulated from 1982 (the last massive write-off of debts). This debt repayable in 1988-1990, was first restructured with a repayment of 10 years, and finally written-off in 1990. The write-offs were carried out through allocation of budget funds to Agroprombank to cover the bad debts of *kolkhozes* and *sovkhozes*. As the 1990 budget resources were not sufficient to finance such an operation, this decision increased the cumulated state debt.

iii) Credit policies in the reform period

In the first year of the reform a new financial market emerged in Russia. However, under conditions of serious macroeconomic instability, the dollarisation of the economy and extremely high inflation rates the banking sector concentrated its activities on short-term loans (several months) at high nominal interest rates. The commercial banks were reluctant to provide any longer-term investment loans.

In the early years of reform, Agroprombank was commissioned by the Federal Government to transfer credits from the state budget to agriculture. Under this scheme the bank carried out purely technical functions (using its extensive network) and did not influence the selection of borrowers, setting of interest rates or other terms of the loan. The bank was allowed only to add its own margin of 3 per cent (maximum) to the interest rate of the loan to cover its operational costs.

Even in the more market oriented credit system, the schemes providing credits to agriculture retained the main features of the centrally administered system and were to a large extent dependent on federal policies and budget sources. Formally the Agroprombank lost its monopolistic position as sole supplier of credit to agriculture, but it retained its dominant position in this sector. The bank is a major agricultural lender in Russia, supplying about 70 per cent of all loans to agriculture, funded mainly by the Central Bank of Russia and the Federal Government, with the rest being provided by other commercial banks with a non-agricultural orientation. Credits extended to agriculture by Agroprombank represented around 50 per cent of its total lending activities.

Direct credits

During the reform period the state administration set up various programmes to support credit flows to agriculture and considerable budget resources were committed under these programmes. In 1992, direct credits from the state budget were granted to agriculture. The rates paid by agricultural enterprises were set at 28 per cent and for individual farmers at 8 per cent. In 1993, a single rate of 28 per cent was set for all farms. The Central Bank rate in this period ranged from 180 to 230 per cent. No clear and transparent rules based on financial evaluation of projects governed the allocation of these direct credits. The allocation of direct soft credits was rationed as in the pre-reform period and decisions were made by the administration. Each region of Russia calculated its needs for direct credits for the year. The Federal Ministry of Finance evaluated the regions' applications and made recommendations to the National Credit Policy Commission (NCPC) who took the final decision. The NCPC was composed of representatives from the Federal Ministry of Finance, Ministry of Economics, Ministry of Agriculture and the Central Bank of Russia. Due to growing budgetary pressure, interest rate subsidies on direct credits were reduced sharply at the end of 1993 and retained only for newly created individual farms. Direct credits provided important support to agriculture in the period between 1992 and 1994 and their share of total budget expenditures to agriculture was 33, 48 and 22 per cent respectively. In 1995 and 1996 their role in support to agriculture dropped to 4 and 1 per cent respectively, and they were replaced by commodity loans and machinery leasing programmes.

The emerging individual farming sector was granted limited credit support from the state budget as early as 1990 when the Government made a first allocation of budget resources to the Russian organisation of farmers (AKKOR) to support family farm development. The bulk of these funds was deposited in Agroprombank to be used as guarantee funds for credits granted to individual farmers. Under this scheme, AKKOR through its regional divisions provided guarantees for the credits extended to individual farmers by the commercial banks. However, this system failed basically due to untransparent rules and opaque criteria which led to fraud and violation of financial rules. The programme was discontinued in 1994.

Long-term credits to finance investments in agriculture and the food industry were also financed from the state budget. In the early years of reform, the administration allocated these credits through different programmes named "Grain for Russia" or "Sugar for Russia" without clearly setting criteria for resource allocation and there was lack of co-ordination between federal and regional administrations. The federal budget often supported small projects of regional or even local importance. In 1995/1996 a more competitive system was created to allocate state support, through the fund of the Ministry of Economy, to medium-term investments. The new system requires a financial participation by the investor and/or the commercial banks (at least 50 per cent) in the projects supported by preferential loans from the fund. However the criteria for the evaluation of the effectiveness of the investment project have not yet been clearly defined. In 1997 the entire programme was frozen due to lack of finance.

Leasing of agricultural machinery

In 1994 the state set up a programme supporting purchases of agricultural machinery. The government financed purchases of agricultural machinery and leased this machinery to farmers under long-term leasing contracts. For this purpose a special leasing fund was created which was financed from the federal budget. Rosagrosnab, the former state-owned monopoly supplier to agriculture was chosen as the supplier of agricultural machinery under the leasing contracts. Rosagrosnab buys the machinery from manufacturers or imports machinery and distributes it to regions according to an agreed scheme. Regional administrations allocate the machinery to producers who have to repay in regular instalments at subsidised interest rates under leasing contracts. Budgetary resources allocated to finance the programme represented, on average, 5 per cent of total budgetary expenditures on agriculture between 1994 and 1996.

In 1997 the leasing system was substantially reformed, in order to make it more transparent and to secure a higher rate of repayment of finances engaged under this programme. Under the new scheme Rosagrosnab lost its monopoly position as supplier of machinery under leasing contracts. The resources from the Federal Leasing Fund are lent to private leasing companies (leasing operators) selected in open tenders. To be selected in a tender the potential leasing operator was to commit its assets to finance leasing operations (Kholod, 1996). The contracts signed with the Federal Leasing Fund provide that all the financial resources from the Fund must be paid back by the leasing operator according to a specified timetable.

Commodity credit

In 1995, the government introduced the concept of "commodity credit" (*tovarny credit*) as a substitute for short-term direct credits for crop producers to finance sowing and harvesting campaigns. These commodity credits were not financial credits but deliveries of fuel to be repaid by deliveries of specific commodities (basically grains) after the harvest. Under this lending scheme the value of the fuel deliveries was offset against the fuel suppliers' fiscal liabilities to the state budget. Like the previous direct short-term credits to agriculture, the commodity credits were still based on administrative decisions and rationing. Moreover, there was no leverage to make farmers repay the credits. In practice the commodity credits were not given to finance seasonal work in general but were closely linked with state procurement. The indicative procurement price (for grains) fixed by the state was also used to fix the terms of trade for the commodity credits.

In 1995, the commodity credits fell short of achieving the expected results. Fuel deliveries were much lower than had been planned, and the terms of trade for such deliveries turned out to be unfavourable for agricultural producers. Immediately after the introduction of the commodity credit schemes the oil companies increased prices of fuels delivered to agriculture by 20 to 30 per cent.[10] Because of unfavourable terms of trade for farmers, and weak enforcement mechanisms, regional governments experienced great difficulties in collecting commodity credit repayments (in general in the form of grain deliveries to state procurement). Additionally, the procedure for allocating commodity credits became more complicated and introduced additional economic agents (fuel producers and suppliers, local authorities) into the credit chain. The introduction of commodity credits further distorted credit markets and markets for the commodities to which these credits were applied. Furthermore, the scheme promoted highly inefficient barter transactions, which limited the scope of possible buyers and suppliers of agricultural products. In addition, in April 1996 a Presidential decree deferred until 1998 the repayment of Rb5.1 trillion of unpaid commodity credits disbursed in 1995. The same decree postponed until 2005 the remaining debts (those not written off in 1995) for direct credits supplied to agricultural producers and food processors from 1992 to 1994.

In 1996, commodity credit remained the main channel of centralised short-term credit for agriculture. Regions received fuel, lubricants and fodder of the total value of Rb12 trillion to be distributed to agricultural producers in the form of this credit. However, the vast majority of agricultural enterprises did not repay the credits and, in turn, regions were not able to repay their debts to the Federal Government. The debt amounted to Rb9.6 trillion at the beginning of 1997. Moreover Rb1.7 trillion of commodity credits were forgiven as disaster relief (the main beneficiaries of this disaster relief were the regions of West Siberia, Northern Caucasus, Ural and Volga Region).

Soft Farm Credit Fund

In 1997 the government created (Government Decree No. 224 of 26 February 1997) a Soft Farm Credit Fund for the Agro-Food Organisations. The Fund was set up to provide preferential short term credits for agricultural producers and to replace the commodity credit. For each *oblast* the upper limits of preferential credits were fixed by the Ministry of Agriculture and Food. The subsidised interest rates were fixed at one quarter of the Central Bank refinance rate. In 1997 two commercial banks (Agroprombank and Alpha Bank) were authorised by the government to provide these credits and for 1998 twelve commercial banks, including CBS-Agro and Alpha Bank, were selected through an open tender.

In 1997 the Fund received Rb2.8 trillion for its operations from the federal budget. The repayments from the 1996 commodity credits were to constitute the second source of finance for the Fund. In May 1997, the government decided to transform the remaining regional debts into three year bonds to be issued by the regions and sold on the stock market. The decision to issue bonds depended on the regions themselves, but for the regions which did not issue these bonds the repayments due for commodity credits were to be deducted from the federal budget transfers to the region. The sales of these bonds were not very successful and by the beginning of 1998 the Fund received less than Rb2 trillion from this source.

2. Input subsidies

i) Pre-reform period

During the Soviet period, input subsidies were one of the most important instruments for lowering agricultural production costs and maintaining low and stable food prices. However, these input subsidies led to widespread price distortions in the whole agro-food chain. Input subsidies were applied mainly to agricultural machinery and fertilisers (chemical plant protection products were not subsidised). In the 1980s input prices paid by agricultural producers were lower than wholesale input prices by 50 per cent for fertilisers, by 40 to 50 per cent for agricultural machinery, by 20 to 40 per cent for tractors and 20 to 30 per cent for trucks. The price gap was paid from the state (USSR) budget to the State agencies supplying the specific inputs to agriculture.

These input subsidies were applied until 1988 and abolished in 1989. They were replaced by differentiated supplementary payments (*i.e.* paid directly to agricultural producers) (Part IV.2.a). In 1989 the input subsidies converted into supplementary payments were estimated at Rb1.3 billion (US$1.4 billion) for fertilisers and Rb1.6 billion (US$1.7 billion) for machinery.

Another subsidised input was electricity which was supplied to industries (including the food industry) at Rb0.038 per Kilowatt-hour while agricultural producers paid only Rb0.010 per kWh. The price gap was compensated by taxing other users (industries) which were paying higher rates.

ii) Reform period

During the first years of price liberalisation the output/input ratios for agricultural producers deteriorated sharply due to the price deregulation and subsidy cuts. The devaluation of the domestic currency and the resulting inflationary pressures added further to the difficulties. Input suppliers reacted to new price and market conditions by raising their prices close to world market levels. The gap between domestic prices and world market levels closed much more slowly in the first years of the reform due to inefficient price transmission along the agro-food chain and falling consumer demand.

In order to make up for the cost price squeeze, government compensated farmers for the rise in input prices by direct payments. Most of these direct payments were to compensate for the costs of inputs such as fuel, gas, fertilisers and chemicals, especially in 1992, when the share of input subsidies in total budgetary expenditures represented above 9 per cent. There were also some smaller input subsidy programmes for fuel and electricity for greenhouses, mixed feed (feedlots and poultry factories), high quality seed (to specialised high quality seed reproduction farms), and pedigree subsidies (to selected pedigree farms) that continued after 1993. In 1993 the subsidies for inputs were relatively low and represented around 2 per cent of total budgetary expenditures in agriculture. However, this share increased again from 5.8 to 11.4 per cent between 1994 and 1996.

Table IV.19. **Electricity prices for agriculture and industry**
Rouble per kilowatt hour (Rb/kWh)

	Industry	Agriculture	Ratio Agriculture/Industry
Up to 1991	0.038	0.010	0.26
1992	1.100	0.204	0.19
1993	12.00	5.33	0.44
1994	59.68	31.87	0.53
1995	137.30	93.70	0.68
1996	244.93	136.38	0.56
1997	264.07	160.56	0.61

Source: Ministry of Agriculture, Ministry of Economy.

From the range of input subsidies used in the pre-reform period, current policies have retained the lower price of electricity supplied to agricultural producers, for which the financing scheme remains the same as in the pre-reform period (*i.e.* not directly from the budget but by transfers from industry which was taxed by higher prices used to compensate for low prices for agriculture). The price gap has been gradually reduced between 1993 and 1995, but increased again in 1996 (Table IV.19). The level of support provided to agricultural producers through lower electricity prices was relatively high, and it represented an equivalent of around 8.5 per cent of budgetary expenditures to agriculture between 1992 and 1994. In 1995 and 1996 this share even increased to 11 and 19 per cent respectively.

3. Tax concessions

i) Pre-reform period

In the 1980s the basic taxes levied under the Soviet system were:

- enterprise profit tax;
- social insurance and maintenance tax;
- tax on fixed assets;
- tax on the use of natural resources;
- turnover tax;
- a personal income tax.

In accordance with the reform of 1986-1987 agricultural enterprises were not subject to the same profit taxes as other enterprises and organisations, but had to pay the so called "payments to the budget" (*platezhi v budzhet*). Under this scheme the *kolkhozes* and *sovkhozes* paid the state for the land and non-land assets allocated to them. These payments were based on the concept of so called production potential which was an equivalent in roubles of the potential of three production factors (land, fixed capital, labour). Using a specific formula for each Soviet Republic, *oblast* and *rayon* the total sum of payments to the budget in proportion to their estimated production potential was determined. Weak and loss-making farms were released from that payment and their production potential was deducted from the calculation for the region.

There was no special treatment for agricultural producers as far as other taxes are concerned. In the *kolkhozes* income tax was paid directly by the enterprise and the members of *kolkhozes* received a net remuneration. In the *sovkhozes*, as in other state enterprises, the employees paid their income tax on an individual basis.

ii) Reform period

The tax reform was one of the key elements of the economic reforms started in 1992. The new tax system was based on the following taxes:

- profit tax, set at 32 per cent of gross profit;
- personal income tax;
- enterprise fixed assets (property) tax, set up and collected by regional authorities but not more than 1 per cent of the accounting value of the property;
- land tax, with rates set at the federal level for the whole territory depending on land quality (production potential); the regional authorities have the right to differentiate the rate of land tax within a range set by the federal law;
- Value added tax (VAT) applied to all goods and services (the basic rate was fixed at 20 per cent, and a lower rate was fixed at 10 per cent for some basic products and services);
- excise tax, set for specific products considered as luxury goods;
- road fund taxes, applied on sales of products such as fuels, cars and other vehicles to provide resources for the road fund established to finance road construction and maintenance.

In addition, a new system of contributions to the Pension Fund, Social Fund, Health Insurance Fund and Employment Fund was introduced (Part IV.8).

Since the introduction of the new tax system agricultural producers have been granted various types of tax concessions. Agricultural producers are exempt from profit taxes on agricultural production and from the fixed assets tax. The rate of payment to the Road Fund is also reduced for agricultural producers. Since 1994 the profit tax exemption has been extended to profits from processing own agricultural production. Since as early as 1991, agricultural producers' contributions to the Pension Fund have been fixed at lower rates.

Since 1990, new individual farmers have been released from land tax payments for the first five years after the establishment of their farms. In 1996 this release was extended for another five years.

E. INFRASTRUCTURAL MEASURES

1. Research and development

i) In the pre-reform period

In the USSR framework Russia had developed an extensive agricultural research system. The system was vertically organised and financed from the state budget. Agricultural research institutions can be classified into four groups based on the system of organisation and financing:

- Most agricultural research was undertaken in the institutions (research institutes and experimental units), organised within the framework of the Lenin All-Union Academy of Agricultural Sciences (*Vsesoyouznaia Akademia Selskokhoziaistvennych Nauk Imeni Lenina* – VASKhNIL). The Academy had 12 scientific branch departments covering specific areas of agricultural science, such as crop selection, livestock breeding, veterinary science, irrigation, economics, etc. The Academy also had eight regional departments. Although the Russian Federation had no specific territorial substructure (it was one Soviet republic) there were four regional divisions. In the years of *perestroika* Soviet republics established their own Academies of agricultural sciences, although those were still formally subordinated to the All-Union Academy. Under this concept the Russian Academy of Agricultural Sciences was established in 1990. Under the supervision of the National Committee for Science and the Ministry of Agriculture and Food (MAF), the Russian Academy of Agricultural Sciences acted as a central administrative body planning and financing the research for subordinated research units. The board of the Academy was the highest administrative body of the Academy. It consisted of elected academics and so called "corresponding members", who were usually the heads of the Academy's research institutes or senior officers of the Academy itself. The board determined the direction of research and assessed the final results. There were

222 research institutes and 28 experimental and selection stations organised under the Russian Academy. There were also 48 breeding centres (39 for crop production, 9 for livestock husbandry) organised under the supervision of research institutes of the Russian Academy.

- Agricultural Research institutions dealing predominantly with applied studies (67 research institutes, 62 experimental and selection stations) were subordinated directly to the MAF. The research activities at these institutes were directed and financed by the Department of Science of MAF. For specific activities – like veterinary control – the institutes were supervised by the corresponding functional department of the MAF.
- Agricultural research was also carried out by the institutions engaged in agricultural education (agricultural universities, teaching academies and colleges). These research activities were also directed and financed through the MAF.
- Some research activities related to agriculture were also carried out at the institutions of the All-Union Academy of Sciences (or republican Academies of Sciences).[11] These research activities were financed through the Academy of Science and in general were not co-ordinated with the research of the Agricultural Academy of Sciences. The Academy of Science was supervised and financed through the State Committee for Science and in general was more independent than the Agricultural Academy.

The results of agricultural research were disseminated through the state institutions at the national or regional levels. Implementation of technological and economic recommendations (based on research results) by farms and agribusiness enterprises was obligatory. In addition to the centrally planned and financed research programmes any research institution could be contracted and financed by the agro-business sector to carry out studies.

ii) During the reform

The beginning of the reforms did not bring about any radical changes in the structure of the institutions engaged in agricultural science. The number of institutions and their organisational structure remained almost the same as before. The abolition of the VASKhNIL late in 1991 was the main structural change of that period. The Russian Academy of Agricultural Sciences (RASKhN) has remained the main Russian research institution since then. In 1996 the post of the president of the Academy was ranked at a level equivalent to a deputy minister of agriculture.

The channels for financing agricultural research also remained the same. Theoretically five year research programmes are elaborated by the RASKhN in co-operation with the MAF and proposed to the institutes together with appropriate finance earmarked for these programmes. However, in reality the programmes are formulated by the institutes, at the basis of available facilities, experience and staff and, after a process of iteration, these programmes, if accepted by the Academy and the MAF, are included in the programme of work to be financed from the state budget. Due to a lack of finance the number of research employees dropped to 39 000 (educational institutions are not included) in 1996. The number of post-graduate students in agricultural sciences fell from 2 300 in 1990 to 1 900 in 1995.

In 1992 and 1993 the government ordered all business enterprises in agriculture and the downstream sector to pay a contribution representing 1.5 per cent of their revenues into a special fund in order to secure the funding of agricultural research from the federal budget. To a limited extent agricultural research related to the problems of family farming was financed from the funds of the Association of the Farmers and Farm Co-operatives of Russia (AKKOR).

Dissemination of research results has been weak. The collapse of the centrally planned economy eliminated the previous extension system. Farms and agrobusinesses, faced with a lack of working capital, tended to use less research intensive technologies and were reluctant to finance the dissemination of information and applied research programmes. Being aware of the problem the government signed a US$170 million loan agreement with the World Bank. The loan was intended to finance the implementation of a project to foster agricultural restructuring. One component of the project was to finance the creation of an extension services network in Russia under the umbrella of the MAF. By 1997, regional extension service offices were set up in 22 regions of the Russian Federation and in 83 *rayons*. Resources were also committed for staff training in the extension offices. A special centre for information dissemination was established near Moscow to collect and disseminate information from a large number of institutions engaged in agricultural research.

2. Education and training

Under the Soviet rule, the system of agricultural education and training was structured into four distinct levels to serve the specialised needs of the agro-industrial complex. Specifically, the system was planned and operated to provide narrowly trained specialists for specific occupations in *kolkhoz*es and *sovkhoz*es and in the food industry enterprises operating under the centrally planned economy. Education programmes focused on crop and livestock production technologies, agricultural mechanisation, and farm accountancy. The economic reforms required the agricultural education and training system to change to meet emerging needs for new educational programmes emphasising on agricultural economics, management, marketing and environment.

In general the agricultural education and training system in Russia consists of four levels:

- initial agricultural vocational schools (SPTUs);
- agricultural colleges;
- agricultural academies and education institutes;
- continuing agricultural education.

Upon completion of primary school students may continue their education in the agricultural vocational schools, which provide on average three-year courses to train qualified workers, or in the agricultural colleges, offering four-year courses to train higher level specialists. General secondary school graduates may continue studying one year in the SPTU and three years in agricultural colleges.

The SPTUs provide courses in six main areas - crop production, animal production, soil improvement, processing, forestry, and more recently individual farming. In the school year 1992/1993, there were 2 180 SPTUs with around 600 000 students. In 1996/1997, the number of SPTUs declined to 1 065 and the number of graduates reached only 144 300.[12] Upper professional agricultural schools and colleges provide higher level education in 14 areas (accounting, veterinary science, agronomy, law, construction, etc.). Colleges also serve as a preparation for higher education (academies, universities). In the school year 1992/1993, there were 300 upper professional agricultural schools and colleges with 289 600 students. In 1996/1997, the number of these schools was slightly lower at 297 with 253 300 students.

Those students who finish full secondary education (colleges) may compete through examinations for entry to the agricultural academies, universities and institutes, which provide 20 specialised programmes in agriculture and forestry (economics, accounting, management, finance, agricultural chemistry, agricultural engineering, agronomy, etc.). Higher education in agriculture normally requires five years of study. In 1996/1997, there were 62 agricultural academies and education institutes in Russia. The main educational centre for agriculture is the Timiryazev's Agricultural Academy in Moscow. Other universities and academies are located throughout the country. Almost every region of the Federation has its own agricultural education institute and several agricultural colleges. Apart from daily courses the education institutes at all levels organise evening and correspondence (distance) courses. New entries to the agricultural academies declined from 29 100 in 1992 to 25 600 in 1997.

Educational activities to train and upgrade specialists in the agro-food sector are carried out by two main institutions -the Agro-Industrial Academy of Russia and the Russian Higher School of Management in the Agro-Industrial Complex. These two major training institutes also provide methodological and pedagogical support and guidance to almost 100 other upgrading and training institutes located throughout Russia.

Initial agricultural vocational schools (SPTUs) are administered and supervised by the Ministry of General and Professional Education (MGPE), which co-ordinates its policy with the MAF. Agricultural colleges, academies, and education institutes are supervised by the MAF although the regional education authority and MGPE play a leading role in setting standards. For that purpose two special departments exist within the MAF: the Department of Secondary Agricultural Education (colleges) and the Department of Higher Education Institutes (universities, academies). The Personnel Department of the MAF supervises the training and upgrading institutes.

Educational and training institutions are mainly financed by the MAF (with the exception of the SPTUs). To some extent these institutions are also funded from the regional and local budgets. Under the 1992 Law "On Education" educational institutes are entitled to earn supplementary income complementary to budget funding. Major sources of such funding are:

- training contracts concluded with official bodies and enterprises;
- research contracts with enterprises, international organisations and other institutions; and
- student fees.[13]

Agricultural education and training institutions have had to face considerable financial difficulties in recent years because of the reduction in public funding. Simultaneously the reforms abolished the system of administrative (mandatory) allocation of graduates to specific farms and other enterprises in the agro-food sector. In this new context many institutions started to offer courses on a fee-paying basis. In fact, almost all institutions, charge fees for "market oriented" courses such as computer courses, extra English tuition, management courses, legal courses, accounting courses, etc. In many cases farms and agro-food enterprises are willing to sponsor students in order to have high-skilled specialists in 4-5 years. In some cases, changes in the curricula are being introduced with technical assistance from international programmes such as Tempus and TACIS.

3. Quality and sanitary control

The system of standards set up in the former USSR, is still valid in Russia and in all the other NIS. The legislation dealing with common NIS markets also deals with the harmonisation of standards among the NIS. In every branch of agriculture and food industry there are hundreds of state standards (GOST). In the meat industry, for example, there are 13 national state standards, 133 technical conditions and 98 regulations (standards describing the rules and methods of specific technological processes). However, these state standards do not cover all varieties of foodstuffs currently delivered to the Russian market (*e.g.* there are no standards for yoghurts and dairy desserts).

Quality control by state institutions is organised directly at processing plants. Every lot of delivered products is sampled and tested.[14] According to the Laws "On protection of consumers' rights" (1994) and "On certification of goods and services" all foodstuffs are subject to certification. The Department of Standards and Certification of Foodstuffs and Raw Material for Foodstuffs in the State Committee on Standards of Russia is the state authority responsible for certification at the federal level.

All imported foodstuffs and raw materials for food production have to be certified.[15] With Ukraine, Moldova and some other NIS there are agreements on the adoption of the former Soviet system of food certification and mutual acceptance of food certificates. The Veterinary Department of the Ministry of Agriculture issues the permit for imports of animal products in accordance with the veterinary situation in the country concerned.[16] Based on a veterinary permit (issued by the Russian Ministry of Agriculture) and a veterinary certificate of the country-supplier a special veterinary licence is issued at the border for each imported lot of animals or animal products.

The certification organisations also issue certificates of conformity, which verify the identity of foodstuffs and of the accompanying documents. For that the importer has to present samples of imported goods and related documents to the filial of the State Committee on Sanitary and Epidemic Control. Since 1 May 1997 all imported foodstuffs must include consumer information (labels, descriptions, and instructions) in Russian. Experts estimate that in early 1997 up to 58 per cent of imported food products in the Moscow region (major consumer of imported foodstuffs) had no Russian labels.

4. Structural policies

No clear priorities regarding structural policy have as yet been set at the federal level. State structural policy for the agro-food sector during the reform period has been limited to two main areas:

- Government support for the establishment of private individual farms, providing credit subsidies to newcomers to rural areas willing to start individual farms. In 1992, credit to individual farmers was more highly subsidised than credit to large farms, but as of 1994 the programme was discontinued (Section III.4.iii).
- Programmes supporting investment in the food industry. However, the programmes supporting investment in various branches of the food industry were not well defined and lacked financing. At the beginning of the reform investment aids were also provided to farms building their own processing capacities, financed mainly from foreign loans and donations. In addition, income from on-farm processing activities has been exempted from taxation since 1993. Nevertheless only small-scale mills and bakeries demonstrated relatively high competitiveness. Dairy and meat processing plants found themselves unable to compete with large-scale enterprises.

5. Agricultural infrastructure

Infrastructure facilities in rural areas have been underdeveloped in Russia. The poorly developed road network and local storage and processing facilities were, for instance, among the major reasons for serious losses of farm production for decades. Almost all the rural areas are supplied with electricity but the gas supply, water supply and waste water management networks were poorly developed. In the pre-reform years and during the reform the already weak infrastructure deteriorated further due to lack of finance for maintenance and development. The division of responsibilities between the different levels of local authorities for the maintenance of Infrastructural networks was not clearly specified and has further contributed to the poor condition of rural infrastructure.

Around 5 to 6 per cent of all agricultural lands in Russia are improved (drained or irrigated). In some areas such as the North-West region, the share is more than 30 per cent. However, the lack of finance has led to a deterioration of drained or irrigated lands.

6. Marketing and promotion

Economic reforms had a major impact on the whole system of marketing agricultural products. In the centrally planned economy the state was the principal distributor both of inputs to agriculture and agricultural and food outputs. The transition to a market oriented economy quickly led to a diminished state role in the agro-food market.

With this reduction in the state's role the shares of other marketing channels (private middlemen, wholesale markets and auctions, producers' own retail outlets) increased. However, direct deliveries to privatised food processing enterprises remained the most important marketing channel. Table IV.20 shows the structure of different marketing channels used for grain and meat in 1996.

The appropriate market infrastructure has been slow to develop in Russia. There is a lack of commodity exchanges, wholesale markets and auctions that would improve market transparency and provide clear market signals to producers. Barter trade remained important even in 1997, and a large part of production is still used for payments in kind for workers on farms. Moreover, governmental programmes on trade credit and leasing were increasingly using barter trade (Part IV.4.) and were an impediment to the development of a functioning market infrastructure. In 1994, the MAF adopted the national programme for the development of agro-food wholesale markets. According to the programme a "Law on agricultural wholesale markets" was to be drafted, and a number of pilot wholesale markets of different types were to be set up in various regions across the country. Mainly due to budget constraints the programme was implemented on a limited scale. Only some regional and local governments endorsed a few activities to develop such markets: by 1997 the markets were reported to exist in 20 regions, while in 30 more regions they were being organised. At the end of 1997, the MAF started to

Table IV.20. **Structure of grain and meat sales in 1996 (survey of 20 regions)**

Per cent

Channels of marketing	Grain	Meat (liveweight)
Direct sales to processors	30.4	64
Direct sales to consumer co-operatives	0.7	1.4
Direct deliveries to consumers	1.1	2.8
Wholesale markets, fairs, commodity exchanges	4	0.6
Kolkhoz (town) markets	12	7.2
Producers own retail outlets	10.9	10.7
Sales through middlemen	1.7	1.3
Barter trade	11.9	3
Payment in kind to workers	16.4	5.1
Sales to other agricultural enterprises	10.9	3.9

Source: Goskomstat.

prepare a new, two-stage programme aiming to create a system of wholesale markets for the whole country by the year 2005.

Some support to develop a market infrastructure and information system was provided to Russia by international institutions. With the technical assistance of the USDA, work on the development of a nation wide market information system in agro-food sector was started in 1994. Also in 1994, the Russian government signed an agreement with the World Bank for a US$240 million loan on Agricultural Restructuring Implementation Support (ARIS). One component of the ARIS programme is to finance market infrastructure development.

Beginning in 1997, in accordance with the ARIS programme, 25 regions in the Russian Federation signed agreements for credits to finance the development of a market information system. As of mid 1997 only preparatory work had been done in this area: five pilot projects for wholesale markets had been drafted for approval by the World Bank experts.

F. RURAL DEVELOPMENT MEASURES

The rural population (39.9 million) represented 27 per cent of the total population of Russia in 1996. The proportion has increased slightly since 1990 (by one percentage point), as the number of people living in rural areas has risen by 3 per cent while the total Russian population declined by 1 per cent between 1990 and 1996. The rural economy is still to a large extent dependent on the agricultural sector as it employs almost 50 per cent of the active population in rural areas. The share of the active population out of the total population in rural areas has been stable at 52 per cent in the 1990s, much lower than the share calculated for the urban population (59 per cent).

The Russian government has so far not formulated a coherent policy for tackling specific problems in rural areas, but has declared its intentions to continue the traditional Soviet approach to closing the gap between the level of services provided in urban and rural areas. However, the limitations of public finance do not allow the maintenance of the existing services, let alone their extension. Support for the rural population is provided almost exclusively through agricultural policy measures. In the 1996 budget there were only two items that may be considered as support for the rural economy: subsidies for housing construction in rural areas; and finance for training programmes. These two programmes combined represented some 6 per cent of total budgetary spending for agriculture.

G. SOCIAL MEASURES

In the Soviet period social policy for rural areas was closely linked with agriculture and agricultural policy. Large scale agricultural enterprises (*kolkhozes* and *sovkhozes*) fulfilled a broad range of social functions (nursery schools, sport, culture and leisure activities, health care, canteens, etc.) in rural areas. Expenditures of agricultural enterprises on social programmes were strongly reduced after 1991, resulting in a decline in the availability of these social services.

There has been no specific social security system for agricultural employees, farmers or their families. The institutional and legislative instruments on which current state social policy is based combine several funds separated from the state budget with clearly defined financing rules:

- Pension fund – the contribution by the employer was set at 31.6 per cent of the wages paid to employees or of the income of an individual entrepreneur. Since 1993 the contribution to the fund has been reduced to 28 per cent. Since 1991, large scale agricultural enterprises and individual farmers pay lower contributions to the Pension fund fixed at 20.6 per cent;
- Social fund – the contribution of employers and entrepreneurs to the social fund is fixed at a flat rate of 5.4 per cent of wages;
- Employment Fund (established in 1992) – payments to the fund were initially fixed at 1 per cent of wages and since the second quarter of 1993 at 2 per cent, and since January 1996 at 1.5 per cent;
- Obligatory Health Insurance Fund (established in 1993) – contributions were set at 3.6 per cent of wages;
- apart from the lower contribution to the Pension fund, the same social policy measures are applied to agriculture and rural areas as to the other sectors of the economy.

H. ENVIRONMENTAL MEASURES

In the Soviet period environmental measures were applied through specialised government institutions designed to plan and control the use of different types of natural resources. More specifically, the following governmental bodies controlled natural resources:

- the Ministry of Water Resources;
- the Ministry of Geology;
- the Ministry of Forestry;
- the Committee on Land of the Ministry of Agriculture;
- the Committee on Hydrometeorology;
- the Committee on Geodesy and Cartography.

These institutions covered all of the USSR and had corresponding branches in the Russian Federation. In addition a special Committee on Environment was established in 1988. In 1990 this committee was transformed into the State Committee on Ecology and Use of Natural Resources.

At the end of 1991 all the Russian branches of the Soviet Union's specialised institutions were grouped together to form the Russian "Ministry of Ecology and Natural Resources", except for the Russian Committee on Land, which as a separate entity, was made responsible mainly for the management of the land reform. The newly established Ministry had since undergone important structural change. Following the adoption of the Federal Law "On mineral resources" (1992) the supervision of geological activities was withdrawn from the Ministry and came under the responsibility of the Ministry of Geology. Similarly, the Committees on Water Resources, Hydrometeorology and Forestry left the Ministry during 1992 and 1993 and were re-established as independent institutions. In December 1993, the Ministry of Ecology and Natural Resources was renamed the "Ministry of Environment and Natural Resource Protection". Finally at the end of 1996 the Ministry was split into two institutions of state administration: the "Ministry of Natural Resources" and the "State Committee for Environment Protection".

The Ministry of Natural Resources is responsible for regulating the use of water and mineral resources and for their environmental protection. The State Committee on Environment Protection is the main governmental body in Russia implementing the state policy on environment, performing ecological control, providing ecological expertise and administering the state-owned natural parks and reserves. The Ministry of Agriculture and Food is responsible for the development and implementation of environment measures in the agriculture and food industry. After the Committee on Fishery merged with the Ministry of Agriculture and Food in the beginning of 1997, the latter took over the allocation of responsibilities for rational use of biological water resources.

The basic documents regulating environmental policies in Russia are:

- Law "On Environment Protection" (1991) amended in 1992 and 1993;
- Law "On Animal World" (1995);
- Law "On Ecological Expertise" (1995);
- "On land improvement (irrigation, drainage, reclamation)" (1996);
- Water Code and Forest Code (1996).

There were also several Russian Federation Government and Presidential decrees which more directly addressed environmental issues related to agriculture. However, their major concern was maintenance of land quality in view of increasing agricultural production rather than environmental issues as such.

I. CONSUMER MEASURES

During the Soviet period the authorities continuously tried to keep retail prices stable and well below the rising costs of food production. To keep food prices low, increasing subsidies were paid to processing enterprises. Moreover, stable food retail prices combined with the increase in nominal wages stimulated demand for food, which could not be matched by supplies from the inefficient farming sector. At the beginning of the 1990s there was increasing evidence of the economic unsustainability of the existing system of consumer support. In April 1991, retail prices in state shops were doubled, nevertheless the state retail system collapsed by autumn 1991.

As part of the general economic reform, price liberalisation was introduced beginning in 1992. From March 1992, retail price controls on food were delegated to the regional authorities and had to be financed from local budgets. By the end of 1992, almost all regions recorded budget deficits and limited their price control activities mostly to bread and milk. However, food prices were regulated differently in regions reflecting local budget conditions and local policy (Annex on Regional Policies).

To a certain extent the social programmes set to support the lower income groups of the population are implicitly directed (at least partly) to food consumption support, insofar as the share of expenditure spent on food for low income households is close to 70 to 80 per cent (the average for Russia was 43 per cent in 1996). Also, the lower rate of VAT applied to foodstuffs (except those liable for excise duties) introduced in 1993 (10 per cent instead of the normal VAT set at 20 per cent) indirectly supports food consumption, but in this case, for the entire population.

J. OVERALL BUDGETARY OUTLAYS ON AGRO-FOOD POLICIES

In the Soviet period, agricultural policies, as described above, led to unsustainable levels of budgetary expenditures. In 1990, the value of food subsidies to consumers corresponded to 25 per cent of total Soviet budget expenditures (8 per cent of GDP), while the share of budgetary transfers to agriculture represented around 10 per cent of total budgetary expenditures (3.3 per cent of GDP).

Some overall budgetary figures (*e.g.* total budget expenditures) for the Russian Federation are available only from 1992, since it was impossible to extract corresponding data from the overall budgetary data of the Soviet Union period. The change in objectives introduced during the reform period was also accompanied by a substantial reduction in subsidies. Between 1992 and 1995, the share of budgetary transfers to agriculture in total budgetary expenditures declined from 13.2 per cent to 4.7 per cent and, then, increased slightly to 5 per cent in 1996 (Table IV.21).

Other (non budgetary) transfers to agriculture were provided in the form of lower electricity prices.[17] The equivalent of support provided through energy prices, expressed as a percentage of total

Table IV.21. **Subsidies and other financial transfers to the agricultural sector – general indicators**

Billion roubles

	1992	1993	1994	1995	1996
Total budgetary transfers to agriculture	790	5 218	13 867	22 982	31 533
Total budgetary expenditures	5 970	57 647	234 840	487 400	652 700
Budgetary transfers to agriculture as a share of total budgetary expenditures %	13.23	9.05	5.90	4.72	4.83
Other transfers to agriculture	63	462	1 113	2 616	6 145
Budgetary and other transfers to agriculture	852	5 679	14 980	25 598	37 678
GDP in nomimal prices	19 006	171 510	610 592	1 630 956	2 256 000
Share of total transfers to agriculture in GDP %	4.48	3.31	2.45	1.57	1.67

Source: Goskomstat, Ministry of Agriculture and Food.

Table IV.22. **Total budgetary expenditures in agriculture**

	1990	1991	1992	1993	1994	1995	1996
Billion roubles (current prices)							
Direct payments	0	11	214	1 029	2 233	5 314	7 432
of which: livestock subsidies	0	0	202	915	2 045	4 272	5 488
Reduction of inputs costs	15	3	338	2 633	4 876	9 214	14 190
of which: input subsidies	1	0	73	117	805	2 493	3 700
credit subsidies	14	3	264	2 517	4 071	6 721	10 490
General services	6	8	91	780	4 480	5 568	6 911
Social services	9	6	147	776	2 278	2 886	3 000
Total budgetary expenditures	30	27	790	5 218	13 867	22 982	31 533
Per cent							
Direct payments	0.0	39.9	27.1	19.7	16.1	23.1	23.6
of which: livestock subsidies	0.0	0.0	25.6	17.5	14.7	18.6	17.4
Reduction of inputs costs	50.2	11.0	42.7	50.5	35.2	40.1	45.0
of which: input subsidies	2.0	0.4	9.3	2.2	5.8	10.8	11.7
credit subsidies	48.1	10.6	33.4	48.2	29.4	29.2	33.3
General services	20.5	28.2	11.6	15.0	32.3	24.2	21.9
Social services	29.3	20.9	18.6	14.9	16.4	12.6	9.5
Total budgetary expenditures	100.0	100.0	100.0	100.0	100.0	100.0	100.0

Source: OECD Secretariat.

budgetary transfers to agriculture, was stable at around 8 per cent between 1992 and 1994, and increased in 1995 and 1996 to 11.4 and 18.9 per cent respectively. As a percentage share in GDP, the sum of budgetary transfers and other transfers to agriculture (through lower electricity prices) declined from 4.44 per cent in 1992 to 1.57 per cent in 1995, but increased to 1.7 per cent in 1996 (Table IV.21).

In the period from 1992 to 1996, agricultural policy measures were limited to *ad hoc* measures addressing the most immediate problems of the agricultural sector. Subsidies and other financial transfers from the state budget to agriculture, were often provided through programmes changing from one year to another. Under these programmes the main budgetary transfers to agriculture consisted of:

- direct payments, mainly to livestock producers;
- financing programmes reducing input costs using both direct inputs subsidies and credit subsidies;
- expenditures on general services; and
- contributions to finance social services provided by large scale farms in rural areas.

During the period from 1992 to 1996, most budget expenditures (40 to 50 per cent) were on programmes reducing input costs. This type of support consisted mainly of credit subsidies. However, the share of direct input subsidies increased steadily between 1993 and 1996. The share of direct payments declined between 1992 and 1994 from 27 to 16 per cent. In 1995 and 1996 this share has been stable at 23 per cent. The share of expenditures on general services increased in the first years of the reforms (1992-1994) and then declined to around 22 per cent in 1996. The share of budgetary contributions to finance social services provided by the large scale farms declined continuously from 18.6 per cent in 1992 to 9.5 per cent in 1996. The level and structure of total budgetary transfers to agriculture in the period from 1990 to 1996 are shown in Table IV.22. These transfers, although related to, should not be confused with the transfers to agriculture as estimated by the PSE calculations (Part V).

NOTES

1. In the rural areas the food supply was mainly based on the production from household plots (autoconsumption).
2. The main sources of budget revenues were the profits from exports of oil and gas, however these revenues were reduced during the 1980s as oil and gas prices dropped significantly.
3. Since 1992, the Russian government has had increasing difficulties carrying out its budgeted commitments due to its inability to collect the taxes it formally claims. The Ministry of Finance and the Central Bank of Russia have adopted the practice of paying budgeted allocations only when the money to cover them is available. In order to secure the most important payments, drafts of state budgets have included a system of priorities, providing for so -called "protected items" which are to be covered first, if necessary by borrowing. Only budgeted items which are "protected" are assured of being allocated.
4. For selected products a premium equivalent of 50 per cent of the administered price was paid for production exceeding the average of the 5 previous years and a premium equivalent to 100 per cent of the administered price, if the delivered production exceeded both the annual target and the average of the 5 year period.
5. The calculated wholesale price took into account the administered prices paid to farmers, which were higher that the wholesale price at which food producers sold their production.
6. At their peak between 1988 and 1991 consumer subsidies represented around 10 per cent of GDP.
7. The minimum guaranteed price is probably not the right term as these prices were offered only for sales to federal reserves and were not used under an open ended system of state intervention. For other market agents the role of these prices was only indicative.
8. The term "mark-up limits" is used in Russian to denote a limitation on price increase relative to the cost of raw materials.
9. Originally these tariffs were planned to be introduced as of March 1994 and for some products at higher rates (sugar 30 per cent); however, the mayors of Moscow and St. Petersburg, the two major consumption centres of imported food, strongly opposed these measures. Thus, the government delayed the introduction of the tariffs and reduced some of them.
10. In autumn 1996, the government decided to introduce white sugar import quotas of 1.5 million tonnes (1.15 million tonnes from Ukraine), but the proposal was vetoed by the president.
11. In fact this price increase was a kind of interest rate for commodity credit. This situation creates a paradoxical situation as the commodity loans are implicitly financed by the state budget (unpaid tax revenues from the oil companies) and the interests on these loans are collected by oil companies in the form of increased prices for deliveries to agriculture under commodity loans.
12. As a rule in the countries of the Soviet block, basic research fall under two academies: The Academy of Sciences (also called the "big Academy") under which was organised all research covering all fields with the exception of agricultural research, whose main activities were organised under the Academy of Agricultural Science (the "small Academy").
13. This substantial reduction is due principally to a change in the statistical classification of the vocational schools in 1994.
14. In principle education is guaranteed and free through university. However, the law stipulates that fee paying students may be taken into schools in addition to the required allotment of students. Each school can have a certain percentage of fee paying students that fall into various categories (*e.g.* those who completed a course and are doing another subject; those taking courses not on the regular curriculum, etc.)
15. For example milk is controlled by organoleptic parameters, temperature, solidity, cleanness, acidity, fat content and efficiency of the temperature treatment. Thermal resistance testing is obligatory for milk to be used for baby food and pasteurisation.

16. Two certificates are required for food products entering Russia: a Gosstandard certificate showing that the product meets all applicable Russian standards, and a Gosanepidnazour certificate of hygiene showing that the product may be imported.

17. In Ireland, the Russian administration has permanent representatives to control beef deliveries to Russia.

18. These transfers were not financed from the budget but by an implicit tax charged on the electricity consumption in industry (Section IV.4.b.ii).

EVALUATION OF SUPPORT TO AGRICULTURE

For the purposes of this study support to Russian agriculture is measured by the Producer Subsidy Equivalent (PSE) and the Consumer Subsidy Equivalent (CSE). PSEs and CSEs, as indicators of government assistance granted for the main agricultural commodities, have been estimated for all OECD countries as well as for several CEECs and used in the analysis of their progress to a more market oriented agriculture. The PSE/CSE concept and related methodology have already been presented in several OECD publications.

In brief, the Producer Subsidy Equivalent indicates the value of monetary transfers to agriculture resulting from agricultural policies in a given year. It includes both transfers from consumers of agricultural products (through domestic market price support) and transfers from taxpayers (through budgetary or tax expenditures). The market price support component is measured by a price gap between domestic and external reference prices multiplied by the quantities produced. Transfers from taxpayers include subsidies paid directly on output, input subsidies, infrastructure subsidies etc. The Consumer Subsidy Equivalent indicates the implicit tax paid by consumers as a result of higher domestic prices maintained by market price support. It also includes other transfers such as subsidies to consumers from government budgets. The methodology used in the calculations continues to evolve through a process of constant review within the OECD.

In any use of PSE and CSE indicators, such as for comparison between countries, it is important to bear in mind the recognised limits of these indicators with respect to policy coverage, commodity coverage, data availability and methodology applied, as well as the specific characteristics of agricultural policies in countries in transition, in particular the macro-economic framework in which agricultural policy measures have been applied in recent years. The PSE and CSE calculations for Russia are, in general, based on actual, official data, albeit often of a preliminary nature, especially for recent years when more approximate methods of estimation have been necessary. Obviously, as actual data may differ from what has been assumed, the calculations may be revised in due course as more reliable data become available. The more detailed description of methodology and the assumptions underlying the PSE and CSE estimates for Russia are presented in Annex I which also includes tables of PSE/CSE calculations and results.

A. AGGREGATE RESULTS

The evolution of total support to Russian agriculture as measured by the Producer Subsidy Equivalent is shown in Table V.1. and Graph V.1. (Annex Tables I.12.i. and I.12.ii for details). The results of the analysis should be interpreted with caution in view of the major changes that took place in Russia between 1986 and 1997. To a large extent, the pattern of support to Russian agriculture followed that

Table V.1. **Aggregate percentage PSEs and CSEs for Russia, 1986-1997**

	1986	1987	1988	1989	1990	1991	1992	1993	1994	1995	1996p	1997e
Percentage PSE	98	97	91	86	80	61	–105	–26	–9	21	32	26
Percentage CSE	–54	–52	–49	–48	–43	–50	172	72	45	–1	–18	–20

e: estimate; p: provisional.
Source: OECD.

observed in other countries in transition. Support, as measured by the PSE, can be divided into three distinct phases: very high during the Soviet period, negative at the beginning of the transition period and increasing in more recent years.

During the Soviet period the market price support was very high, and the net percentage PSE was on average 90 per cent between 1986 and 1990. The high level of market price support was due to the fact that domestic prices in Russia of most products covered by the PSE calculations were substantially above the border reference prices used in the estimates and calculated at official exchange rates. Between 1986 and 1991, these high administered prices were maintained by the centrally planned system and were isolated from changes in world market prices by the state monopoly on all trade. Therefore, the price gap measurement (which forms the basis for the calculation of Market Price Support: MPS) embraces all measures applied in this period, including the setting of fixed prices and margins at various points in the food chain as well as the setting of exchange rates (Parts I and IV). Budgetary support in the Soviet period was also very high, in particular through soft credits and subsidies reducing the costs of inputs.

During 1992-1993, the dramatic macroeconomic developments and rather chaotic adjustments resulted in a sharp fall in the net percentage PSE from 61 per cent in 1991 to minus 105 per cent in 1992 and minus 26 per cent in 1993, reflecting an implicit taxation of the agricultural sector. This sharp drop in PSEs coincided with a substantial real depreciation of the currency and reflects the predominant role of macroeconomic developments much more than specific agricultural policies. Indeed, the implicit taxation, resulting largely from the depreciation of the Rouble, was not specific to agriculture, as most other sectors were also facing transitional adjustment difficulties to the new macroeconomic conditions. A strongly undervalued Rouble, particularly in 1992 and 1993, could be seen as a form of "macroeconomic protection" of the Russian economy as a whole, including agriculture, against imports and by increasing export competitiveness on foreign markets. However, this "macroeconomic protection" was not exploited by the farming sector due to the inefficiencies in the whole agro-food economy, particularly in the downstream sector, which was unable to transmit higher world prices for many products to the farmgate. Moreover, taxes on exports of agricultural products (70 per cent for grains, meat and meat products, and vegetable oil; 15 per cent for milk and dairy products and oilseeds; 70 ECU per tonne for sugar) imposed by the Russian government at the beginning of the transition period in an attempt to prevent the outflow of these products from the domestic market and maintain low prices for consumers, as well as local restrictions on free movements of agricultural commodities between regions, contributed to keeping producer prices low compared to world market prices, thus maintaining the implicit taxation of producers.

From 1994, most export restriction measures were progressively reduced and completely abolished by March 1996. Strong real appreciation of the Rouble in 1994 combined with the introduction of border protection against imports of many products, in particular of animal products, contributed to the increase in the measured level of support to agriculture up to 1996. The estimated percentage PSE in aggregate terms became positive at 21 per cent in 1995 and increased to 32 per cent in 1996. In 1997, the fall in prices of several agricultural commodities (primarily grains and oilseeds), the decrease in budgetary support (in nominal and real terms), together with stabilised border protection and stabilised real exchange rate led to a fall in the average aggregate PSE to 26 per cent.

During the Soviet period the level of support in Russia was significantly higher than the average in (24) OECD countries (which declined from 47 per cent in 1986 to 42 per cent in 1991) and also higher than in the other CEECs for which OECD has calculated PSEs (Hungary, Poland, Czech Republic, Slovakia and the three Baltic countries) (OECD 1997*c* and OECD 1998, forthcoming). In 1997, the percentage PSE was below the (24) OECD average of 35 per cent, but already higher than the levels estimated in 1997 for the Baltic countries (8 per cent in Latvia, 9 per cent in Estonia and 18 per cent in Lithuania), Slovakia (25 per cent) and for the Czech Republic[1] (11 per cent), Hungary[1] (16 per cent), and Poland[1] (22 per cent) (Annex Tables I.16 and I.17).

In the 1986-1991 period, consumption subsidies compensated for part of the market transfers (about one third of the total value of market transfers) from consumers to producers resulting from Soviet market price support policies (Annex Table I.12.ii). Hence, the implicit tax on consumers as measured by the percentage CSE ranged from minus 54 to minus 43 per cent, *i.e.*, much lower than the levels of net percentage PSE. During the early years of the reform, consumer subsidies were substantially reduced and were mostly granted to milk and cereal products. In 1995, consumer subsidies had become almost negligible. As of 1992 and up to 1994, negative levels of market price support resulted in an implicit subsidy for consumers. As market transfers accounted for the major part of total CSE, changes in the total value of CSE mirrored developments in market price support. The level of the implicit consumer subsidy decreased gradually from 172 per cent in 1992 to 45 per cent in 1994. In 1995, market transfers became negative, and in 1997 the implicit tax on consumers was 20 per cent.

B. EXCHANGE RATE SENSITIVITY

Since the main component of support to Russian agriculture is market price support, PSE estimates are very sensitive to the exchange rate applied. The basic set of PSEs/CSEs presented in the study is calculated at official exchange rates, on the assumption that these rates reflect the economic conditions in which the economic agents made their decisions. However a second series of PSEs/CSEs was calculated with an adjustment made to the official exchange rate and the results are presented in Annex Table I.21. The adjusted exchange rate used in the study is the "Atlas Conversion Factor" calculated by the World Bank (for the definition see Annex I, section C.5). The ratio of adjusted to official exchange rates presented in Annex Table I.20 shows to what extent the Rouble was overvalued in the Soviet period (the adjusted exchange rate was on average 1.7 times the official one) and undervalued between 1992 and 1994. The currency was most undervalued in 1992 when the adjusted exchange rate was only 22 per cent of the official exchange rate. Afterwards, very high inflation rates combined with much lower depreciation in 1993 and 1994 led to a real appreciation of the currency which returned to a more market-related equilibrium. In 1995 and 1996, the ratio of the adjusted to the official exchange rate was

◆ Graph V.1. ***PSEs for Russia at official and adjusted exchange rates***

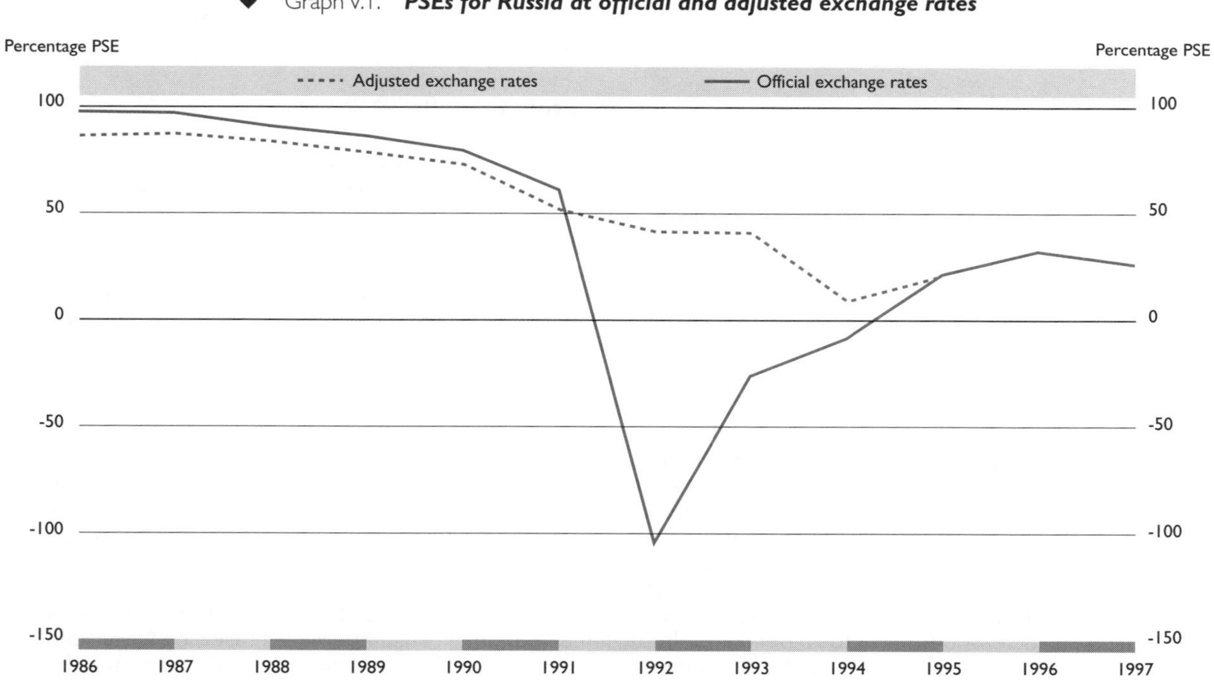

Note: When 1997 calculations were made, an estimate of the 1997 Atlas Conversion Factor was not available.
Source: OECD.

close to one. When 1997 calculations were made, an estimate of the 1997 "Atlas Conversion Factor" was not available.

The results of this adjustment were relatively small for the period from 1986 to 1991, since the level of support as measured by the percentage PSE at adjusted exchange rates followed the same pattern as for the percentage PSE at official exchange rates (Graph V.3.). During that period, percentage PSEs at adjusted exchange rates were only 9 per cent lower on average. The greatest disparity took place in 1992 when the currency sharply depreciated (by 10 924 per cent), while the rate of inflation was 4 times lower, resulting in an undervaluation in market terms. Subsequent very high inflation rates combined with much lower depreciation between 1992 and 1994 led to a real appreciation of the currency which returned to a more market-related equilibrium. Given the magnitude of changes in real exchange rates, the adjustment had a significant effect on PSEs, especially between 1992 and 1994. Percentage PSEs at adjusted exchange rates fell from 51 per cent in 1991 to 41 per cent in 1992, stabilised in 1993, decreased to 9 per cent in 1994, and increased then gradually to respectively 21 and 32 per cent in 1995 and 1996. On the other hand, PSEs at official exchange rates dropped sharply to minus 105 per cent in 1992, but thereafter recovered steadily and reached 32 per cent in 1996. In effect, the adjustment deferred the impact of the policy induced depreciation of 1992, spreading it out over several years when the adjusted exchange rate depreciated in line with inflation (Graph V.1).

C. DECOMPOSITION ANALYSIS OF SUPPORT

1. Composition of support

Shares of the different components of support in the net total PSE are set out in Annex Table I.14. Market price support was by far the most important component. It increased from between 83 and 99 per cent in the pre-reform period to over 100 per cent between 1991 and 1994. A share greater than 100 per cent was possible because the strong negative values of MPS were partly compensated by budget expenditures and the feed cost adjustment component which became positive as of 1992. In

Box V.1. **Decomposition results**

The decomposition of the total PSE and total CSE helps to identify the relative importance of changes in the various PSE and CSE components in explaining the overall year-to-year changes in PSEs and CSEs. The decomposition analysis is presented in a graphical form using a "tree diagram" to illustrate the contribution of changes in each component of support, and the overall yearly change in support. The lower value in brackets is the approximate contribution to the total change (*i.e.*, the effect on the total PSE or total CSE of the change in the component, on the assumption that no other change had taken place) (Annex 1). Figure V.1. below shows the changes of the various PSE and CSE components between 1996 and 1997.

In 1997, the net total PSE fell by over 15 per cent. As the level of production slightly increased by 0.4 per cent, the fall in PSE was associated with a decrease of nearly 16 per cent in unit PSE. Among the components of the decrease in the unit PSE, nominal decreases of respectively 37 and 43 per cent in other support and direct payments were the main elements contributing to the decline in total assistance. The fall in the feed adjustment component to a negative value contributed also to the reduction in unit PSE while the increase in the unit market price support of 20 per cent partially reduced the fall in unit PSE. The unit market price support increase resulted from higher producer prices and lower border prices expressed in roubles. The slight decrease in border prices was the result of a fall of 12 per cent in dollar prices mostly attenuated by some nominal currency depreciation.

The total CSE in 1997 increased by 13 per cent in absolute terms over 1996 and the implicit tax on consumers increased from 18 to 20 per cent. Although the volume of consumption increased by 2 per cent, the rise in the absolute value of CSE was almost entirely attributable to higher unit CSE, reflecting the increase in market transfers (as there were no consumption subsidies). The increase in market transfers, in turn, was mainly due to higher consumer prices.

◆ Figure V.1. ***Decomposition of PSE and CSE changes from 1996 to 1997***

Decomposition of PSE changes

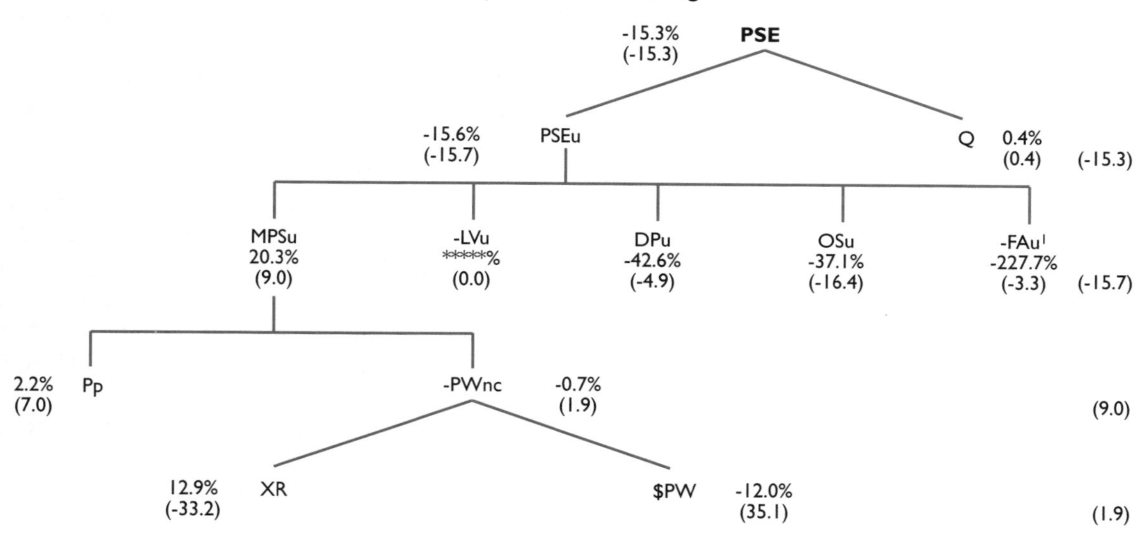

1. FAu was positive in 1996 and negative in 1997.

Decomposition of CSE changes

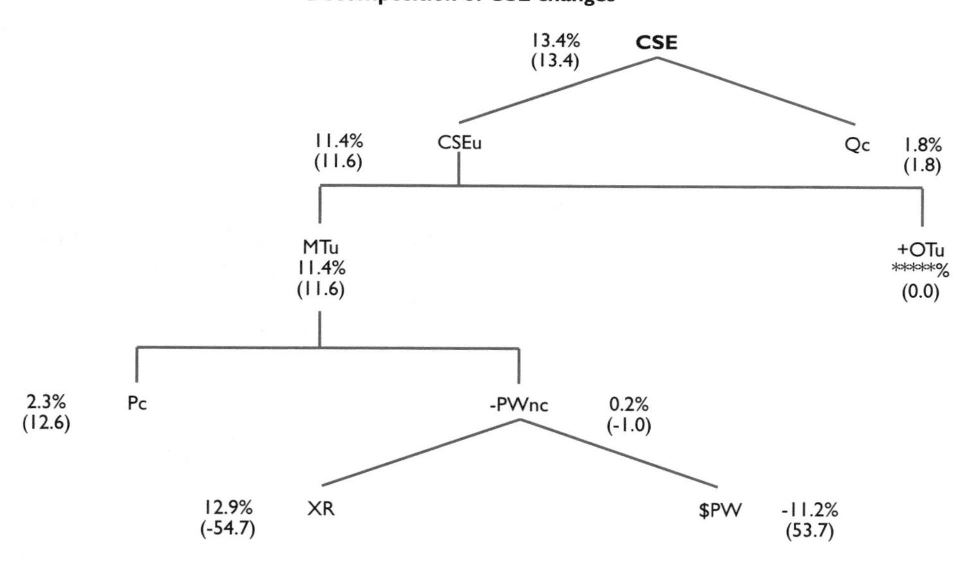

PSE	: net total PSE	**MPS**	: market price support	**u**	: per unit
CSE	: net total CSE	**MT**	: market transfers	**nc**	: in national currency
Q	: quantity produced	**LV**	: levies on output	**$**	: in US$
Qc	: quantity consumed	**DP**	: direct payments	**...%**	: percentage change
Pp	: production price	**FA**	: feed adjustment	**(...)**	: contribution to total change in PSE or CSE in percentage points
Pc	: consumption price	**OS**	: other support		
PW	: world price	**OT**	: other transfers		
XR	: exchange rate/US$			***	: undefined

1. The upper value shown for each element is the annual percentage change in the Fisher Ideal index. The lower value in brackets is the approximate percentage points contribution to the total change (*i.e.,* the effect on the total PSE or total CSE of the change in the component if no other change had taken place).

Source: OECD Secretariat 1998.

1995, the MPS component was negative but was more than offset by the feed adjustment effect and transfers from the budget, which explains the positive net PSE value. In 1996, the net PSE increased to 32 per cent due to a positive market price support and large budget transfers which accounted for 55 per cent of total net PSE. In 1997, MPS accounted for 61 per cent of total support, budget expenditures for 41 per cent while the feed adjustment component became again negative (minus 2 per cent). Direct payments were not an important component of agricultural budgetary support during the Soviet period but their importance increased in the reform period, from 11 per cent in 1986-1987 to an average of 23 per cent in 1991-1997. As of 1992, direct payments consisted mainly of subsidies paid directly to livestock producers (before 1992, premiums and supplementary price payments were channelled indirectly to producers through the purchase prices paid by the procurement agencies and were included in the average prices used in the calculations; hence, they were not treated as separate direct payments). The share of reduction of input costs in total PSE was about 15 per cent between 1986 and 1990. This type of support was continued after 1990 and contributed significantly to offsetting the negative value of market price support up to 1995. The share of general services remained stable at 5 per cent during the Soviet period and increased in recent years (14 per cent on average in 1995-1997).

Annex Table I.14 also shows the shares of CSE components. The most important component is market transfers which are the corollary on the consumer side of market price support for producers. Other transfers consist of consumer subsidies which partly compensated for the negative market transfers during the Soviet period and which, as of 1992, were additional to positive market transfers. In 1986-1991, the ratio of other transfers to market transfers, indicating the relative importance of consumer subsidies in offsetting the tax on consumers due to producer price support varied from 33 to 46 per cent. As from 1992, consumer subsidies were substantially reduced and their share in the total CSE was on average less than 3 per cent between 1992 and 1994. Between 1995 and 1997, consumer subsidies were negligible.

2. Commodity composition

The PSE/CSE calculations do not cover all agricultural products.[2] During the period under review, the products covered by the PSE estimates accounted for on average 73 per cent of the total value of agricultural production in Russia. The share of livestock products was much higher (95 per cent) than that of crops (48 per cent). In OECD countries, the share of agricultural products covered by the PSE calculations varies from 40 per cent in Turkey to 94 per cent in Finland.

Shares of various commodities in the total net PSE are shown in Annex Table I.15.i. During the pre-1991 period, support to livestock products dominated total support in absolute terms and on average accounted for 79 per cent of the net total PSE. The high share of livestock products in total support was due to their dominance in the total value of agricultural production. In effect, the share of livestock production in total gross agricultural output was on average 68 per cent in 1986-1990. There was less difference between percentage PSE for livestock and crop products during that period. Between 1986 and 1990, the percentage PSE for crops and for livestock averaged 90 per cent. In 1992, PSE fell sharply for both livestock and crop products but the fall was steeper for livestock (minus 183 per cent) than for crops (minus 46 per cent). Thereafter, the percentage PSE for livestock increased steadily to 33 and 39 per cent in 1995 and 1996 respectively. In the case of crops, PSEs were almost at the same level, at respectively minus 25 and minus 27 per cent in 1993 and 1994, and started to increase as of 1995, but at a slower pace than for livestock commodities. In 1996, the percentage PSE for crops became positive at 16 per cent. In 1997, the level of support decreased for both crop and livestock commodities, but whilst crop PSEs declined only 2 percentage points from 16 per cent in 1996 to 14 per cent in 1997, the fall was more significant for livestock commodities, with PSEs decreasing from 39 to 32 per cent (Graphs V.2, V.3.*i* and V.3.*ii*).

◆ Graph V.2. ***Evolution of PSEs for crops and livestock commodities***

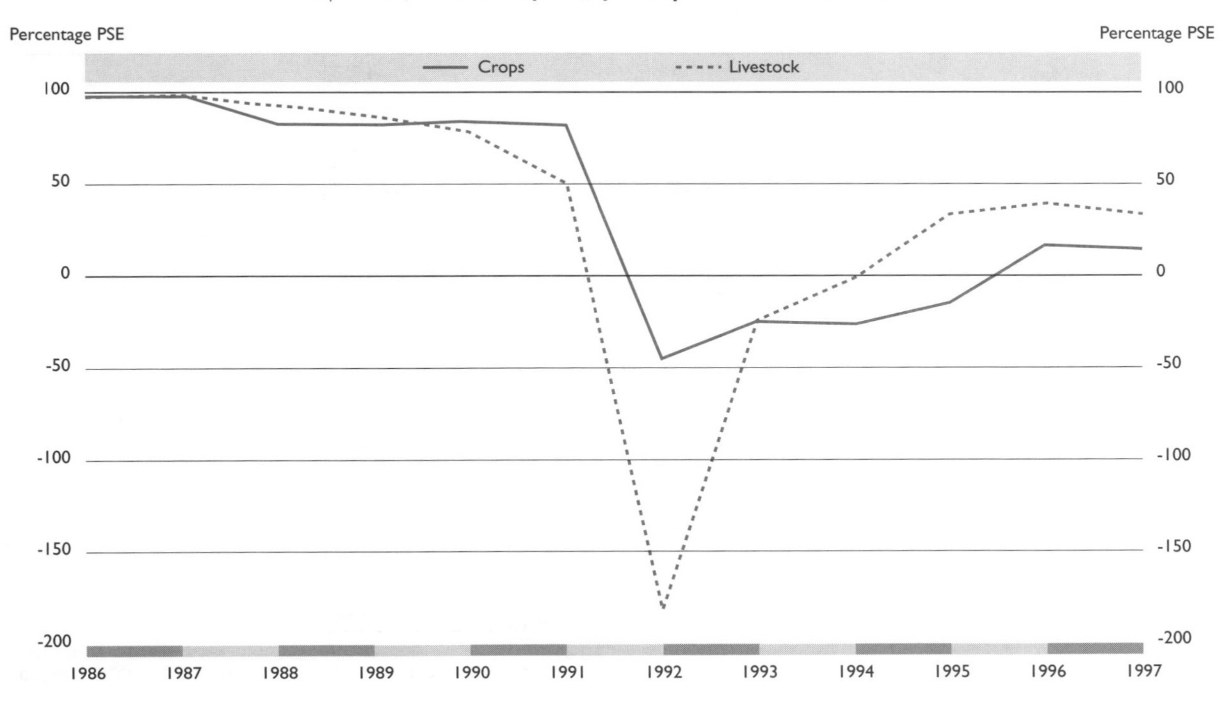

Source: OECD.

◆ Graph V.3*i*. ***Russian percentage PSE in 1997, by commodity***

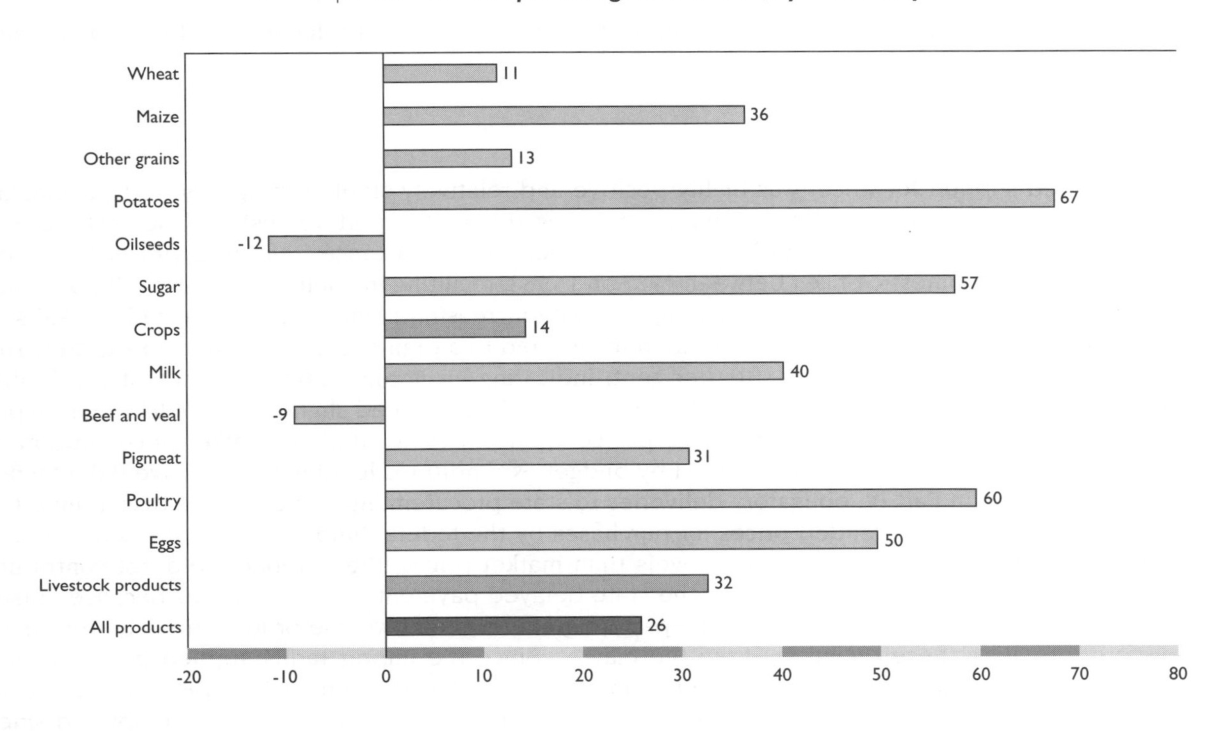

Source: OECD.

◆ Graph V.3ii. ***Russian percentage CSE in 1997, by commodity***

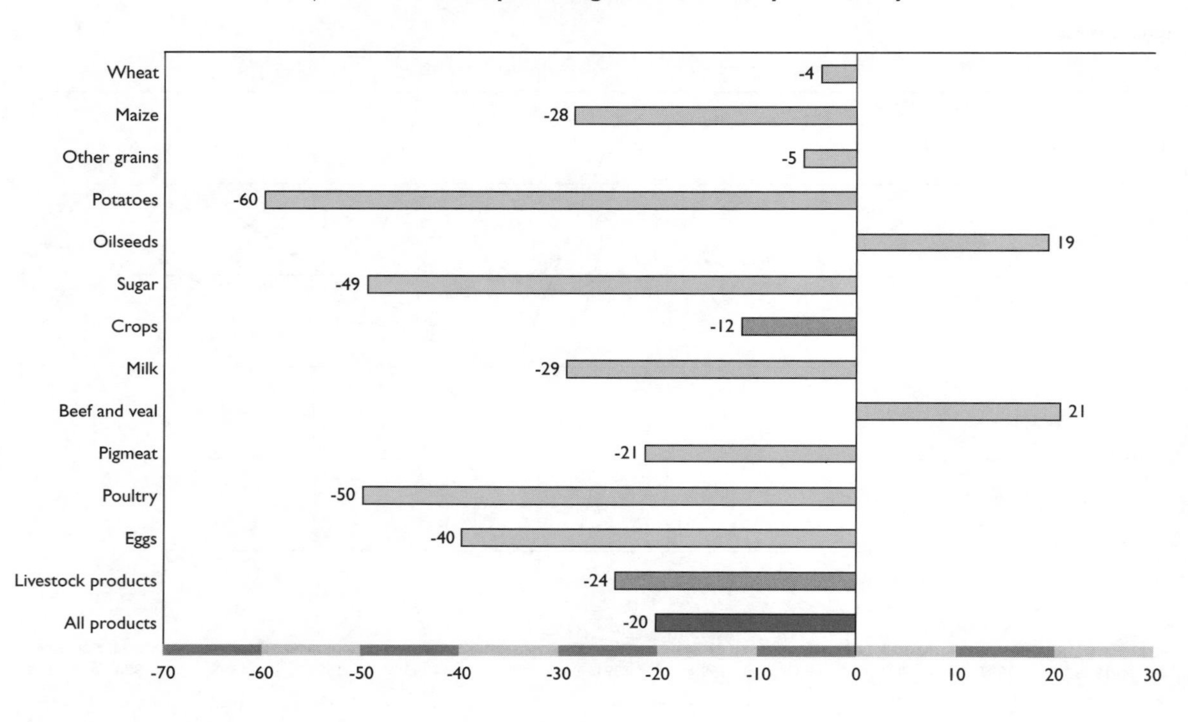

Source: OECD.

D. ANALYSIS OF SUPPORT BY COMMODITY

Detailed PSE results, on a product by product basis, are given in Tables V.2.i and V.2.ii below and Annex Tables I.1 to I.11.

1. Wheat (Annex Table I.1)

Measured support for wheat was highly positive and relatively stable during 1986-1991, averaging 83 per cent. Developments in border reference prices (EU unit export values) and domestic prices explain the small fluctuations recorded during that period. For instance, reference prices increased more rapidly than domestic prices between 1987 and 1988, resulting in a fall of the PSE, while between 1989 and 1991 decreasing reference prices together with increasing domestic prices caused the PSEs to increase again. The sharp currency depreciation in 1992 led to a dramatic fall in measured support. The percentage PSE dropped to minus 101 per cent, indicating an implicit taxation of wheat producers. Following an increase between 1992 and 1993, percentage PSE declined slightly in 1994 to minus 36 per cent, remained negative in 1995 at minus 22 per cent, and in 1996, although market price support of wheat was still negative, it was compensated by budget expenditures leading to a positive PSE of 9 per cent. As described in Part IV, obligatory deliveries to state procurement agencies were discontinued in 1993 and so-called recommended prices for purchases by the federal funds replaced the procurement prices. Although they were set at higher levels than market prices, they probably did not contribute much to an increase in domestic prices as the state delayed payment for its purchases because of lack of resources. Moreover, farms tended to keep more grain for their own use or for various barter deals, therefore insulating these quantities from the market. Since the import tariffs applied on grains and grain products are relatively low (1 per cent for grains from 15 March 1994 to 14 May 1996 and 5 per cent from 15 May 1996; moreover, a 10 per cent duty on imports of grain products has been applied since 1995), their influence on producer prices is rather limited. On the other hand, barriers imposed on inter-

Table V.2.*i*. **Russian percentage PSE, by commodity**

	1986	1987	1988	1989	1990	1991	1992	1993	1994	1995	1996p	1997e
Wheat	90	88	74	76	82	86	–101	–31	–36	–22	9	11
Maize	111	106	103	100	104	90	–62	–4	50	31	46	36
Other grains	98	96	74	76	84	79	–16	–34	–34	–44	16	13
Potatoes (not included in the aggregation)	100	85	83	91	87	87	–67	32	1	–25	71	67
Oilseeds	98	107	88	94	84	73	–55	–150	–36	6	3	–12
Sugar	117	116	106	98	91	75	16	59	37	39	72	57
Crops	**98**	**98**	**83**	**82**	**85**	**82**	**–46**	**–25**	**–27**	**–15**	**16**	**14**
Milk	106	103	98	95	91	64	–155	–44	–18	41	46	40
Beef and Veal	102	102	97	93	85	64	–202	–64	–62	–27	4	–9
Pigmeat	78	84	80	69	55	20	–236	11	35	39	40	31
Poultry	90	93	89	84	55	29	–131	40	63	72	69	60
Eggs	80	76	68	60	40	33	–207	–28	35	59	54	50
Livestock products	**98**	**97**	**92**	**87**	**78**	**50**	**–183**	**–27**	**–1**	**33**	**39**	**32**
All products	**98**	**97**	**91**	**86**	**80**	**61**	**–105**	**–26**	**–9**	**21**	**32**	**26**

e: estimate; p: provisional.
Source: OECD.

Table V.2.*ii*. **Russian percentage CSE, by commodity**

	1986	1987	1988	1989	1990	1991	1992	1993	1994	1995	1996p	1997e
Wheat	–35	–40	–21	–30	–50	–73	127	67	65	36	6	–4
Maize	–81	–66	–78	–78	–84	–83	82	29	–27	–18	–32	–28
Other grains	–52	–49	–35	–45	–58	–68	36	61	58	55	–1	–5
Potatoes (not included in the aggregation)	–67	–52	–56	–68	–66	–78	85	–9	20	35	–59	–60
Oilseeds	–39	–32	8	–19	–24	–36	78	173	57	4	10	19
Sugar	–52	–53	–39	–22	–33	–54	4	–37	–15	–26	–57	–49
Crops	**–51**	**–47**	**–41**	**–41**	**–54**	**–71**	**70**	**49**	**44**	**26**	**–9**	**–12**
Milk	–56	–53	–53	–50	–53	–35	297	112	65	–20	–27	–29
Beef and veal	–52	–55	–46	–49	–25	–40	389	127	116	53	15	21
Pigmeat	–52	–55	–49	–49	–27	–35	440	51	11	–12	–22	–21
Poultry	–33	–27	–38	–44	–22	–25	333	23	–15	–41	–49	–50
Eggs	–81	–72	–76	–77	–61	–60	347	85	1	–38	–37	–40
Livestock products	**–54**	**–53**	**–51**	**–51**	**–38**	**–38**	**353**	**90**	**46**	**–10**	**–23**	**–24**
All products	**–54**	**–52**	**–49**	**–48**	**–43**	**–50**	**172**	**72**	**45**	**–1**	**–18**	**–20**

e: estimate; p: provisional.
Source: OECD.

regional trade such as temporary bans or restrictions for grain "exports" from one region to another may have had more distortive effects on domestic prices. In 1997, percentage PSE increased to 11 per cent, due to a fall in reference prices which was steeper than a fall in domestic wheat prices, following a good wheat crop in Russia.

Wheat CSEs were negative between 1986 and 1991 mainly reflecting negative market transfers during this period and mirroring developments on the PSE side. However, consumer subsidies offset to a large extent these negative market transfers. In particular between 1986 and 1989, the ratio of consumer subsidies to market transfers was about 50 per cent. In 1990 and 1991, consumer subsidies declined and this ratio was only 25 and 9 per cent respectively; consumer subsidies were still applied after 1992 but their share in total CSE decreased substantially (on average 10 per cent in 1992-1994). Between 1992 and 1996, CSEs were positive, with transfers to consumers peaking at 127 per cent in 1992, exceeding 60 per cent in 1993 and 1994, still high at 36 per cent in 1995 but declining to 6 per cent in 1996. In 1997, market transfers became negative and wheat consumers were implicitly taxed at minus 4 per cent.

2. Maize (Annex Table I.2)

The volume of maize production is relatively low in Russia compared to other grains. The general level of support to maize was higher in percentage terms than that for wheat, with percentage PSE averaging 102 per cent between 1986 and 1991 and 41 per cent in 1994-1997. Support to maize fell sharply in 1992 to minus 62 per cent due to strong world prices in Rouble terms but increased rapidly to minus 4 per cent in 1993, 50 per cent in 1994 and remained high at 36 per cent in 1997.

The CSE for maize was mostly influenced by market transfers as consumer subsidies were relatively low and did not compensate for highly negative market transfers during the Soviet period. The percentage CSE was negative over the whole period, except for 1992 and 1993 when the reference prices were higher than the domestic prices. In 1997, the implicit tax on consumers was 28 per cent.

3. Other grains (Annex Tables I.3, a, b, c)

Other grains included in the PSE calculations for Russia are rye, barley, and oats. Production of these grains is dominated by barley, which represents about 50 per cent of the output of these three grains, while rye and oats account for about 25 per cent each. During the Soviet era, support to other grains in percentage terms was higher than that for wheat but lower than that for maize. Percentage PSEs followed the same pattern as those for wheat between 1986 and 1991, varying from 74 to 98 per cent. Fluctuations during that period mainly reflect changes in border reference prices. For instance, support for rye, barley and oats producers fell by about 20 per cent between 1987 and 1989 due to strong border reference prices. The fall in PSEs between 1990 and 1991 from 84 to 79 per cent was principally due to a reduction in budget expenditures. Between 1992 and 1995, PSEs were negative, meaning an implicit tax on producers. The policy of taxing grain exports during the transition period also kept domestic prices well below world prices. The export duty imposed on grains was set at 70 per cent up to 1 October 1993 and at 10-25 per cent from 1 October 1993. Export duties were discontinued as of 1 September 1995 and in 1996 domestic prices of rye, barley and oats rose more rapidly than the reference prices. In 1996, the aggregate percentage PSE for these three grains became positive at 16 per cent, and decreased slightly to 13 per cent in 1997.

Between 1986 and 1991, support for rye was higher than that for barley which in turn was higher than that for oats. In 1992, PSEs for all three grains decreased sharply due to the depreciation of the Rouble. While PSEs for rye and oats were negative at minus 39 and minus 57 respectively, PSE for barley was positive at 7 per cent, due to a rise in domestic prices between 1991 and 1992. However, between 1992 and 1995 domestic prices for barley increased less than for other grains and percentage PSEs fell steadily, reaching a low of minus 77 per cent in 1995. In 1996, the rapid increase in domestic prices of barley (by about 150 per cent compared to 100 per cent for prices of rye and oats) led to a recovery in the level of market price support, which added to budget expenditures resulted in a positive PSE of one per cent. PSEs for oats increased slightly between 1992 and 1995, still remaining negative in 1995 but in 1996 domestic prices rose more rapidly than reference prices resulting in a PSE of 27 per cent. On the other hand, rye PSEs recovered faster and in 1995 rye PSEs had already reached 17 per cent, and continued to increase to 33 per cent in 1996. In 1997, both domestic and world reference prices decreased for the three grains. In the case of barley, the rate of decline was greater for reference prices than for domestic prices, and the PSEs increased slightly from 1 per cent in 1996 to 4 per cent in 1997. For rye and oats, the decline in domestic and reference prices was almost the same but reduced budget expenditures lowered PSEs to 26 per cent for rye and to 19 per cent for oats.

The development of CSEs for coarse grains can also be divided into two distinct periods. In the Soviet period, negative market transfers, only partly compensated by consumer subsidies, indicated that consumers were implicitly taxed. The level of implicit taxation peaked in 1991 when the CSE reached minus 68 per cent. Consumer subsidies were maintained between 1992-1994, representing 4 per cent of total CSE in 1992-1993, but only 2 per cent in 1994. Between 1992 and 1995 CSEs were positive with transfers to consumers of about 60 per cent in 1993 and 1994, 55 per cent in 1995, but they declined in 1996 and 1997 to respectively minus 1 and minus 5 per cent, reflecting the recovery in the level of support on the producer side.

4. Potatoes (Annex Table I.4)

Potatoes are an important crop in Russia with an average share in total value of crops of about 25 per cent and of 12 per cent in total value of agricultural production for the period 1986-1996. Individual results of PSE estimates for potatoes are presented in tables I.4.i, ii and iii but due to methodological problems related to reference prices and negligible amounts of potatoes traded internationally, potatoes are not included in the calculation of aggregate PSE. However since potatoes are such an important agricultural commodity in Russia, an attempt has been made to estimate the level of support by using the standard PSE methodology. During the period under review reference German farm-gate prices fluctuated widely and domestic prices of potatoes, which are in fact a non tradable good, adjusted very slowly (if at all) to world reference prices.

Measured support for potatoes was high and relatively stable between 1986 and 1991, ranging from 83 to 100 per cent and dropping sharply to minus 67 per cent in 1992. Since then, measured support has been very volatile: it increased to 32 per cent in 1993, fell back to 1 per cent in 1994 and plunged to minus 25 per cent in 1995. These large fluctuations were mostly due to reference price movements: between 1992 and 1993, reference prices in US$ declined, causing increasing domestic prices to rise above reference prices. On the other hand, the sharp drop in percentage PSE between 1993 and 1995 was due to a sharp increase in reference prices, not matched by the increase in the domestic farmgate prices for potatoes. In 1996, domestic prices increased only slightly but as reference prices dropped by two -thirds, PSE soared to plus 71 per cent. In 1997, PSE decreased slightly to 67 per cent mainly due to a reduction in budget expenditures.

In the years of highly negative market transfers (1986-1991) only partially compensated by consumer subsidies, consumers were implicitly taxed with CSEs varying from minus 52 to minus 78 per cent. As consumer subsidies were discontinued in 1992, developments in CSE have mirrored developments on the PSE side since 1992. Transfers to consumers were positive in 1994 and 1995, but in 1996 and 1997 consumers were implicitly taxed at 60 per cent.

5. Oilseeds (Annex Table I.5)

Sunflower is the major oil crop in Russia, accounting for 90 per cent of total oilseed output. Contrary to most other agricultural products, production of sunflower did not fall between 1986 and 1996 and the sunflower area has been steadily increasing between 1991 and 1995. Measured support was very high during the 1986-1991 period, with the PSE averaging 91 per cent. In 1992, the PSE dropped to minus 55 per cent reflecting the unstable macroeconomic environment. It continued to fall to minus 150 per cent in 1993, because of a sharp rise in world prices. Moreover a 15 per cent export tax imposed on oilseeds up to 1993 contributed to keeping producer prices well below world prices. When this export duty was removed in 1994, 40 per cent of the oilseed production was exported. As for other crops, direct market intervention was discontinued in 1992. Since 1995 the market has been protected by tariffs of 15 per cent for oils and 10 per cent for sunflower seed. In 1994 and 1995, world reference prices decreased slightly (in ECU terms) and domestic prices continued their upward trend, which almost closed the gap with reference prices. In 1995 and 1996, market price support was still negative but offsetting budget expenditures resulted in positive PSEs of respectively 6 and 3 per cent. In 1997 domestic prices decreased while reference prices increased, and as budget expenditures did not offset the negative market price support as was the case in 1995 and in 1996, PSE became negative at minus 12 per cent.

In the Soviet era consumer subsidies for sunflower offset about two-thirds of the negative market transfers on average, which led to a lower taxation of consumers of sunflower oil than of other crops. Consumer subsidies were abolished in 1993 and since then, CSEs have mirrored PSEs indicating positive transfers to consumers. The implicit tax on producers in 1997 implied an implicit subsidy of 19 per cent on the consumer side.

6. Sugar (Annex Table I.6)

Sugar was one of the most heavily supported crops in Russia, with PSEs ranging from 75 per cent to 117 per cent in 1986-1991. The high positive PSEs during the Soviet period were mainly due to market price support. As for the other crops, percentage PSEs decreased sharply in 1992, due to the currency depreciation but the negative market price support was more than offset by budgetary expenditures and the resulting PSE was positive at 16 per cent. As for grains and oilseeds, recommended prices set in 1993 and minimum guaranteed prices announced in 1994 and 1995 had probably a limited impact on market prices because of the lack of state resources. Since 1994 however, the market has been supported by import tariffs (20 per cent for refined sugar in 1994 which increased to 25 per cent in 1995, and 1 per cent for imports of raw cane sugar) and import quotas, which have contributed to the maintenance of high domestic prices. Thus, contrary to other crops, transfers to producers continued at a high level in the period from 1993 to 1997, with the producer subsidy peaking at 72 per cent in 1996 and decreasing to 57 per cent in 1997.

The percentage CSE for sugar was negative over the whole period under review (except in 1992, when market price support was negative), meaning that there was an implicit tax on consumption. However, the effect of high market price support was moderated by substantial consumer subsidies up to 1991, and the ratio of consumer subsidies to (negative) market transfers averaged 50 per cent during that period. In 1992, consumer subsidies for sugar were discontinued and the negative percentage CSEs mirrored negative market transfers.

7. Milk (Annex Table I.7)

Together with beef and veal, milk was the most heavily supported livestock product in the pre-reform period (until 1991), in both absolute and percentage terms. Net percentage PSE decreased from 106 per cent in 1986 to 91 per cent in 1990. In 1991, the percentage PSE decreased further to 64 per cent due to the combined effects of an increase in the reference price, a fall in the level of budgetary expenditures and a higher negative feed adjustment component. As for the other products, the milk PSE dropped to minus 155 per cent in 1992 in line with the depreciation of the local currency. In 1992, price liberalisation led to a substantial fall in demand, which motivated the introduction of livestock subsidies for milk and meat producers. These subsidies were linked with the level of production delivered to state procurement and are considered in the PSE estimates as direct payments. The share of subsidies granted for milk was substantial, on average 36 per cent of direct payments during 1992-1997. Since 1993, the milk market has also been supported by border protection (import tariffs in the range of 15-20 per cent) which contributed to increased market price support. As a result, the high level of implicit taxation of milk producers has been progressively reduced since 1992 and in 1995 domestic prices were above world prices resulting in positive market price support. Moreover, the net PSE was increased by the strongly positive feed adjustment and reached 41 per cent. In 1996, domestic prices continued to rise more rapidly than world reference prices, increasing net PSE to 46 per cent, but in 1997 milk PSE decreased to 40 per cent mainly as a result of lower budget expenditures and a negative feed adjustment component.[3]

In the Soviet period, milk CSEs reflected high consumption aids, which to a large extent (43 per cent) offset negative market transfers. Following price liberalisation consumer subsidies were maintained for milk and milk products on a regional basis up to 1994, albeit at a much lower level. In 1992, the high implicit taxation of milk producers resulted in huge market transfers to consumers which were substantially reduced in the following years. As of 1995, consumers were again taxed and the percentage CSE was minus 29 per cent in 1997.

8. Beef and veal (Annex Table I.8)

Measured support for beef and veal followed a similar pattern to that for milk. During the Soviet period, beef and veal was under strong government control and highly subsidised both on the producer and the consumer side. As explained in Part IV, there was no specialised beef meat production in

Russia, dual-purpose cattle were dominant and therefore beef and veal production is closely linked to that of milk. The net percentage PSEs for beef and veal were extremely high during the 1986-1990 period (96 per cent on average) and declined to 64 per cent in 1991 as in the case of milk. Support to beef production became strongly negative between 1992 and 1994, reflecting a sharp drop in domestic prices (beef prices were only 85 per cent of the level of pork prices). The high export duties (70 per cent) imposed on meat and meat products during the transition period until October 1993 contributed to maintaining domestic prices well below world prices. Following price liberalisation in 1992, substantial subsidies for production delivered to state procurement were introduced (they represented 17 per cent of all direct payments between 1992 and 1997) and since March 1994, meat products (beef and pigmeat) have also been protected by increasing import tariffs (8 per cent from March 1994 to July 1995; 15 per cent from July 1995). Though this led to a rise in the level of support, PSEs remained negative during 1995-1997 (except in 1996 where offsetting budget expenditures and a positive feed adjustment component resulted in a slightly positive PSE). In 1997 the implicit taxation on producers was 9 per cent, and beef and veal was the only livestock product with a negative PSE.

High consumer subsidies in the period from 1986 to 1991 offset to a large extent high transfers from consumers. CSEs ranged from minus 55 per cent to minus 25 per cent in that period. In 1992, as in the case of milk, consumer subsidies were granted at regional level. Thereafter their amount became negligible. Consumer subsidies were granted to meat products in 1992 but were completely eliminated in 1993. Since then, high positive CSEs reflect developments in market price support and indicate that consumers are implicitly subsidised.

9. Pigmeat (Annex Table I.9)

Percentage PSEs for pigmeat were high during the period 1986-1990, ranging from 55 per cent to 84 per cent. This rate fell to 20 per cent in 1991, mainly due to the highly negative feed adjustment component (reflecting high support to grains) and slumped to minus 236 per cent due to the developments in the exchange rate. In contrast to beef and veal, the PSE for pigmeat was negative only in 1992. Market price support was still significantly negative in 1993 (due to strong reference prices), but was more than offset by a positive feed adjustment component, resulting in a positive percentage PSE. Since March 1994, domestic prices of pigmeat have also been supported by the same border measures applied for beef and veal (import tariffs in the range of 8-15 per cent since 1994), contributing to a further increase in PSEs to 40 per cent in 1996. In 1997, PSE decreased to 31 per cent, mainly due to a reduction in budget expenditures.

As in the case of milk and beef and veal, CSEs for pigmeat in the Soviet period reflected the high level of consumer aids which to a large (and increasing extent) offset market transfers. In 1992, the extremely high level of market transfers reflected the very low producer price. Between 1992 and 1997, percentage CSEs declined steadily and turned negative in 1995, with an implicit tax on consumers of 21 per cent in 1997.

10. Poultry (Annex Table I.10)

The trend in measured support to poultrymeat is similar to that for pigmeat. Between 1986 and 1989, percentage PSEs were high and averaged 89 per cent, decreased to 55 and 29 per cent in 1990 and 1991 respectively, because of strong world prices and more negative feed adjustment. PSEs became negative in 1992, indicating an implicit taxation of producers. After 1992, poultry producers received subsidies on production delivered to state procurement as for the other livestock products. However, these subsidies were smaller than in the case of milk and beef. Since March 1994, poultry prices have been supported by increasing import tariffs which have been also substantially higher than for beef and pigmeat (20 per cent from March 1994 to June 1995; 25 per cent from July 1995 to May 1996, 30 per cent since May 1996 but not less than 0.3 ECU per kg). This contributed to a rapid rise in domestic prices, and PSEs for poultry were significantly positive between 1993 and 1997. They increased from 40 per cent in 1993 to 72 per cent in 1995 due to a rapid rise in domestic prices and a positive feed adjustment component, but decreased slightly in 1996 to 69 per cent (due to lower positive feed adjustment) and to 60 per cent in 1997 principally due to lower budget expenditures.

In the Soviet period, the market transfer component of the CSE was substantially offset by consumption aids. The implicit tax on consumers was in the order of 22 to 44 per cent during that period. In 1992 and 1993, domestic prices fell below reference prices, leading to positive CSEs of 333 and 23 per cent respectively. Thereafter, the CSE turned negative, in parallel with the increase in the poultry PSE due to higher domestic prices.

11. Eggs (Annex Table I.11)

Measured support for eggs was consistently high during the Soviet period, although it was lower than for poultry. It decreased from 80 per cent in 1986 to 60 per cent in 1989, and as in the case of poultry, declined further in 1990 and 1991. The sharp currency depreciation led to a slump in market price support, with percentage PSEs of minus 207 and 28 per cent in 1992 and 1993 respectively. Between 1993 and 1995, price recovery in the sector resulted in increased support for egg producers. In 1995, the rise in domestic prices combined with a fall in the reference price increased the level of support to 59 per cent. PSE decreased then to 50 per cent in 1997.

As no consumer subsidies were applied to eggs, the CSE consisted only of market transfers and followed closely the development of market price support throughout the period under review.

NOTES

1. The Czech Republic became a member of the OECD in 1995. Hungary and Poland joined OECD in 1996.
2. The products covered in the PSE/CSE calculations for Russia are: wheat, maize, other grains (rye, barley and oats), sunflower, sugarbeet, milk, beef and veal, pigmeat, poultrymeat and eggs. The PSE/CSE calculations are also made for potatoes, but results are not included in the overall evaluation of support policies in Russia for the reasons explained in the section on potatoes (see below).
3. It should be noted that in 1997 direct subsidies to livestock producers derived from local budget reports were underestimated as they only referred to subsidies granted in 32 regions out of 89.

BIBLIOGRAPHY

Agra Europe. East Europe Agriculture and Food, various issues.

Agropromizdat (1989), Otsenivanie v agropromyshlennom komplekse, Moscow.

Berkowitz, D.M., D.N. DeJong and S. Husted (1996), Transition in Russia: It's Happening, Working Paper No. 306, University of Pittsburgh, Pittsburgh.

Blasi, J., M. Kroumova and D. Kruse (1997), Kremlin Capitalism: Privatizing the Russian Economy, Cornell University Press, Ithaca NY.

Brooks, K. and Z. Lerman (1994), "Land Reform and Farm Restructuring in Russia", World Bank Discussion Papers No. 233, Washington, D.C.

Brooks, K., E. Krylatykh, Z. Lerman, A. Petrikov and V. Uzun (1996) "Agricultural Reform in Russia. A View from the Farm Level", World Bank Discussion Papers No. 327, Washington, D.C.

Butsykin P.I. (1997), "Obshtchii agrarnyi rynok stran SNG", Agrarnaia Nauka, No. 1.

Butterfield, J. (1995), Agricultural Enterprise Restructuring in Russia, paper presented to the Ad Hoc Group on East/West Economic Relations in Agriculture, 4-6 October, Paris.

CEA (1993), Structure of Retail Prices for Staple Foodstuffs in the First Half of 1993. Survey Results, Moscow.

CNFA (The Citizens Network for Foreign Affairs) (1996), "Issues of Food Security in Contemporary Russia", Food Systems Policy Roundtable Journal No. 1, May/June, Moscow.

Cook, E.C., W.M. Liefert, R. Koopman (1991) "Government Intervention in Soviet Agriculture: Estimates of Consumer and Producer Subsidy Equivalents", *Staff reports* No. AGES 9146.

De Masi, P. and V. Koen (1996), "Relative Price Convergence in Russia," IMF Staff Papers 43: 97-122.

Economic Research Service/USDA (1997), "NIS and Baltic Countries Look to Join the WTO", Agricultural Outlook, November.

EEM Moscow Bulletin, various issues, Moscow.

Figes, O. (1996), A People's Tragedy. The Russian Revolution 1891-1924, London.

Gardner, B.L. and K.M. Brooks (1994), "Food Prices and Market Integration in Russia 1992-93, "American Journal of Agricultural Economics 76: 641-646.

Goodwin, B.K., T. Grennes and C. McCurdy (1996), Spatial Price Dynamics and Integration in Russian Food Markets, Working Paper, North Carolina State University, Raleigh.

Gordeev, AV., I.G. Ushatchev (1997), *Sostaianie i mery po razvitiu agropromyshlennovo proizvodstva Rossiiskoi Federatsii, Ezhegodnoi doklad,* MAF, Mowcow.

Goskomstat (1994), Agropromyshlennyi kompleks Rossiiskoi Federatsii. Statitistcheskii sbornik, Moscow.

Goskomstat (1995), Selskoie khoziaistvo Rossii, Moscow.

Goskomstat (1996), Rossiiskii statistitcheskii ezhegodnik, Moscow.

Goskomstat (1997*a*), Sotsyalno-ekonomitcheskoye polozhenie Rossii: 1996, Moscow.

Goskomstat (1997*b*), 1997: Russia in Figures, Moscow.

Goskomstat (1997*c*), "O sostaianii prodovolstvennovo rinka w Rossiiskoi Federatsii (po materialam Goskomstata Rossii", Voprosi statistiki 9/1997, Moscow.

Goskomstat (1997*d*), Osnovnyie pokazateli agropromyshlennovo kompleksa Rossiiskoi Federatsii v 1996 godu, Moscow.

Goskomstat (1997*e*), Proizvodstvenno-ekonomitcheskiie pokazateli razvitiia agropromyshlennovo kompleksa Rossii v 1996 godu, Parts I and II, Moscow.

Goskomstat (1998), Sotsyalno-ekonomitcheskoye polozhenie Rossii: 1997, Moscow.

Government of the Russian Federation (1996), Program for Stabilisation and Development of Agricultural Production in the Russian Federation for 1996-2000, Moscow.

Harley, M. (1996), "Use of the Producer Subsidy Equivalent as a Measure of Support to Agriculture in Transition", Amer. J. Agr. Econ. 78 (August 1996): 799-804, Proceedings of the ASSA winter meeting, San Francisco, CA, 5-7 January 1996.

IMF, World Bank, OECD, EBRD (1991), A Study of the Soviet Economy, Washington, D.C., Paris, London.

Interfax. Food and Agriculture Report, various issues.

International Finance Corporation (IFC) and The Overseas Development Administration (ODA) (1995), Land Privatisation and Farm Reorganization in Russia, Washington, D.C.

Kalinin, N.I. (1997), "Prava na zemliu v Rossiiskoi Federatsii", Agrarnaia Nauka, No. 4.

Khramova, I. (1997), "O pravie sobstvennosti na zemliu", Ekonomika i Zhizn, No. 24.

Kholod, L.I. (1996), Mechanism to Implement Financial Credit Policy Concept in Agriindustrial Complex in 1997, in Agricultural and Agribusiness Credit in Russia (Conference Materials), Moscow.

Klugman, J. (1996), Poverty in Russia during the Transition, Washington D.C., World Bank.

Kopsidis, M. (1997), Marktintegration und Landwirtschaftliche Entwicklung: Lehren aus der Wirtschaftsgeschichte und Entwicklungsökonomie für den Russischen Getreidemarkt im Transformationsproze, Discussion Paper No. 5/97, Institute of Agricultural Development in Central and Eastern Europe, Halle/Saale.

Le Houeroue, P. and M. Rutkowski (1996), "Federal Transfers in Russia: Their Impact on Regional Revenues and Incomes," Comparative Economic Studies 38(2/3): 21-44.

Lesik, B. (1996), "Reformy: orlovskii opyt. Z zemelnoi dolei - v dorogu", Selskaia Zhizn', 31 October.

Libert, B. (1995), The Environment Heritage of Soviet Agriculture, CAB International, Wallingford.

Liefert, W.M., R.B. Koopman, E.C. Cook (1993), "Agricultural Reform in the Former USSR", *Comparative Economic Studies,* Vol. 35, No. 4, Winter 1993.

Liefert, W.M., D. Sedik, R. Koopman, E. Serova and O. Melyukhina (1996), "Producer Subsidy Equivalents for Russian Agriculture: Estimation and Interpretation", Amer. J. Agr. Econ. 78 (August 1996): 792-798, Proceedings of the ASSA winter meeting, San Francisco, CA, 5-7 January 1996.

Loy, J.-P. and P. Wehrheim (1996), Spatial Food Market Integration in Russia, Discussion Paper Series "The Russian Food Economy in Transition" No. 6, Institute for Food Economics and Consumption Studies, University of Kiel, Kiel.

MAF (1996), Production and economic indicators of the development of the agri-food sector of Russia in 1995, Ministry of Agriculture and Food, Moscow.

MAF (1997), Sostaianie i miery po razvitiu agropromyshlennovo proizvodstva Rossiiskoi Federatsii. Ezhegodnyi doklad, Moscow.

Melyukhina, O. and P. Wehrheim (1996), Russian Agricultural and Food Policies in the Transition Period: Federal and Regional Responsibilities in Flux, Discussion Paper Series "The Russian Food Economy in Transition" No. 5, Institute for Food Economics and Consumption Studies, University of Kiel, Kiel.

Melyukhina, O., M. Qaim and P. Wehrheim (1997), Measuring Regional Protection Rates for Food Commodities in Russia: Producer and Consumer Perspectives, Discussion Paper Series "The Russian Food Economy in Transition" No. 11, Institute for Food Economics and Consumption Studies, University of Kiel, Kiel.

Nazarenko, V.I. (1995), Problems of the Agricultural Economy in Russia following the Disintegration of the USSR, Food and Agricultural Policy Research Center, Tokyo.

Nazarenko, V.I. (1997), Agrarian Reform and Agricultural Situation in Russia, All-Russia Research Institute of Information and Technical-Economic Studies of Agroindustrial Complex, Moscow.

OECD (1987), National Policies and Agricultural Trade, Paris.

OECD (1991), The Soviet Agro-Food System and Agricultural Trade. Prospects for Reform, Paris.

OECD (1995), OECD Economic Surveys: The Russian Federation 1995, Paris.

OECD (1996*a*), Review of Agricultural Policies: Estonia, Paris.

OECD (1996*b*), Agro-Food Sector Policy in OECD Countries and the Russian Federation, Proceedings from the Seminar organised by the OECD/CCET with the Co-operation of the Ministry of Agriculture and Food Supplies of the Russian Federation, Paris.

OECD (1996*c*), The Changing Social Benefits in Russian Enterprises, Paris.

OECD (1997*a*), Agricultural Policies in Transition Economies. Monitoring and Evaluation 1997, Paris.

OECD (1997*b*), OECD Economic Surveys: The Russian Federation 1997, Paris.

OECD (1997*c*), Agricultural Policies in OECD Countries. Monitoring and Evaluation 1997, Paris.

OECD (1998), *Agricultural Policies in non-OECD Economies: Monitoring and Evaluation 1998,* Paris, forthcoming.

Prosterman, R.L., R.G. Mitchell and B.J. Rorem (1997), "Prospects for Peasant Farming in Russia", Rural Development Reports on Foreign Aid and Development #92, Report on Fieldwork Conducted in Collaboration with the Agrarian Institute in Moscow, Seattle, WA.

Roskhleboproduct Corporation (1995), Marketing Margins in the Bread Chain. Unpublished material received by IET.

Roskomzem (1995), The State National Report on the Status and Use of Land in the Russian Federation (in Russian), ed. N.V. Komov et al., Russian Land Committee and Ministry of Environment, Moscow.

Russian Federation Customs Committee (1997), Tamozhennaia statistika vneshnei torgovli Rossiiskoi Federatsii, Moscow.

Sedik, D., C. Foster and W. Liefert (1996), "Economic Reforms and Agriculture in the Russian Federation, 1992-95" in Communist Economies & Economic Transformation, Vol. 8, No. 2.

Semiletov, L.M. (1991), "Purchasing prices for agricultural products in Russian Federation", Planning and accounting in agricultural enterprises, No. 7, Moscow.

Serova, E.V. and O. Melyukhina (1995), "Finance Subsidies and Pricing in the Russian Food and Agriculture Sector in Transition", Series: The Russian Food Economy in Transition, Discussion Paper No. 2 Institute for Food Economics and Consumption Studies, Kiel: University of Kiel.

Serova, E.V. (1996), Land and Property in Russian Agriculture in the Transition, mimeo, Institute for Economy in Transition, Moscow.

Serova, E.V. and R.G. Yanbykh (1996), "Kreditovanie selskhovo khoziaistva v Rossii: Sostoianie i perspektivi", Voprosy Ekonomiki, No. 8, Moscow.

Shleifer, A. (1994), Establishing Property Rights, paper presented to the World Bank Annual Conference on Development Economics, Washington, D.C., April 28-29.

Sinelnikov, S. (1995), Budget crisis in Russia: 1985-1995, Eurasia, Moscow.

Strokov S. and W.H. Meyers (1996), Producer Subsidy Equivalents and Evaluation of Support to Russian Agricultural Producers, Working Paper 96-WP 168, November 1996, Center for Agricultural and Rural Development, Iowa State University.

Tabata, S. (1994), "The anatomy of Russian Foreign Trade Statistics", Post-Soviet Geography, No. 8.

Udatchin, A.A. (1996), "Pravo na dolevuiu sobstvennost", Agrarnaia Nauka, No. 6.

United States Department of Agriculture (USDA), Foreign Attache Report, various issues.

Ushatchev I.G. (1997), "Agrarnaia reforma v Rossii: integratsia v mirovoi rynok", Agrarnaia Nauka No. 1.

Uzun, V., N. Shagaida, M. Garadzha, G.A. Rodionova, G.M. Antonov and E.F. Kunina (1996), Ekonomicheskie i sotsialnyie posledstvia reformirovania selkhozpredpriiatsii po Nizhegorodskoi modeli (predvaritelnyie resultaty), Agrarian Institute, Moscow.

Van Atta, D. (1997), "Agrarian Reform in Post-Soviet Russia", in J.F.M. Swinnen (ed.), Political Economy of Agrarian Reform in Central and Eastern Europe, Ashgate.

Von Braun, J., M. Qaim and H. Tho Seeth (1997), "Food Consumption in Russia: Econometric Analyses with Household Data", Series: *The Russian Food Economy in Transition,* Discussion Paper No. 8, Institute for Food Economics and Consumption Studies, Kiel: University of Kiel.

Wädekin K.E. (1982), Agrarian Policies in Communist Europe: A Critical Introduction, London.

Wichern, R. (1997), The Agricultural Situation in Russia: An Overview, paper prepared for the Commission of the EU, Institute of Agricultural Economics, Kiel.

World Bank (1992), Food and Agricultural Policy Reforms in the Former USSR. An Agenda for the Transition, Washington, D.C.

World Bank (1993), "Foreign Trade Statistics in the USSR and Successor States", Studies of Economies in Transformation, M.V. Belkindas, O.V. Ivanova (eds.), Washington, D.C.

World Bank (1995), "Workers in an integrating world", World Development Report 1995.

World Bank (1996*a*), Russian Federation, Toward Medium-Term Viability, Washington, D.C.

World Bank (1996*b*), Fiscal Management in Russia, Washington, D.C.

WT/ACC/4 (1996), *Accession to the World Trade Organization,* Information to be Provided on Domestic Support and Export Subsidies in Agriculture. WTO, Technical Note by the Secretariat, 18 March.

Annex I

ASSISTANCE TO RUSSIAN AGRICULTURE

INTRODUCTION

In this Annex, section A briefly explains the concepts of Producer Subsidy Equivalents (PSE) and Consumer Subsidy Equivalents (CSE). The methodology used to analyse changes in PSEs and CSEs at an aggregate Russian level by reference to changes in the components of PSEs or CSEs is described in section B. Some particular methodological issues concerning the estimation of PSEs and CSEs for the Russian Federation are discussed in section C. Section D contains the main PSE and CSE results and related data in tabular and graphical form.

A. CONCEPTS AND METHODOLOGY

There are two sources of support or assistance to agriculture. In the first form, called **Market Price Support**, consumers give support to agriculture in the form of higher prices. Taxpayers have also to contribute when product surpluses generated by the high prices have to be disposed of by export subsidies, for example. The higher consumer prices are maintained by domestic supply restrictions such as supply quotas and by foreign trade barriers such as import quotas, import tariffs and export subsidies. The consumer transfer per tonne of product is measured by subtracting an external world reference price for a commodity from a domestic reference price for the same commodity. This price gap, if it is positive, is the support per unit of product for the agricultural commodity in question. (If it is negative, it is a tax on agriculture and thus benefits consumers). For measuring the price gap, it is important to use world and domestic reference prices for like commodities so that the price gap only reflects a difference in price and not differences in quality, variety or degree of processing. This basic model assumes that the average tax per tonne paid by consumers is the same as the average subsidy per tonne transferred to producers (*i.e.* that MPSu = MTu). Although this is the most common assumption, it does not always apply, for example to the case of pooling systems (Section C).

The second form of support is **budget transfers** or subsidies which do not directly affect consumer or market prices. Such support could consist of deficiency payments, input subsidies, infrastructure subsidies, etc. Some such programmes are commodity or commodity group specific. Some are very general, such as agricultural research funding. Some subsidies that should be included in PSEs may not appear explicitly in government budgets or accounts. Subsidies may be concessionary, such as tax concessions affording farmers some additional allowances that reduce their tax bills. The value of such concessions would not usually appear in a budget statement. Another example is concessionary energy (*e.g.* electricity) charges. Again, the value of these may not be explicit. Sometimes they will be estimated when, for example, the government has to pay compensation to the electricity company for charging less to farmers.

CSEs consist of the consumer transfer, which is usually measured through the same price gap calculation made for PSEs. In addition, there may be direct budgetary subsidies to consumers that are often employed to reduce the impact of high market prices that consumers pay to producers.

The PSE is intended to measure the value of the monetary transfers to producers from consumers of agricultural products and from taxpayers resulting from a given set of agricultural policies in a given year. The denomination "agricultural" is interpreted in a broad sense to include policies that tend to target agriculture.

Five categories of agricultural policy measures are included in the calculations of PSEs. The first is market price support, while the other four cover the second form of support which does not directly influence market prices:

a) all measures that transfer money to producers through affecting producer and consumer prices simultaneously, such as support prices and trade measures (***Market Price Support***);

b) all measures that transfer money directly from taxpayers to producers, without raising prices to consumers (***Direct Payments***);

c) all measures that lower input costs with no distinction being made between subsidies to capital and those to other inputs (***Reduction in Inputs Costs***);

d) measures that in the long term reduce costs but which are not directly received by producers (***General Services***);

e) finally, other indirect support, the main elements of which are sub-national subsidies (*i.e.* regional subsidies) and tax concessions (***Other Indirect Support***).

The PSE calculations include all the transfers that specifically result from agricultural policies. However, they exclude certain budget outlays on policy measures, such as subsidies specific to food processing and distribution sectors, outlays that are not specific to the agricultural sector (*e.g.* certain transport subsidies), outlays for stockholding, and budgetary payments associated with measures that result in the permanent withdrawal of resources from agriculture. In those cases in which it is not possible to allocate budgetary subsidies in categories *ii*) to *v*) to individual commodities, this is done according to the share of each commodity in total agricultural output.

In order to avoid double-counting in the calculation of aggregate and average PSEs, an estimate is made to account for the effects of the market price support element of agricultural policies in redistributing transfers within the agricultural sector. This is the **feed adjustment** to the PSE calculations. It is carried out for all livestock products to account for the effects of market price support for feed grains and oilseeds and of taxes on processed feedstuffs in increasing the costs of animal feed to livestock producers. Transfers to producers before deduction of the feed adjustment is called the **Gross PSE**. Transfers to producers after deduction of the feed adjustment is called the **Net PSE**.

The PSE is expressed in three ways:

- **Total PSE**: the total value of transfers to producers;
- **Percentage PSE**: the total value of transfers as a percentage of the total value of production (valued at domestic prices), adjusted to include direct payments and to exclude levies;
- **Unit PSE**: the total value of transfers per tonne.

In algebraic form, these PSE expressions are written as:

Gross Total PSE: $Q^*(P-PWnc) + DP - LV + OS$
Net Total PSE: $Q^*(P-PWnc) + DP - LV + OS - FA$
Unit PSE: $PSEu = PSE/Q$
Percentage PSE = $100^* \; PSE/(Q^*P+DP-LV)$
where

Q = volume of production;
P = domestic producer reference price (in practice, this may be a farm gate price suitably adjusted to include handling margins to the frontier where it is compared to the world or external reference price. (In which case, the rate of assistance at the farm gate is assumed to be the same as that measured at the frontier and the above formulae for the Net and Gross PSE are simplifications);
PWnc = world price (reference price) at the border in domestic currency;
DP = direct payments;
LV = levies on production;

OS = all other budgetary-financed support;
FA = feed adjustment (only for livestock products).

The CSE measures the value of monetary transfers from domestic consumers to producers and taxpayers resulting from a given set of agricultural policies in a given year. In other words, it measures the implicit tax imposed on consumers by agricultural policies.

Two categories of agricultural policy measures are included in the OECD calculations of CSEs:

- transfers from consumers to producers due to market price support (***Market Transfers***);
- budgetary transfers to consumers resulting from agricultural policies (***Other Transfers***).

The CSE is also expressed in three ways:

- **Total CSE**: the total value of transfers from consumers;
- **Percentage CSE**: the total value of transfers as a percentage of the total value of consumption (valued at the implicit consumer price, see below) including transfers such as consumer subsidies;
- **Unit CSE**: the total value of transfers per tonne.

In algebraic form, these CSE expressions, as measured by the OECD, are written as:

- Total CSE: $CSE = -Q_c*(Pc-PWnc) + OT$
- Unit CSE: $CSE_u = CSE/Q_c$
- Percentage $CSE = 100* CSE/(Q_c*Pc)$

where,

Pc = implicit consumer price (defined as the farm gate producer price, less unit market price support, plus unit market transfers. As these unit values are usually equal, the implicit consumer price usually equals the farm gate producer price. This is always the case in the Russian Federation);
Q_c = volume of consumption;
OT = budgetary subsidies to consumers resulting from agricultural policies.

There is thus a very close relationship between the values of PSEs and CSEs. All market price support policies that create a wedge between domestic and world prices raise consumer prices: a positive (negative) transfer to producers from consumers is equivalent to a subsidy (tax) to producers and a tax (subsidy) on consumers. Specific consumer subsidies from government budgets, such as food subsidies, paid in implementing agricultural policies, partly offset consumer taxes. Direct payments and other support, which are implemented through the budget, raise the effective price received by producers but they do not raise the price paid by consumers.

B. METHODOLOGY FOR PSE AND CSE DECOMPOSITION

The purpose of decomposing the PSE/CSE is to provide a means of analysing year-to-year changes in total PSEs and total CSEs at an aggregate (all commodity) level, by reference to the changes in the main components of the total PSEs and total CSEs. The approach has a number of advantages. Firstly, it identifies the relative importance of changes in the various PSE and CSE components in explaining the overall year-to-year change in total PSEs and CSEs for the Russian Federation. Secondly, it allows for the condensation of a large volume of data into a concise form. The approach is briefly summarised here. (see OECD 1992, for a more detailed description).

The methodology is based on expressing a net total PSE for a given commodity in terms of its components, a ***production volume*** component and a ***unit PSE*** component. The unit PSE is in turn broken down into a series of ***unit value*** components: ***market price support, output levies, direct payments, other support*** (*i.e.* the categories described as "Reduction in Input Costs", "General Services" and "Other Indirect Support" under "Concepts" above) ***and feed adjustment***. Unit market price support is itself further decomposed into a ***domestic producer price*** component and a ***border price in***

domestic currency. The latter in turn is made up of an ***exchange rate*** and a ***border price in US dollars*** component.

Likewise, the CSE is broken down into a ***consumption volume*** component and a ***unit CSE***. The unit CSE has two unit value components: ***market transfers*** and ***other transfers***. As they are the mirror image of market price support, unit market transfers consist of a ***domestic consumer price*** (as consumption is usually valued at the farm gate, this is often equal to the producer price) and a *border price in domestic currency*. The latter is broken down into an ***exchange rate*** and *a border price in* US *dollars* component.

For each component, two indicators are calculated: the percentage change in that component and the contribution, in terms of percentage points, of that change to the overall change in the total PSE. The contribution can also be interpreted as the change that would have occurred in the total PSE if nothing else had changed. The sum of the contributions from all components equals the change in total PSE. Similarly, the changes in CSEs can be expressed in terms of shares in total CSE and changes in its components. CSE indices are constructed and contributions estimated as for PSEs.

For the total PSE and for each of the components of the unit PSE, year-to-year percentage changes in Fisher Ideal indices[1] are calculated to provide a country index aggregated across all commodities.

Algebraically, the decomposition analysis for PSE, in terms of percentage changes, is presented as follows:

$$\mathring{PSE} = \mathring{PSE}_u + \mathring{Q} + \mathring{PSE}_u * \mathring{Q} \quad (1)$$

$$\mathring{PSE}_u = \mathring{S}_{mps} * \mathring{MPS}_u - \mathring{S}_{lv} * \mathring{LV}_u + \mathring{S}_{dp} * \mathring{DP}_u + \mathring{S}_{os} * \mathring{OS}_u - \mathring{S}_{fa} * \mathring{FA}_u \quad (2)$$

$$\mathring{S}_{mps} * \mathring{MPS}_u = \mathring{S}_p * \mathring{P} - \mathring{S}_{pwnc} * \mathring{PW}_{nc} \quad (3)$$

$$\mathring{PW}_{nc} = \mathring{XR} + \$\mathring{PW} + \mathring{XR} * \$\mathring{PW} \quad (4)$$

where

° indicates the percentage change in the nominated variable;

MPS_u is unit market price support (per tonne);

LV_u is unit levies on output (per tonne);

DP_u is unit direct payments (per tonne);

OS_u is unit other support (per tonne);

FA_u is feed adjustment per unit (per tonne);

XR is the exchange rate in units of domestic currency per US$;

S_{mps}, S_{lv}, S_{dp}, S_{os}, S_{fa} are the shares of market price support, levies, direct payments, other support and feed adjustment, respectively, in the total PSE;

$PW is the ***implicit*** border price in $US; it is calculated as the difference between domestic prices and unit market price support.[2]

Equation (2) shows that the change in the unit PSE is equal to the sum of the percentage changes in its components weighted by the shares of those components in the base year. However, as the changes are expressed by Fisher Ideal Indices, the above expressions are not exact. Thus, approximation techniques are used to preserve the additivity of the decomposition formulae.

The decomposition analysis is based on the assumption that components of assistance are independent of one another, which is a useful simplification but needs to be interpreted carefully. In many cases the components are related; for instance, market price support and direct payments may both be influenced by border price changes. In the case where market price support is provided solely by a tariff, changes in the internal price would be a direct consequence of changes in the world price.

Results of the decomposition are presented in graphical form using a "tree diagram" to clarify the relationships between the components of assistance. A "tree diagram" for the decomposition of annual change in the total PSE and total CSE for the Russian Federation (all commodities) between 1995 and 1996 is presented in Box V.1. The upper value shown for each element is the annual percentage change in the Fisher Ideal Index. The lower value in brackets is the approximate percentage points contribution to the total change (*i.e.* the effect on the total PSE or total CSE of the change in the component if no other change had taken place).

The choice of the *numéraire* currency to be used for international comparison is arbitrary from a technical point of view. By convention, the US dollar has been used predominantly in OECD PSE work and is therefore used in this study. However, it can be shown that the use of an alternative *numéraire* currency affects only the values and contributions of the exchange rate and border price in the *numéraire* currency indices. All other PSE components are unaffected and the differences are fully determined by the change in the exchange rate between the "old" and "new" *numéraire* currencies. Likewise, the contributions are determined using the share weights, which remain unaltered by any change in the choice of *numéraire* currency.

C. ESTIMATION OF RUSSIAN PSEs AND CSEs FROM 1986 TO 1997

1. Calculation of Market Price Support

The detachment of the Russian economy from the former Soviet Union presents an obvious difficulty for ensuring consistency of measuring PSEs and CSEs for Russia between the two periods: the period of Soviet policy between 1986-1991 and the transition period after 1991. The policies determining MPS for Russia during the 1986-1991 period were set at the all Soviet Union level. In the centrally planned system, these policies were essentially the setting of fixed prices and margins at various points in the food chain, the state monopoly in agro-food trade and the setting of exchange rates. As explained above, market price support (for producers) and market transfers (from consumers) per tonne of product should be measured by calculating the gap between a domestic reference price for the product and a suitable external reference or border price. This price gap is then weighted by total production or total consumption to produce total market price support and total market transfers respectively. Therefore, the method adopted is to measure the price gaps or protection levels resulting from the Soviet policy and to weigh them by Russian production and consumption figures to arrive at total MPS and MT. This process is facilitated by the fact that production and consumption figures are available at the republican level.

The method described above assumes that the tax per tonne paid by consumers as a result of price support measures and border restrictions is equal to the subsidy per tonne transferred to producers (*i.e.* MPSu=MTu). In some countries this is not always the case. Some OECD countries operate a pooling system whereby consumers pay a price which is an average of a high domestic price paid to producers and a lower import price. This "pooling" system requires that the state controls trade in the product. Imports are purchased by a state monopoly at a low price and sold to consumers at an intermediate price that allows the state to pay a high support price to producers, possibly without any taxpayer contribution. This type of system could be appropriate for the calculation of Market Price Support and Market Transfers in economies in transition. Given that the state operated a trade monopoly and fixed internal prices, it was certainly feasible. In the Russian case, this was not the procedure followed. When a product was imported, a process of equalisation took place whereby the price of the imported product, whether higher or lower than the domestic equivalent in Rouble at the corrected exchange rate, was "sold" at the fixed domestic price. This means, in effect, that consumers (before considering consumer subsidies) paid the full producer price, so the basic model, in which MPSu=MTu, is appropriate to the Russian case.

2. Budget expenditures

As far as budget subsidies are concerned, much data is available at the republican level. However, for many types of budgetary expenditures paid from the all Soviet Union budget, in particular input subsidies and consumer subsidies, estimates are based on Russia's share of All-Union expenditures. For instance subsidies reducing the costs of various inputs (purchases of agricultural machinery, fertilisers, mixed feed) are estimated according to Russia's share in All-Union volume of sales of these various inputs, and consumer subsidies for Russia are estimated on the basis of Russia's share in

All-Union value of state purchases of those commodities for which consumer subsidies were given, some consumer subsidies being directly financed by the Russian budget.

As mentioned in Section A, the calculation of the PSE consists of allocating budgetary expenditures among the various commodities. For most of the subsidies in Russia, data was available only on the total amount of funds distributed by the state throughout the agricultural economy, not on the amount of funds allocated for production of specific commodities. The distribution method is related to the way each measure is applied (and to which commodities it is related), but if there is no specific basis for allocating the expenditure, allocation is done according to the share of each commodity in total value of agricultural production.

Transfers from taxpayers to agriculture include also other (non budgetary) transfers to agriculture, such as subsidies lowering the electricity prices for agricultural producers. They also include the write-offs of debts accumulated in Russia between 1983 and 1990, and 1992 and 1994. This subsidy is included in the PSE calculation by dividing the total amount of written-off credits acquired in each period by the number of years in the period. For the period 1992 to 1994, an additional adjustment to take account of the rate of inflation has been made.

3. Reference prices

Two key reference prices are used in the measurement of a price gap due to policies: external reference price and domestic producer reference price.

i) External reference price

The external price is in principle the unit export value or average export price for a product for which the country is a net exporter and the unit import value or average import price for a product for which the country is a net importer. The trade prices should, as far as possible, be those of the country being assessed to ensure a comparison of 'like with like". There are a number of problems about using Russian trade data. First of all, during the Soviet period a large part of trade was with other republics inside the USSR and the related "trade prices" prices are not appropriate because they consist essentially of domestic trade prices fixed in roubles by central authorities. Also official data on agricultural trade between the RSFSR and other republics inside the USSR as well as between the RSFSR and other countries are incomplete. Moreover, a large part of the trade between Russia and other countries, in particular NIS, took place on a barter basis. With just a few exceptions, every year during the period under review, Russia was a net importer of all commodities covered by the PSE estimates. Russian unit import values are available since 1992, but because it was not possible to find consistent trade series for the whole period under review and for the reasons mentioned above, unit import values were not used as reference prices.

In the case of many OECD countries, unit trade values have also proven to be unreliable and quoted trade prices have been used as reference prices (*e.g.* the annual average of a regularly quoted export price of a specific commodity at a specific location). The chosen price is one that, as far as possible, is representative of the product produced domestically. When a country's own unit trade values are not available or deemed to be unreliable and no suitable quoted trade prices are available, previous practice has been to use the trade prices of a third country. This practice does carry the risk of poor comparability between the domestic product and the reference product. If this can be demonstrated, an adjustment for quality differences can be attempted. For many countries in transition, the OECD practice has been to use EU reference prices when problems with the country's own trade prices are found. This is a useful approach for a number of reasons. The EU is a major trader in the region and as such tends to determine trade prices for the region. Hence, its reference prices are a good indicator of the alternative price that would have faced Russia in the absence of its own trade barriers or systemic failures. In addition, as exported products would be competing with the EU export price on any third market, the same EU export price can be used when the country is a net exporter of the product in question.

This was the approach adopted, therefore. EU reference prices (as used for the calculation of the EU's own PSEs and those of some neighbouring countries) were used for most products. Exceptions were rye and potatoes where a German price was used (as no EU price was available and as Germany is a major producer of rye and potatoes); beef and pigmeat, where Hungarian export prices were used (Hungary is a major exporter of these products at a quality more comparable to the Russian product); and milk where the OECD methodology applies the adjusted New Zealand export price for all countries.

ii) Domestic producer reference price

In principle, when a price comparison is made between the two reference prices (domestic and external), the comparison should be for the identical product in terms of quality and stage of processing. If the external reference price is for a quality of product very different from the average product produced, unit value of production would not be the appropriate domestic price for measurement of the price gap. In the case of the Russian Federation, the choice of domestic price was limited by the availability of data. Before 1991, the domestic producer prices used were the average state purchases prices (received by large-scale producers and household plots), as other marketing channels were insignificant. Between 1991 and 1996, as new marketing channels started to develop, the prices used are weighted average producer prices received by large-scale farms, private farms, and households with weights based on marketed output. In 1997, prices received by large-scale farms were applied, as the price information for households and private farms was not available at the time when 1997 calculations were made. It should be noted that during the Soviet period average prices included several supplementary payments which covered measures such as premiums to stimulate production, payments to low-profitable farms, payments compensating expenses of some inputs and investments and also some social measures (for example, costs of maintenance of kindergartens) which are usually not included in the PSE calculations. However, because all these premiums and supplementary price payments were channelled indirectly to producers through the purchase prices paid by the procurement agencies, it was not possible to treat them as distinct budgetary expenditures. The domestic reference prices are then adjusted by technical coefficients and margins to create, as near as possible, a product comparable to the product to which the external reference price applied.

4. Farm gate comparison

It is OECD practice to measure support to agriculture as near as practicable to the farm gate. However, external reference prices (if for traded goods) are always applied to a product to which some value has been added after the farm gate. Hence, comparison of a farm gate domestic price with an external reference price will exclude this value-added and tend to understate the price gap. In the case of the Russian Federation, however, due to inefficiencies in the food chain, the cost of transport and handling from farm to port and processing margins can be high relative to farm prices. Ignoring these margins could therefore produce significant underestimates.

This problem has led to two practices. First, the external reference price should be sought for a product that is as little transformed as possible. An export price for salami as a reference price for pork would create severe problems in identifying both the technical and economic margins involved between the farm gate product and the highly processed product. The same problem would arise in using an external price for flour to measure the price gap for milling wheat. The errors in such a procedure are likely to be very large. It is for this reason that for meats generally external reference prices for a carcass with minimal processing or value-added are preferred, while for grains an export price for the grain in its rawest form is preferred. The second practice involves making technical and value-added adjustments to the prices on which the comparison is based. The first practice of choosing a product with minimal transformation minimises the errors in making these adjustments. The appropriate margin can be added to the farm gate price to bring it to the frontier for comparison or, alternatively, the margin can be subtracted from the external reference price to bring it back to the farm gate for comparison.

The margins for Russia are based on price estimates of the Ministry of Agriculture (1996), surveys of the Centre for Economic Analysis (CEA, 1993), Goskomstat (1995) and Roskhleboproduct Corporation (1995). Although they have fluctuated over the period studied, in recent years they have been between 40 and 50 per cent of the average farmgate price. Margins were not applied for potatoes as German farmgate prices were used as the external reference price.

5. Exchange rate in PSE estimation

Currency exchange rates enter into the calculation of the PSE in two ways: firstly, when an external reference price is used that is expressed in a foreign currency, and secondly, when total PSEs are converted to some *numéraire* currency such as the US dollar for comparison with other country values. It is obvious that the exchange rate used for this purpose should be some economically meaningful figure. Since the official exchange rate seem to reflect in the most adequate way the macroeconomic conditions in which economic agents in Russia have been making decisions, these rates were selected for the calculations of basic series of PSEs and CSEs, except in 1990 and 1991 for which estimates based on commercial exchange rates were used. In effect, a commercial exchange rate was introduced in November 1990 at a depreciated level compared with the official exchange rate, and as it was applied to most trade transactions, the exchange rate used in the PSE/CSE calculations in 1990 is not the official exchange rate still fixed at 0.6 Rb per US$ but the weighted average between the official exchange rate for the first 10 months and the commercial exchange rate (1.7462Rb/US$) for November and December 1990. In 1991 the exchange rate used is a commercial exchange rate based on foreign trade of Russia in roubles and US dollars. In order to take into account some significant overvaluation and undervaluation in the period from 1986 to 1997, a second set of PSEs and CSEs was calculated with adjusted exchange rates.

The adjusted exchange rate used in the study is the "Atlas Conversion Factor" calculated by the World Bank. This "Atlas Conversion Factor" represents a three-year average exchange rate, with exchange rates of the current year and the two preceding years adjusted for differences in the rates of inflation between the country for which the atlas conversion factor is calculated and the G-5 countries (France, Germany, Japan, the United Kingdom, and the United States). The inflation rate for G-5 countries is represented by changes in the Special Drawing Right (SDR) deflators. The ratio of adjusted to official exchange rates is presented in Annex Table I.20. This ratio indicates that the Rouble was overvalued during the Soviet period, with the adjusted exchange rate being on average 1.7 times higher than the official one. In 1992 and 1993 on the contrary the adjusted exchange rate reflects a strong undervaluation linked with the sharp depreciation of the Rouble. In effect in 1992 the Rouble depreciated by 110 times against the US dollar, while prices increased by about 25 times. Afterwards very high inflation rates combined with much lower depreciation between 1992 and 1994 led to a real appreciation of the currency which returned to a more market-related equilibrium. In 1995 and 1996, the ratio of the adjusted to the official exchange rate was close to one.

As an alternative, the purchasing power parity (PPP) could be applied. However, since the PPP reflects to a large extent a wide range of non-tradeable goods (such as services, housing rents and charges, etc.) the exchange rate based on PPP does not reflect adequately the price gap for tradeable goods such as agricultural products. For this reason it has not been applied in the case of Russia.

NOTES

1. The Laspeyres price index is a weighted average of price changes between year 1 and year 0 with the weights being the quantity for year 0:

   $$L = \frac{\Sigma P_1 * Q_0}{\Sigma P_0 * Q_0}$$

   The Paasche price index is a weighted average of price changes between year 1 and year 0 with the weights being the quantity for year 1:

   $$L = \frac{\Sigma P_1 * Q_1}{\Sigma P_0 * Q_1}$$

   The Fisher Ideal Index is the geometric average of the Laspeyres and Paasche indices:

   $$F = (L * P)^{1/2} = \frac{\Sigma P_1 * Q_0}{\Sigma P_0 * Q_0} * \frac{\Sigma P_1 * Q_1}{\Sigma P_0 * Q_1}$$

2. It may not therefore equate exactly with the actual reference price used in estimating the PSE, as transport costs, quality adjustment factors etc., are all reflected in this implicit price.

D. TABLES OF PSE AND CSE RESULTS AND RELATED DATA

Annex Table I.1.*i.* **Producer Subsidy Equivalents: Wheat**

	Units	1986	1987	1988	1989	1990	1991	1992	1993	1994	1995	1996p	1997e
I. Level of production	'000 t	47 434	36 868	39 864	44 004	49 596	38 899	46 167	43 547	32 129	30 119	34 917	44 188
II. Production price (farm gate)	Rb/t	122	125	130	198	279	689	8 181	48 249	106 027	365 260	639 872	621 843
III. Value of production	Rb mn	5 787	4 609	5 182	8 713	13 837	26 801	377 692	2 101 099	3 406 541	11 001 266	22 342 411	27 477 998
IV. Levies	Rb mn	0	0	0	0	0	0	0	0	0	0	0	0
V. Direct payments	Rb mn	96	61	0	0	0	1 100	2 890	39 472	25 015	55 143	180 073	2 592
VI. Adjusted value of production	Rb mn	5 883	4 669	5 182	8 713	13 837	27 901	380 582	2 140 571	3 431 557	11 056 409	22 522 483	27 480 590
VII. Gross total PSE	Rb mn	5 318	4 127	3 852	6 632	11 298	23 987	–384 265	–655 553	–1 238 613	–2 485 388	1 950 122	3 141 640
A. Market price support	Rb mn	3 636	2 845	2 890	5 258	9 252	21 502	–457 912	–1 171 037	–2 036 037	–3 940 120	–1 261 552	987 993
B. Direct payments	Rb mn	96	61	0	0	0	1 100	2 890	39 472	25 015	55 143	180 073	2 592
C. Reduction of input costs	Rb mn	1 339	1 011	758	1 031	1 507	472	56 893	389 906	476 448	1 064 563	2 347 074	1 254 143
D. General services	Rb mn	246	210	204	344	539	913	13 864	86 106	295 961	335 027	684 527	896 912
VIII. Gross unit PSE	Rb/t	112	112	97	151	228	617	–8 323	–15 054	–38 551	–82 519	55 850	71 097
IX. Gross percentage PSE	**%**	**90**	**88**	**74**	**76**	**82**	**86**	**–101**	**–31**	**–36**	**–22**	**9**	**11**

e: estimate; p: provisional.
Source: OECD.

Annex Table I.1.*ii.* **Consumer Subsidy Equivalents: Wheat**

	Units	1986	1987	1988	1989	1990	1991	1992	1993	1994	1995	1996p	1997e
I. Level of consumption	'000 t	52 573	50 981	48 840	51 960	57 260	53 823	56 617	48 945	42 616	39 420	37 814	38 814
II. Consumption price (farm gate)	Rb/t	122	125	130	198	279	689	8 181	48 249	106 027	365 260	639 872	621 843
III. Value of consumption	Rb mn	6 414	6 373	6 349	10 288	15 976	37 084	463 184	2 361 547	4 518 447	14 398 549	24 196 120	24 136 214
IV. Total CSE	Rb mn	–2 260	–2 518	–1 333	–3 078	–7 986	–27 171	589 565	1 575 516	2 934 233	5 156 863	1 366 221	–867 836
A. Market transfers	Rb mn	–4 030	–3 933	–3 541	–6 209	–10 681	–29 751	561 561	1 316 197	2 700 606	5 156 863	1 366 221	–867 836
B. Other transfers	Rb mn	1 770	1 415	2 209	3 131	2 695	2 580	28 003	259 319	233 627	0	0	0
V. Unit CSE	Rb/t	–43	–49	–27	–59	–139	–505	10 413	32 190	68 853	130 818	36 130	–22 359
VI. Percentage CSE	**%**	**–35**	**–40**	**–21**	**–30**	**–50**	**–73**	**127**	**67**	**65**	**36**	**6**	**–4**

e: estimate; p: provisional.
Source: OECD.

RUSSIA

Definitions and notes to Table 1

PSE: Wheat

I. **Level of production:** total production of all wheat at farm level, calendar year.[1]

II. **Production price (farm gate):** before 1991, these prices are the average state purchases prices (received by large-scale producers and household plots), as other marketing channels were insignificant. As of 1991, average prices received by large-scale agricultural producers reflecting all marketing channels in the Goskomstat classification (procurement agencies, private traders, barter deals, payments in kind to farm workers, free market...).[1]

III. **Value of production:** (I)*(II).

V. **Direct payments**

VI. **Adjusted value of production:** III + V.

A. **Market price support:** for calculation, see following table.

B, C, D See notes to Table 12.

CSE: Wheat

I. **Level of consumption:** total domestic use of wheat (sum of feed, food, seed, and industrial uses), marketing year (July/June).[2]

II. **Consumption price (farm gate):** implicit price measured at the farm gate; equal to the production price minus the sum of unit market price support and unit market transfers [= Pp – (MPSu + MTu)]. As MPSu = –MTu, consumption price = average production price (see following table).

III. **Value of consumption:** (I)*(II).

A. **Market transfers:** for calculation, see following table.

B. **Other transfers:** 1986-1991: consumer subsidies paid to food processing industries, designed to maintain relatively low and stable prices for food. In 1992-1994, consumer subsidies were maintained for bakery and cereal products. Since 1992, consumer subsidies were mostly financed from local budgets. In 1995, they became negligible.

Sources:

1. Goskomstat database. 1991-1995 production data published by Goskomstat in *Agriculture of the Russian Federation*, and *Russian Statistical Yearbook*.
2. USDA database (Production, Supply, and Distribution database).

Annex Table I.1.*iii.* **Market Price Support and Market Transfers: Wheat**

	Units	1986	1987	1988	1989	1990	1991	1992	1993	1994	1995	1996p	1997e
PSE – MARKET PRICE SUPPORT (MPS)													
1. Total consumption	'000 t	52 573	50 981	48 840	51 960	57 260	53 823	56 617	48 945	42 616	39 420	37 814	38 814
2. Total production	'000 t	47 434	36 868	39 864	44 004	49 596	38 899	46 167	43 547	32 129	30 119	34 917	44 188
3. Average price at farm level	**Rb/t**	**122**	**125**	**130**	**198**	**279**	**689**	**8 181**	**48 249**	**106 027**	**365 260**	**639 872**	**621 843**
4. Handling margin	%	20	20	20	20	20	25	40	45	50	50	50	50
5. Adjusted domestic price	Rb/t	146	150	156	238	335	861	11 453	69 961	159 041	547 890	959 808	932 765
6. World reference price	ECU/t	92	83	97	143	112	79	102	100	97	125	156	136
7. Official exchange rate	Rb/ECU	0.59	0.69	0.71	0.66	0.99	2.16	249	1 092	2 614	5 953	6 504	6 627
8. World reference price	Rb/t	54	57	69	94	111	170	25 339	108 953	254 097	744 118	1 014 003	899 226
9. Price difference	Rb/t	92	93	87	143	224	691	–13 886	–38 992	–95 056	–196 228	–54 195	33 538
10. Percentage MPS	**%**	**63**	**62**	**56**	**60**	**67**	**80**	**–121**	**–56**	**–60**	**–36**	**–6**	**4**
11. Unit market price support	Rb/t	77	77	73	119	187	553	–9 919	–26 891	–63 371	–130 818	–36 130	22 359
12. Consumer transfers	Rb mn	3 636	2 845	2 890	5 258	9 252	21 502	–457 912	–1 171 037	–2 036 037	–3 940 120	–1 261 552	867 836
13. Budget transfers	Rb mn	0	0	0	0	0	0	0	0	0	0	0	120 156
14. Market price support	Rb mn	3 636	2 845	2 890	5 258	9 252	21 502	–457 912	–1 171 037	–2 036 037	–3 940 120	–1 261 552	987 993
CSE – MARKET TRANSFERS (MT)													
15. Unit market transfers	Rb/t	–77	–77	–73	–119	–187	–553	9 919	26 891	63 371	130 818	36 130	–22 359
16. Consumer transfers	Rb mn	–3 636	–2 845	–2 890	–5 258	–9 252	–21 502	457 912	1 171 037	2 036 037	3 940 120	1 261 552	–867 836
17. Budget transfers	Rb mn	–394	–1 089	–651	–951	–1 430	–8 249	103 649	145 159	664 569	1 216,742	104 669	0
18. Market transfers	Rb mn	–4 030	–3 933	–3 541	–6 209	–10 681	–29 751	561 561	1 316 197	2 700 606	5 156 863	1 366 221	–867 836
19. Consumption price (farm level)	Rb/t	122	125	130	198	279	689	8 181	48 249	106 027	365 260	639 872	621 843

e: estimate; p: provisional.

Notes:

5 = 3*(1 + (4/100))
6 = EU export price of commercial quality wheat.
8 = 6*7
9 = 5 – 8
10 = 9/5*100
11 = 10*3/100
12 = If 1 > 2 then 11*2; if 2 > 1 then 11*1
13 = If 1 > 2 then 0; if 2 > 1 then 11*(2 – 1)
14 = 12 + 13 or 11*2
15 = –11
16 = –12
17 = If 1 > 2 then 15*(1 – 2); if 2 > 1 then 0
18 = 16 + 17 or 15*1
19 = Consumption price (farm gate) = average producer price –(MPSu + MTu); as MPSu = –MTu, consumption price = average producer price.

Source: OECD.

Annex Table I.2.*i.* **Producer Subsidy Equivalents: Maize**

	Units	1986	1987	1988	1989	1990	1991	1992	1993	1994	1995	1996p	1997e
I. Level of production	'000 t	1 708	3 844	3 814	4 663	2 451	1 969	2 135	2 441	892	1 738	1 088	2 671
II. Production price (farm gate)	Rb/t	313	222	421	404	734	1 114	8 838	57 805	250 746	538 504	919 365	730 822
III. Value of production	Rb mn	535	853	1 606	1 884	1 799	2 193	18 869	141 102	223 665	935 920	1 000 269	1 952 026
IV. Levies	Rb mn	0	0	0	0	0	0	0	0	0	0	0	0
V. Direct payments	Rb mn	9	11	0	0	0	90	144	2 651	1 642	4 691	8 062	184
VI. Adjusted value of production	Rb mn	544	865	1 606	1 884	1 799	2 283	19 014	143 753	225 308	940 611	1 008 331	1 952 210
VII. Gross total PSE	Rb mn	604	917	1 661	1 878	1 863	2 059	–11 725	–5 860	113 416	296 129	460 460	709 113
A. Market price support	Rb mn	448	679	1 363	1 580	1 597	1 856	–15 404	–40 478	61 059	172 370	316 674	556 119
B. Direct payments	Rb mn	9	11	0	0	0	90	144	2 651	1 642	4 691	8 062	184
C. Reduction of input costs	Rb mn	124	187	235	223	196	39	2 842	26 185	31 282	90 566	105 078	89 094
D. General services	Rb mn	23	39	63	74	70	75	693	5 783	19 432	28 502	30 646	63 716
VIII. Gross unit PSE	Rb/t	353	239	436	403	760	1 046	–5 492	–2 401	127 148	170 385	423 217	265 486
IX. Gross percentage PSE	**%**	**111**	**106**	**103**	**100**	**104**	**90**	**–62**	**–4**	**50**	**31**	**46**	**36**

e: estimate; p: provisional.
Source: OECD.

Annex Table I.2.*ii.* **Consumer Subsidy Equivalents: Maize**

	Units	1986	1987	1988	1989	1990	1991	1992	1993	1994	1995	1996p	1997e
I. Level of consumption	'000 t	10 981	7 208	15 667	15 673	6 121	10 238	6 214	5 771	2 154	1 800	1 300	2 700
II. Consumption price (farm gate)	Rb/t	313	222	421	404	734	1 114	8 838	57 805	250 746	538 504	919 365	730 822
III. Value of consumption	Rb mn	3 437	1 600	6 596	6 332	4 493	11 405	54 919	333 593	540 107	969 307	1 195 175	1 973 219
IV. Total CSE	Rb mn	–2 780	–1 063	–5 115	–4 932	–3 780	–9 468	44 834	95 697	–147 445	–178 518	–378 378	–562 157
A. Market transfers	Rb mn	–2 882	–1 274	–5 600	–5 312	–3 987	–9 650	44 834	95 697	–147 445	–178 518	–378 378	–562 157
B. Other transfers	Rb mn	102	211	485	380	207	181	0	0	0	0	0	0
V. Unit CSE	Rb/t	–253	–148	–326	–315	–618	–925	7 215	16 582	–68 452	–99 177	–291 060	–208 206
VI. Percentage CSE	**%**	**–81**	**–66**	**–78**	**–78**	**–84**	**–83**	**82**	**29**	**–27**	**–18**	**–32**	**–28**

e: estimate; p: provisional.
Source: OECD.

RUSSIA

Definitions and notes to Table 2

PSE: Maize

I. **Level of production:** production of grain maize (maize for silage is excluded), calendar year.[1]

II. **Production price (farm gate):** before 1991, these prices are the average state purchases prices (received by large-scale producers and household plots), as other marketing channels were insignificant. As of 1991, average prices received by large-scale agricultural producers reflecting all marketing channels in the Goskomstat classification (procurement agencies, private traders, barter deals, payments in kind to farm workers, free market...).[1]

III. **Value of production:** (I)*(II).

V. **Direct payments**

VI. **Adjusted value of production:** III + V.

A. **Market price support:** for calculation, see following table.

B, C, D See notes to Table 12.

CSE: Maize

I. **Level of consumption:** total domestic use of maize (sum of feed, food, seed, and industrial uses), marketing year (July/June).[2]

II. **Consumption price (farm gate):** implicit price measured at the farm gate; equal to the production price minus the sum of unit market price support and unit market transfers transfers [= Pp – (MPSu + MTu)]. As MPSu = –MTu, consumption price = average production price (see following table).

III. **Value of consumption:** (I)*(II).

A. **Market transfers:** for calculation, see following table.

B. **Other transfers:** 1986-1991: consumer subsidies paid to food processing industries, designed to maintain relatively low and stable prices for food.

Sources:

1. Goskomstat database. 1991-1995 production data published by Goskomstat in *Agriculture of the Russian Federation*, and *Russian Statistical Yearbook*.
2. USDA database (Production, Supply, and Distribution database).

Annex Table I.2.*iii.* **Market Price Support and Market Transfers: Maize**

	Units	1986	1987	1988	1989	1990	1991	1992	1993	1994	1995	1996p	1997e
PSE – MARKET PRICE SUPPORT (MPS)													
1. Total consumption	'000 t	10 981	7 208	15 667	15 673	6 121	10 238	6 214	5 771	2 154	1 800	1 300	2 700
2. Total production	'000 t	1 708	3 844	3 814	4 663	2 451	1 969	2 135	2 441	892	1 738	1 088	2 671
3. Average price at farm level	**Rb/t**	**313**	**222**	**421**	**404**	**734**	**1 114**	**8 838**	**57 805**	**250 746**	**538 504**	**919 365**	**730 822**
4. Handling margin	%	20	20	20	20	20	25	40	45	50	50	50	50
5. Adjusted domestic price	Rb/t	376	266	505	485	881	1 393	12 373	83 817	376 119	807 756	1 379 048	1 096 233
6. World reference price	ECU/t	103	79	108	118	100	99	90	99	105	111	145	118
7. Official exchange rate	Rb/ECU	0.59	0.69	0.71	0.66	0.99	2.16	249	1 092	2 614	5 953	6 504	6 627
8. World reference price	Rb/t	61	54	76	78	99	214	22 474	107 862	273 441	658 991	942 457	783 924
9. Price difference	Rb/t	315	212	429	407	782	1 178	–10 101	–24 045	102 678	148 765	436 591	312 309
10. Percentage MPS	**%**	**84**	**80**	**85**	**84**	**89**	**85**	**–82**	**–29**	**27**	**18**	**32**	**28**
11. Unit market price support	Rb/t	262	177	357	339	651	943	–7 215	–16 582	68 452	99 177	291 060	208 206
12. Consumer transfers	Rb mn	448	679	1 363	1 580	1 597	1 856	–15 404	–40 478	61 059	172 370	316 674	556 119
13. Budget transfers	Rb mn	0	0	0	0	0	0	0	0	0	0	0	0
14. Market price support	Rb mn	448	679	1 363	1 580	1 597	1 856	–15 404	–40 478	61 059	172 370	316 674	556 119
CSE – MARKET TRANSFERS (MT)													
15. Unit market transfers	Rb/t	–262	–177	–357	–339	–651	–943	7 215	16 582	–68 452	–99 177	–291 060	–208 206
16. Consumer transfers	Rb mn	–448	–679	–1 363	–1 580	–1 597	–1 856	15 404	40 478	–61 059	–172 370	–316 674	–556 119
17. Budget transfers	Rb mn	–2 434	–595	–4 237	–3 732	–2 391	–7 794	29 430	55 220	–86 386	–6 149	–61 705	–6 038
18. Market transfers	Rb mn	–2 882	–1 274	–5 600	–5 312	–3 987	–9 650	44 834	95 697	–147 445	–178 518	–378 378	–562 157
19. Consumption price (farm level)	Rb/t	313	222	421	404	734	1 114	8 838	57 805	250 746	538 504	919 365	730 822

e: estimate; p: provisional.

Notes:

5 = 3*(1 + (4/100))
6 = EU import price.
8 = 6*7
9 = 5 – 8
10 = 9/5*100
11 = 10*3/100
12 = If 1 > 2 then 11*2; if 2 > 1 then 11*1
13 = If 1 > 2 then 0; if 2 > 1 then 11*(2 – 1)
14 = 12 + 13 or 11*2
15 = –11
16 = –12
17 = If 1 > 2 then 15*(1 – 2); if 2 > 1 then 0
18 = 16 + 17 or 15*1
19 = Consumption price (farm gate) = average producer price –(MPSu + MTu); as MPSu = –MTu, consumption price = average producer price.

Source: OECD.

RUSSIA

Definitions and notes to Table 3

PSE: Coarse grains

CSE: Coarse grains

These tables are the aggregation of tables 3A, 3B and 3C concerning respectively PSEs and CSEs for rye, barley and oats.

Annex Table I.3.*i.* **Producer Subsidy Equivalents: Other grains**

	Units	1986	1987	1988	1989	1990	1991	1992	1993	1994	1995	1996p	1997e
I. Level of production	'000 t	50 990	49 469	42 551	46 771	55 992	43 185	52 117	47 565	43 800	28 446	30 213	37 635
II. Production price (farm gate)	Rb/t	130	129	141	188	262	486	11 317	41 968	87 273	235 109	534 179	520 435
III. Value of production	Rb mn	6 620	6 372	6 021	8 810	14 663	20 977	589 813	1 996 185	3 822 564	6 687 920	16 139 151	19 586 583
IV. Levies	Rb mn	0	0	0	0	0	0	0	0	0	0	0	0
V. Direct payments	Rb mn	110	84	0	0	0	861	4 513	37 501	28 070	33 522	130 076	1 848
VI. Adjusted value of production	Rb mn	6 730	6 456	6 021	8 810	14 663	21 838	594 326	2 033 686	3 850 634	6 721 442	16 269 227	19 588 430
VII. Gross total PSE	Rb mn	6 563	6 195	4 474	6 702	12 321	17 155	–94 307	–684 823	–1 305 816	–2 948 080	2 555 230	2 519 880
A. Market price support	Rb mn	4 640	4 422	3 357	5 312	10 153	15 210	–209 316	–1 174 568	–2 200 625	–3 832 445	235 261	984 740
B. Direct payments	Rb mn	110	84	0	0	0	861	4 513	37 501	28 070	33 522	130 076	1 848
C. Reduction of input costs	Rb mn	1 531	1 398	881	1 042	1 597	370	88 846	370 437	534 634	647 172	1 695 421	893 966
D. General services	Rb mn	282	291	237	347	571	715	21 650	81 807	332 105	203 670	494 472	639 327
VIII. Gross unit PSE	Rb/t	129	125	105	143	220	397	–1 810	–14 398	–29 813	–103 638	84 574	66 956
IX. Gross percentage PSE	**%**	**98**	**96**	**74**	**76**	**84**	**79**	**–16**	**–34**	**–34**	**–44**	**16**	**13**

e: estimate; p: provisional.
Source: OECD.

Annex Table I.3.*ii.* **Consumer Subsidy Equivalents: Other grains**

	Units	1986	1987	1988	1989	1990	1991	1992	1993	1994	1995	1996p	1997e
I. Level of consumption	'000 t	50 751	52 663	46 347	48 923	57 768	48 055	52 931	47 601	41 166	33 502	30 651	33 900
II. Consumption price (farm gate)	Rb/t	132	129	140	187	261	482	11 374	41 871	87 728	237 359	534 050	521 347
III. Value of consumption	Rb mn	6 676	6 819	6 511	9 170	15 078	23 176	602 046	1 993 119	3 611 401	7 951 992	16 369 166	17 673 680
IV. Total CSE	Rb mn	–3 443	–3 367	–2 310	–4 091	–8 695	–15 725	217 829	1 221 824	2 109 854	4 368 693	–225 927	–944 439
A. Market transfers	Rb mn	–4 744	–4 762	–3 639	–5 538	–10 471	–16 819	208 319	1 170 756	2 068 124	4 368 693	–225 927	–944 439
B. Other transfers	Rb mn	1 301	1 395	1 329	1 447	1 776	1 094	9 510	51 068	41 730	0	0	0
V. Unit CSE	Rb/t	–68	–64	–50	–84	–151	–327	4 115	25 668	51 252	130 401	–7 371	–27 860
VI. Percentage CSE	**%**	**–52**	**–49**	**–35**	**–45**	**–58**	**–68**	**36**	**61**	**58**	**55**	**–1**	**–5**

e: estimate; p: provisional.
Source: OECD.

Annex Table I.3.A.*i.* **Producer Subsidy Equivalents: Rye**

	Units	1986	1987	1988	1989	1990	1991	1992	1993	1994	1995	1996p	1997e
I. Level of production	'000 t	9 717	11 080	12 530	12 593	16 431	10 639	13 887	9 166	5 989	4 098	5 934	7 480
II. Production price (farm gate)	Rb/t	155	169	161	201	278	424	9 236	45 142	101 598	310 388	632 540	588 041
III. Value of production	Rb mn	1 506	1 873	2 017	2 531	4 568	4 511	128 260	413 772	608 470	1 271 970	3 753 492	4 398 547
IV. Levies	Rb mn	0	0	0	0	0	0	0	0	0	0	0	0
V. Direct payments	Rb mn	25	25	0	0	0	185	981	7 773	4 468	6 376	30 252	415
VI. Adjusted value of production	Rb mn	1 531	1 897	2 017	2 531	4 568	4 696	129 242	421 545	612 939	1 278 346	3 783 744	4 398 962
VII. Gross total PSE	Rb mn	1 583	1 991	1 733	1 819	3 749	4 081	–50 819	–211 651	–98 798	216 064	1 263 843	1 130 976
A. Market price support	Rb mn	1 145	1 470	1 359	1 420	3 074	3 663	–75 829	–313 166	–241 233	47 868	724 287	786 231
B. Direct payments	Rb mn	25	25	0	0	0	185	981	7 773	4 468	6 376	30 252	415
C. Reduction of input costs	Rb mn	348	411	295	299	498	80	19 320	76 785	85 102	123 085	394 305	200 757
D. General services	Rb mn	64	85	79	100	178	154	4 708	16 957	52 864	38 736	115 000	143 573
VIII. Gross unit PSE	Rb/t	163	180	138	144	228	384	–3 659	–23 091	–16 497	52 724	212 983	151 200
IX. Gross percentage PSE	**%**	**103**	**105**	**86**	**72**	**82**	**87**	**–39**	**–50**	**–16**	**17**	**33**	**26**

e: estimate; p: provisional.
Source: OECD.

Annex Table I.3.A.*ii.* **Consumer Subsidy Equivalents: Rye**

	Units	1986	1987	1988	1989	1990	1991	1992	1993	1994	1995	1996p	1997e
I. Level of consumption	'000 t	12 939	12 586	12 727	12 357	15 129	11 894	13 339	9 791	6 160	5 650	6 000	7 050
II. Consumption price (farm gate)	Rb/t	155	169	161	201	278	424	9 236	45 142	101 598	310 388	632 540	588 041
III. Value of consumption	Rb mn	2 005	2 127	2 049	2 484	4 206	5 043	123 199	441 985	625 844	1 753 692	3 795 240	4 145 689
IV. Total CSE	Rb mn	–1 087	–1 023	–662	–636	–1 760	–3 586	82 346	385 587	289 851	–65 996	–732 342	–741 033
A. Market transfers	Rb mn	–1 525	–1 670	–1 380	–1 393	–2 830	–4 095	72 837	334 519	248 121	–65 996	–732 342	–741 033
B. Other transfers	Rb mn	438	647	719	758	1 070	509	9 510	51 068	41 730	0	0	0
V. Unit CSE	Rb/t	–84	–81	–52	–51	–116	–302	6 173	39 382	47 054	–11 681	–122 057	–105 111
VI. Percentage CSE	**%**	**–54**	**–48**	**–32**	**–26**	**–42**	**–71**	**67**	**87**	**46**	**–4**	**–19**	**–18**

e: estimate; p: provisional.
Source: OECD.

RUSSIA

Definitions and notes to Table 3A

PSE: Rye

I. **Level of production:** total production of rye at farm level, calendar year.[1]

II. **Production price (farm gate):** before 1991, these prices are the average state purchases prices (received by large-scale producers and household plots), as other marketing channels were insignificant. As of 1991, average prices received by large-scale agricultural producers reflecting all marketing channels in the Goskomstat classification (procurement agencies, private traders, barter deals, payments in kind to farm workers, free market...).[1]

III. **Value of production:** (I)*(II).

V. **Direct payments**

VI. **Adjusted value of production:** III + V.

A. **Market price support:** for calculation, see following table.

B, C, D See notes to Table 12.

CSE: Rye

I. **Level of consumption:** total domestic use of rye (sum of feed, food, seed, and industrial uses), marketing year (July/June).[2]

II. **Consumption price (farm gate):** implicit price measured at the farm gate; equal to the production price minus the sum of unit market price support and unit market transfers transfers [= Pp – (MPSu + MTu)]. As MPSu = –MTu, consumption price = average production price (see following table).

III. **Value of consumption:** (I)*(II).

A. **Market transfers:** for calculation, see following table.

B. **Other transfers:** 1986-1991: consumer subsidies paid to food processing industries, designed to maintain relatively low and stable prices for food. In 1992-1994, consumer subsidies were maintained for bakery and cereal products. Since 1992, consumer subsidies were mostly financed from local budgets. In 1995, they became negligible.

Sources:

1. Goskomstat database. 1991-1995 production data published by Goskomstat in *Agriculture of the Russian Federation*, and *Russian Statistical Yearbook*.
2. USDA database (Production, Supply, and Distribution database).

Annex Table I.3.A.*iii.* **Market Price Support and Market Transfers: Rye**

	Units	1986	1987	1988	1989	1990	1991	1992	1993	1994	1995	1996p	1997e
PSE – MARKET PRICE SUPPORT (MPS)													
1. Total consumption	'000 t	12 939	12 586	12 727	12 357	15 129	11 894	13 339	9 791	6 160	5 650	6 000	7 050
2. Total production	'000 t	9 717	11 080	12 530	12 593	16 431	10 639	13 887	9 166	5 989	4 098	5 934	7 480
3. Average price at farm level	**Rb/t**	**155**	**169**	**161**	**201**	**278**	**424**	**9 236**	**45 142**	**101 598**	**310 388**	**632 540**	**588 041**
4. Handling margin	%	20	20	20	20	20	25	40	45	50	50	50	50
5. Adjusted domestic price	Rb/t	186	203	193	241	334	530	12 930	65 456	152 397	465 582	948 810	882 062
6. World reference price	ECU/t	76	63	89	160	110	46	83	105	81	75	118	109
7. Official exchange rate	Rb/ECU	0.59	0.69	0.71	0.66	0.99	2.16	249.2	1 091.7	2 614.2	5 952.9	6 504.2	6 626.6
8. World reference price	Rb/t	45	44	63	106	109	100	20 575	114 997	212 816	448 061	765 724	724 395
9. Price difference	Rb/t	141	159	130	135	224	430	–7 645	–49 541	–60 419	17 521	183 086	157 667
10. Percentage MPS	**%**	**76**	**78**	**67**	**56**	**67**	**81**	**–59**	**–76**	**–40**	**4**	**19**	**18**
11. Unit market price support	Rb/t	118	133	108	113	187	344	–5 460	–34 166	–40 279	11 681	122 057	105 111
12. Consumer transfers	Rb mn	1 145	1 470	1 359	1 393	2 830	3 663	–72 837	–313 166	–241 233	47 868	724 287	741 033
13. Budget transfers	Rb mn	0	0	0	27	244	0	–2 992	0	0	0	0	45 198
14. Market price support	Rb mn	1 145	1 470	1 359	1 420	3 074	3 663	–75 829	–313 166	–241 233	47 868	724 287	786 231
CSE – MARKET TRANSFERS (MT)													
15. Unit market transfers	Rb/t	–118	–133	–108	–113	–187	–344	5 460	34 166	40 279	–11 681	–122 057	–105 111
16. Consumer transfers	Rb mn	–1 145	–1 470	–1 359	–1 393	–2 830	–3 663	72 837	313 166	241 233	–47 868	–724 287	–741 033
17. Budget transfers	Rb mn	–380	–200	–21	0	0	–432	0	21 354	6 888	–18 129	–8 056	0
18. Market transfers	Rb mn	–1 525	–1 670	–1 380	–1 393	–2 830	–4 095	72 837	334 519	248 121	–65 996	–732 342	–741 033
19. Consumption price (farm level)	Rb/t	155	169	161	201	278	424	9 236	45 142	101 598	310 388	632 540	588 041

e: estimate; p: provisional.

Notes:

5 = 3*(1 + (4/100))
6 = German unit export value to non-EU member countries.
8 = 6*7
9 = 5 – 8
10 = 9/5*100
11 = 10*3/100
12 = If 1 > 2 then 11*2; if 2 > 1 then 11*1
13 = If 1 > 2 then 0; if 2 > 1 then 11*(2 – 1)
14 = 12 + 13 or 11*2
15 = –11
16 = –12
17 = If 1 > 2 then 15*(1 – 2); if 2 > 1 then 0
18 = 16 + 17 or 15*1
19 = Consumption price (farm gate) = average producer price –(MPSu + MTu); as MPSu = –MTu, consumption price = average producer price.

Source: OECD.

Annex Table I.3.B.*i.* **Producer Subsidy Equivalents: Barley**

	Units	1986	1987	1988	1989	1990	1991	1992	1993	1994	1995	1996p	1997e
I. Level of production	'000 t	25 589	26 100	19 417	22 201	27 235	22 174	26 989	26 843	27 054	15 786	15 933	20 774
II. Production price (farm gate)	Rb/t	122	114	127	169	253	454	12 672	37 751	76 621	204 025	503 143	499 571
III. Value of production	Rb mn	3 122	2 975	2 466	3 752	6 890	10 067	342 005	1 013 350	2 072 905	3 220 739	8 016 577	10 378 088
IV. Levies	Rb mn	0	0	0	0	0	0	0	0	0	0	0	0
V. Direct payments	Rb mn	52	39	0	0	0	413	2 617	19 037	15 222	16 144	64 611	979
VI. Adjusted value of production	Rb mn	3 174	3 015	2 466	3 752	6 890	10 480	344 622	1 032 387	2 088 126	3 236 882	8 081 188	10 379 067
VII. Gross total PSE	Rb mn	3 134	2 997	1 874	2 914	5 972	8 056	24 913	–195 200	–780 008	–2 492 951	97 755	453 976
A. Market price support	Rb mn	2 227	2 169	1 416	2 322	4 953	7 123	–41 775	–443 816	–1 265 247	–2 918 839	–1 054 611	–359 429
B. Direct payments	Rb mn	52	39	0	0	0	413	2 617	19 037	15 222	16 144	64 611	979
C. Reduction of input costs	Rb mn	722	653	361	444	751	177	51 517	188 050	289 922	311 662	842 143	473 674
D. General services	Rb mn	133	136	97	148	268	343	12 554	41 529	180 094	98 083	245 612	338 752
VIII. Gross unit PSE	Rb/t	122	115	96	131	219	363	923	–7 272	–28 832	–157 922	6 135	21 853
IX. Gross percentage PSE	**%**	**99**	**99**	**76**	**78**	**87**	**77**	**7**	**–19**	**–37**	**–77**	**1**	**4**

e: estimate; p: provisional.
Source: OECD.

Annex Table I.3.B.*ii.* **Consumer Subsidy Equivalents: Barley**

	Units	1986	1987	1988	1989	1990	1991	1992	1993	1994	1995	1996p	1997e
I. Level of consumption	'000 t	26 269	27 744	22 928	24 508	30 530	25 635	28 368	27 041	24 711	18 002	16 251	18 150
II. Consumption price (farm gate)	Rb/t	122	114	127	169	253	454	12 672	37 751	76 621	204 025	503 143	499 571
III. Value of consumption	Rb mn	3 205	3 163	2 912	4 142	7 724	11 638	359 479	1 020 825	1 893 382	3 672 858	8 176 577	9 067 214
IV. Total CSE	Rb mn	–1 721	–1 739	–1 265	–2 095	–5 010	–7 843	43 909	447 089	1 155 670	3 328 579	1 075 659	314 029
A. Market transfers	Rb mn	–2 286	–2 306	–1 672	–2 563	–5 552	–8 234	43 909	447 089	1 155 670	3 328 579	1 075 659	314 029
B. Other transfers	Rb mn	565	567	408	468	542	391	0	0	0	0	0	0
V. Unit CSE	Rb/t	–66	–63	–55	–85	–164	–306	1 548	16 534	46 767	184 900	66 190	17 302
VI. Percentage CSE	**%**	**–54**	**–55**	**–43**	**–51**	**–65**	**–67**	**12**	**44**	**61**	**91**	**13**	**3**

e: estimate; p: provisional.
Source: OECD.

RUSSIA

Definitions and notes to Table 3B

PSE: Barley

I.	**Level of production:** total production of barley at farm level, calendar year.[1]
II.	**Production price (farm gate):** before 1991, these prices are the average state purchases prices (received by large-scale producers and household plots), as other marketing channels were insignificant. As of 1991, average prices received by large-scale agricultural producers reflecting all marketing channels in the Goskomstat classification (procurement agencies, private traders, barter deals, payments in kind to farm workers, free market...).[1]
III.	**Value of production:** (I)*(II).
V.	**Direct payments**
VI.	**Adjusted value of production:** III + V.
A.	**Market price support:** for calculation, see following table.
B, C, D	See notes to Table 12.

CSE: Barley

I.	**Level of consumption:** total domestic use of barley (sum of feed, food, seed, and industrial uses), marketing year (July/June).[2]
II.	**Consumption price (farm gate):** implicit price measured at the farm gate; equal to the production price minus the sum of unit market price support and unit market transfers transfers [= Pp – (MPSu + MTu)]. As MPSu = –MTu, consumption price = average production price (see following table).
III.	**Value of consumption:** (I)*(II).
A.	**Market transfers:** for calculation, see following table.
B.	**Other transfers:** 1986-1991: consumer subsidies paid to food processing industries, designed to maintain relatively low and stable prices for food.

Sources:

1. Goskomstat database. 1991-1995 production data published by Goskomstat in *Agriculture of the Russian Federation*, and *Russian Statistical Yearbook*.
2. USDA database (Production, Supply, and Distribution database).

Annex Table I.3.B.*iii.* **Market Price Support and Market Transfers: Barley**

	Units	1986	1987	1988	1989	1990	1991	1992	1993	1994	1995	1996p	1997e
PSE – MARKET PRICE SUPPORT (MPS)													
1. Total consumption	'000 t	26 269	27 744	22 928	24 508	30 530	25 635	28 368	27 041	24 711	18 002	16 251	18 150
2. Total production	'000 t	25 589	26 100	19 417	22 201	27 235	22 174	26 989	26 843	27 054	15 786	15 933	20 774
3. Average price at farm level	**Rb/t**	**122**	**114**	**127**	**169**	**253**	**454**	**12 672**	**37 751**	**76 621**	**204 025**	**503 143**	**499 571**
4. Handling margin	%	20	20	20	20	20	25	40	45	50	50	50	50
5. Adjusted domestic price	Rb/t	146	137	152	203	304	568	17 741	54 739	114 932	306 038	754 715	749 357
6. World reference price	ECU/t	71	54	92	117	86	77	80	72	71	98	131	117
7. Official exchange rate	Rb/ECU	0.59	0.69	0.71	0.66	0.99	2.16	249	1 092	2 614	5 953	6 504	6 627
8. World reference price	Rb/t	42	37	65	77	85	166	19 908	78 713	185 083	583 388	854 000	775 309
9. Price difference	Rb/t	104	100	88	126	218	402	–2 167	–23 974	–70 151	–277 351	–99 286	–25 953
10. Percentage MPS	**%**	**71**	**73**	**57**	**62**	**72**	**71**	**–12**	**–44**	**–61**	**–91**	**–13**	**–3**
11. Unit market price support	Rb/t	87	83	73	105	182	321	–1 548	–16 534	–46 767	–184 900	–66 190	–17 302
12. Consumer transfers	Rb mn	2 227	2 169	1 416	2 322	4 953	7 123	–41 775	–443 816	–1 155 670	–2 918 839	–1 054 611	–314 029
13. Budget transfers	Rb mn	0	0	0	0	0	0	0	0	–109 576	0	0	–45 400
14. Market price support	Rb mn	2 227	2 169	1 416	2 322	4 953	7 123	–41 775	–443 816	–1 265 247	–2 918 839	–1 054 611	–359 429
CSE – MARKET TRANSFERS (MT)													
15. Unit market transfers	Rb/t	–87	–83	–73	–105	–182	–321	1 548	16 534	46 767	184 900	66 190	17 302
16. Consumer transfers	Rb mn	–2 227	–2 169	–1 416	–2 322	–4 953	–7 123	41 775	443 816	1 155 670	2 918 839	1 054 611	314 029
17. Budget transfers	Rb mn	–59	–137	–256	–241	–599	–1 112	2 134	3 274	0	409 739	21 049	0
18. Market transfers	Rb mn	–2 286	–2 306	–1 672	–2 563	–5 552	–8 234	43 909	447 089	1 155 670	3 328 579	1 075 659	314 029
19. Consumption price (farm level)	Rb/t	122	114	127	169	253	454	12 672	37 751	76 621	204 025	503 143	499 571

e: estimate; p: provisional.

Notes:

5 = 3*(1 + (4/100))
6 = EU export price for feed barley.
8 = 6*7
9 = 5 – 8
10 = 9/5*100
11 = 10*3/100
12 = If 1 > 2 then 11*2; if 2 > 1 then 11*1
13 = If 1 > 2 then 0; if 2 > 1 then 11*(2 – 1)
14 = 12 + 13 or 11*2
15 = –11
16 = –12
17 = If 1 > 2 then 15*(1 – 2); if 2 > 1 then 0
18 = 16 + 17 or 15*1
19 = Consumption price (farm gate) = average producer price –(MPSu + MTu); as MPSu = –MTu consumption price = average producer price.

Source: OECD.

Annex Table I.3.C.*i.* **Producer Subsidy Equivalents: Oats**

	Units	1986	1987	1988	1989	1990	1991	1992	1993	1994	1995	1996p	1997e
I. Level of production	'000 t	15 684	12 289	10 604	11 977	12 326	10 372	11 241	11 556	10 757	8 562	8 346	9 381
II. Production price (farm gate)	Rb/t	127	124	145	211	260	617	10 635	49 244	106 088	256 390	523 494	512 733
III. Value of production	Rb mn	1 992	1 524	1 538	2 527	3 205	6 400	119 548	569 064	1 141 189	2 195 211	4 369 081	4 809 948
IV. Levies	Rb mn	0	0	0	0	0	0	0	0	0	0	0	0
V. Direct payments	Rb mn	33	20	0	0	0	263	915	10 691	8 380	11 003	35 213	454
VI. Adjusted value of production	Rb mn	2 025	1 544	1 538	2 527	3 205	6 662	120 463	579 754	1 149 569	2 206 214	4 404 294	4 810 402
VII. Gross total PSE	Rb mn	1 847	1 207	867	1 969	2 600	5 018	–68 401	–277 972	–427 009	–671 194	1 193 631	934 928
A. Market price support	Rb mn	1 268	784	581	1 570	2 126	4 424	–91 712	–417 587	–694 145	–961 474	565 585	557 937
B. Direct payments	Rb mn	33	20	0	0	0	263	915	10 691	8 380	11 003	35 213	454
C. Reduction of input costs	Rb mn	461	334	225	299	349	113	18 008	105 603	159 610	212 425	458 973	219 534
D. General services	Rb mn	85	70	60	100	125	218	4 388	23 321	99 147	66 852	133 860	157 002
VIII. Gross unit PSE	Rb/t	118	98	82	164	211	484	–6 085	–24 054	–39 696	–78 392	143 018	99 662
IX. Gross percentage PSE	**%**	**91**	**78**	**56**	**78**	**81**	**75**	**–57**	**–48**	**–37**	**–30**	**27**	**19**

e: estimate; p: provisional.
Source: OECD.

Annex Table I.3.C.*ii.* **Consumer Subsidy Equivalents: Oats**

	Units	1986	1987	1988	1989	1990	1991	1992	1993	1994	1995	1996p	1997e
I. Level of consumption	'000 t	11 544	12 333	10 692	12 058	12 109	10 526	11 224	10 769	10 295	9 850	8 400	8 700
II. Consumption price (farm gate)	Rb/t	127	124	145	211	260	617	10 635	49 244	106 088	256 390	523 494	512 733
III. Value of consumption	Rb mn	1 466	1 529	1 550	2 544	3 148	6 495	119 367	530 309	1 092 176	2 525 442	4 397 350	4 460 777
IV. Total CSE	Rb mn	–635	–605	–384	–1 360	–1 925	–4 296	91 573	389 148	664 333	1 106 111	–569 244	–517 435
A. Market transfers	Rb mn	–934	–786	–586	–1 581	–2 089	–4 490	91 573	389 148	664 333	1 106 111	–569 244	–517 435
B. Other transfers	Rb mn	298	181	203	221	164	194	0	0	0	0	0	0
V. Unit CSE	Rb/t	–55	–49	–36	–113	–159	–408	8 159	36 136	64 530	112 295	–67 767	–59 475
VI. Percentage CSE	**%**	**–43**	**–40**	**–25**	**–53**	**–61**	**–66**	**77**	**73**	**61**	**44**	**–13**	**–12**

e: estimate; p: provisional.
Source: OECD.

RUSSIA

Definitions and notes to Table 3C

PSE: Oats

I. **Level of production:** total production of oats at farm level, calendar year.[1]

II. **Production price (farm gate):** before 1991, these prices are the average state purchases prices (received by large-scale producers and household plots), as other marketing channels were insignificant. As of 1991, average prices received by large-scale agricultural producers reflecting all marketing channels in the Goskomstat classification (procurement agencies, private traders, barter deals, payments in kind to farm workers, free market...).[1]

III. **Value of production:** (I)*(II).

V. **Direct payments**

VI. **Adjusted value of production:** III + V.

A. **Market price support:** for calculation, see following table.

B, C, D See notes to Table 12.

CSE: Oats

I. **Level of consumption:** total domestic use of oats (sum of feed, food, seed, and industrial uses), marketing year (July/June).[2]

II. **Consumption price (farm gate):** implicit price measured at the farm gate; equal to the producer price minus the sum of unit market price support and unit market transfers transfers [= Pp – (MPSu + MTu)]. As MPSu = –MTu, consumption price = average production price (see following table).

III. **Value of consumption:** (I)*(II).

A. **Market transfers:** for calculation, see following table.

B. **Other transfers:** 1986-1991: consumer subsidies paid to food processing industries, designed to maintain relatively low and stable prices for food.

Sources:

1. Goskomstat database. 1991-1995 production data published by Goskomstat in *Agriculture of the Russian Federation*, and *Russian Statistical Yearbook*.
2. USDA database (Production, Supply, and Distribution database).

Annex Table I.3.C.*iii.* **Market Price Support and Market Transfers: Oats**

	Units	1986	1987	1988	1989	1990	1991	1992	1993	1994	1995	1996p	1997e
PSE – MARKET PRICE SUPPORT (MPS)													
1. Total consumption	'000 t	11 544	12 333	10 692	12 058	12 109	10 526	11 224	10 769	10 295	9 850	8 400	8 700
2. Total production	'000 t	15 684	12 289	10 604	11 977	12 326	10 372	11 241	11 556	10 757	8 562	8 346	9 381
3. Average price at farm level	**Rb/t**	**127**	**124**	**145**	**211**	**260**	**617**	**10 635**	**49 244**	**106 088**	**256 390**	**523 494**	**512 733**
4. Handling margin	%	20	20	20	20	20	25	40	45	50	50	50	50
5. Adjusted domestic price	Rb/t	152	149	174	253	312	771	14 889	71 404	159 132	384 585	785 241	769 100
6. World reference price	ECU/t	94	105	153	145	106	110	106	113	98	93	105	103
7. Official exchange rate	Rb/ECU	0.59	0.69	0.71	0.66	0.99	2.16	249	1 092	2 614	5 953	6 504	6 627
8. World reference price	Rb/t	55	72	108	96	105	238	26 311	123 801	255 926	553 028	683 590	679 887
9. Price difference	Rb/t	97	77	66	157	207	533	–11 422	–52 397	–96 794	–168 443	101 651	89 213
10. Percentage MPS	**%**	**64**	**51**	**38**	**62**	**66**	**69**	**–77**	**–73**	**–61**	**–44**	**13**	**12**
11. Unit market price support	Rb/t	81	64	55	131	173	427	–8 159	–36 136	–64 530	–112 295	67 767	59 475
12. Consumer transfers	Rb mn	934	784	581	1 570	2 089	4 424	–91 573	–389 148	–664 333	–961 474	565 585	517 435
13. Budget transfers	Rb mn	335	0	0	0	37	0	–139	–28 439	–29 813	0	0	40 503
14. Market price support	Rb mn	1 268	784	581	1 570	2 126	4 424	–91 712	–417 587	–694 145	–961 474	565 585	557 937
CSE – MARKET TRANSFERS (MT)													
15. Unit market transfers	Rb/t	–81	–64	–55	–131	–173	–427	8 159	36 136	64 530	112 295	–67 767	–59 475
16. Consumer transfers	Rb mn	–934	–784	–581	–1 570	–2 089	–4 424	91 573	389 148	664 333	961 474	–565 585	–517 435
17. Budget transfers	Rb mn	0	–3	–5	–11	0	–66	0	0	0	144 637	–3 659	0
18. Market transfers	Rb mn	–934	–786	–586	–1 581	–2 089	–4 490	91 573	389 148	664 333	1 106,111	–569 244	–517 435
19. Consumption price (farm level)	Rb/t	127	124	145	211	260	617	10 635	49 244	106 088	256 390	523 494	512 733

e: estimate; p: provisional.

Notes:

5 = 3*(1 + (4/100))
6 = EU import price.
8 = 6*7
9 = 5 – 8
10 = 9/5*100
11 = 10*3/100
12 = If 1 > 2 then 11*2; if 2 > 1 then 11*1
13 = If 1 > 2 then 0; if 2 > 1 then 11*(2 – 1)
14 = 12 + 13 or 11*2
15 = –11
16 = –12
17 = If 1 > 2 then 15*(1 – 2); if 2 > 1 then 0
18 = 16 + 17 or 15*1
19 = Consumption price (farm gate) = average producer price –(MPSu + MTu); as MPSu = –MTu, consumption price = average producer price.

Source: OECD.

Annex Table I.4.*i.* **Producer Subsidy Equivalents: Potatoes**

	Units	1986	1987	1988	1989	1990	1991	1992	1993	1994	1995	1996p	1997e
I. Level of production	'000 t	12 827	38 028	33 692	33 760	30 848	34 330	38 330	37 650	33 828	39 909	38 652	37 015
II. Production price (farm gate)	Rb/t	183	157	192	242	295	1 103	9 205	57 316	247 129	909 998	1 134 733	1 020 788
III. Value of production	Rb mn	2 347	5 970	6 469	8 170	9 100	37 882	352 827	2 157 945	8 359 897	36 317 128	43 859 695	37 784 468
IV. Levies	Rb mn	0	0	0	0	0	0	0	0	0	0	0	0
V. Direct payments	Rb mn	39	79	0	0	0	1 554	1 090	4 785	8 556	182 036	360 265	3 564
VI. Adjusted value of production	Rb mn	2 386	6 049	6 469	8 170	9 100	39 436	353 917	2 162 730	8 368 453	36 499 164	44 219 960	37 788 032
VII. Gross total PSE	Rb mn	2 376	5 120	5 367	7 433	7 889	34 132	–236 239	683 939	101 174	–9 081 544	31 408 590	25 490 668
A. Market price support	Rb mn	1 694	3 459	4 166	6 144	6 543	30 619	–299 053	204 973	–1 677 494	–12 772 355	25 836 885	22 595 190
B. Direct payments	Rb mn	39	79	0	0	0	1 554	1 090	4 785	8 556	182 036	360 265	3 564
C. Reduction of input costs	Rb mn	543	1 310	946	966	991	668	48 772	385 745	1 043 804	2 402 793	3 867 665	1 658 588
D. General services	Rb mn	100	272	254	322	354	1 291	12 951	88 436	726 309	1 105 983	1 343 774	1 233 326
VIII. Gross unit PSE	Rb/t	185	135	159	220	256	994	–6 163	18 166	2 991	–227 556	812 599	688 658
IX. Gross percentage PSE	**%**	**100**	**85**	**83**	**91**	**87**	**87**	**–67**	**32**	**1**	**–25**	**71**	**67**

e: estimate; p: provisional.
Source: OECD.

Annex Table I.4.*ii.* **Consumer Subsidy Equivalents: Potatoes**

	Units	1986	1987	1988	1989	1990	1991	1992	1993	1994	1995	1996p	1997e
I. Level of consumption	'000 t	40 887	38 742	36 514	33 966	32 212	33 829	37 194	38 405	36 558	37 406	38 015	38 015
II. Consumption price (farm gate)	Rb/t	183	157	192	242	295	1 103	9 205	57 316	247 129	909 998	1 134 733	1 020 788
III. Value of consumption	Rb mn	7 482	6 082	7 011	8 220	9 503	37 329	342 368	2 201 224	9 034 585	34 039 402	43 136 870	38 805 256
IV. Total CSE	Rb mn	–5 031	–3 167	–3 922	–5 627	–6 261	–28 958	290 187	–209 084	1 812 877	11 971 302	–25 411 084	–23 205 623
A. Market transfers	Rb mn	–5 398	–3 524	–4 515	–6 182	–6 833	–30 172	290 187	–209 084	1 812 877	11 971 302	–25 411 084	–23 205 623
B. Other transfers	Rb mn	367	357	593	554	571	1 213	0	0	0	0	0	0
V. Unit CSE	Rb/t	–123	–82	–107	–166	–194	–856	7 802	–5 444	49 589	320 037	–668 449	–610 433
VI. Percentage CSE	**%**	**–67**	**–52**	**–56**	**–68**	**–66**	**–78**	**85**	**–9**	**20**	**35**	**–59**	**–60**

e: estimate; p: provisional.
Source: OECD.

RUSSIA

Definitions and notes to Table 4

PSE: Potatoes

I. **Level of production:** total production of potatoes (seed and food potatoes) at farm level, calendar year.[1]

II. **Production price (farm gate):** before 1991, these prices are the average state purchases prices (received by large-scale producers and household plots), as other marketing channels were insignificant. Between 1991 and 1996, these prices are weighted average producer prices received by large-scale farms, private farms, and households with weights based on their marketed output. In 1997, prices received by large-scale farms were applied, as the price information for households and private farms was not available at the time when 1997 calculations were made (for calculation, see following table).[1]

III. **Value of production:** (I)*(II).

V. **Direct payments**

VI. **Adjusted value of production:** III + V.

A. **Market price support:** for calculation, see following table.

B, C, D See notes to Table 12.

CSE: Potatoes

I. **Level of consumption:** total domestic use of potatoes, defined as production plus imports minus exports minus net change in stocks (balance sheets), calendar year.[2]

II. **Consumption price (farm gate):** implicit price measured at the farm gate; equal to the production price minus the sum of unit market price support and unit market transfers transfers [= Pp – (MPSu + MTu)]. As MPSu = –MTu, consumption price = average production price (see following table).

III. **Value of consumption:** (I)*(II).

A. **Market transfers:** for calculation, see following table.

B. **Other transfers:** 1986-1991: consumer subsidies paid to food processing industries, designed to maintain relatively low and stable prices for food.

Sources:

1. Goskomstat database. 1991-1995 production data published by Goskomstat in *Agriculture of the Russian Federation*, and *Russian Statistical Yearbook*.
2. Goskomstat database. 1991-1995 data published by Goskomstat in *Consumption of basic food products by population of the Russian Federation*.

Annex Table I.4.*iii.* **Market Price Support and Market Transfers: Potatoes**

	Units	1986	1987	1988	1989	1990	1991	1992	1993	1994	1995	1996p	1997e
PSE – MARKET PRICE SUPPORT (MPS)													
1. Total consumption	'000 t	40 887	38 742	36 514	33 966	32 212	33 829	37 194	38 405	36 558	37 406	38 015	38 015
2. Total production	'000 t	12 827	38 028	33 692	33 760	30 848	34 330	38 330	37 650	33 828	39 909	38 652	37 015
3. Production marketed by large-scale farms	'000 t						4 075	2 730	2 131	1 350	1 162	1 157	
4. Production marketed by households	'000 t						9 019	9 389	9 564	5 319	4 646	4 271	
5. Production marketed by private farms	'000 t						0	21	23	10	13	12	
6. Producer prices for large-scale farms	Rb/t	183	157	192	242	295	1 027	8 824	60 555	257 243	864 141	964 091	1 020 788
7. Producer prices for households	Rb/t						1 138	9 312	56 486	243 632	921 269	1 180 646	
8. Producer prices for private farms	Rb/t						0	10 905	102 314	742 126	979 277	1 245 437	
9. Average price at farm level	**Rb/t**	**183**	**157**	**192**	**242**	**295**	**1 103**	**9 205**	**57 316**	**247 129**	**909 998**	**1 134 733**	**1 020 788**
10. Handling margin	%	0	0	0	0	0	0	0	0	0	0	0	0
11. Adjusted domestic price	Rb/t	183	157	192	242	295	1 103	9 205	57 316	247 129	909 998	1 134 733	1 020 788
12. World reference price	US$/t	85	110	114	100	106	121	88	56	135	270	91	71
13. Official exchange rate	Rb/US$	0.60	0.60	0.60	0.60	0.78	1.75	193	932	2 204	4 554	5 124	5 785
14. World reference price	Rb/t	51	66	68	60	83	212	17 007	51 872	296 718	1 230 035	466 284	410 355
15. Price difference	Rb/t	132	91	124	182	212	892	–7 802	5 444	–49 589	–320 037	668 449	610 433
16. Percentage MPS	**%**	**72**	**58**	**64**	**75**	**72**	**81**	**–85**	**9**	**–20**	**–35**	**59**	**60**
17. Unit market price support	Rb/t	132	91	124	182	212	892	–7 802	5 444	–49 589	–320 037	668 449	610 433
18. Consumer transfers	Rb mn	1 694	3 459	4 166	6 144	6 543	30 172	–290 187	204 973	–1 677 494	–11 971 302	25 411 084	22 595 190
19. Budget transfers	Rb mn	0	0	0	0	0	447	–8 865	0	0	–801 052	425 802	0
20. Market price support	Rb mn	1 694	3 459	4 166	6 144	6 543	30 619	–299 053	204 973	–1 677 494	–12 772 355	25 836 885	22 595 190
CSE – MARKET TRANSFERS (MT)													
21. Unit market transfers	Rb/t	–132	–91	–124	–182	–212	–892	7 802	–5 444	49 589	320 037	–668 449	–610 433
22. Consumer transfers	Rb mn	–1 694	–3 459	–4 166	–6 144	–6 543	–30 172	290 187	–204 973	1 677 494	11 971 302	–25 411 084	–22 595 190
23. Budget transfers	Rb mn	–3 705	–65	–349	–37	–289	0	0	–4 111	135 383	0	0	–610 433
24. Market transfers	Rb mn	–5 398	–3 524	–4 515	–6 182	–6 833	–30 172	290 187	–209 084	1 812 877	11 971 302	–25 411 084	–23 205 623
25. Consumption price (farm level)	Rb/t	183	157	192	242	295	1 103	9 205	57 316	247 129	909 998	1 134 733	1 020 788

e: estimate; p: provisional.

Notes:

9 = (3*6 + 4*7 + 5*8)/(3 + 4 + 5)
11 = 9*(1 + (10/100))
12 = German price at farm gate.
14 = 12*13
15 = 11–14
16 = 15/11*100
17 = 16*9/100
18 = If 1 > 2 then 17*2; if 2 > 1 then 17*1
19 = If 1 > 2 then 0; if 2 > 1 then 17*(2 – 1)
20 = 18 + 19 or 17*2
21 = –17
22 = –18
23 = If 1 > 2 then 21*(1 – 2); if 2 > 1 then 0
24 = 22 + 23 or 21*1
25 = Consumption price (farm gate) = average producer price –(MPSu + MTu); as MPSu = –MTu consumption price = average producer price.

Source: OECD.

Annex Table I.5.*i.* **Producer Subsidy Equivalents: Sunflower**

	Units	1986	1987	1988	1989	1990	1991	1992	1993	1994	1995	1996p	1997e
I. Level of production	'000 t	2 363	3 067	2 958	3 789	3 427	2 896	3 110	2 765	2 553	4 200	2 765	2 824
II. Production price (farm gate)	Rb/t	272	363	375	507	428	837	17 799	72 062	260 457	829 762	791 645	784 303
III. Value of production	Rb mn	643	1 113	1 109	1 921	1 467	2 424	55 355	199 251	664 947	3 485 000	2 188 898	2 214 872
IV. Levies	Rb mn	0	0	0	0	0	0	0	0	0	0	0	0
V. Direct payments	Rb mn	11	15	0	0	0	99	171	442	681	17 468	17 642	209
VI. Adjusted value of production	Rb mn	653	1 128	1 109	1 921	1 467	2 523	55 526	199 693	665 627	3 502 469	2 206 540	2 215 081
VII. Gross total PSE	Rb mn	640	1 205	973	1 797	1 236	1 849	–30 282	–299 156	–236 478	217 150	59 513	–259 443
A. Market price support	Rb mn	453	895	767	1 494	1 019	1 624	–40 952	–344 097	–380 195	–148 675	–224 565	–429 962
B. Direct payments	Rb mn	11	15	0	0	0	99	171	442	681	17 468	17 642	209
C. Reduction of input costs	Rb mn	149	244	162	227	160	43	8 467	36 333	85 266	242 226	199 373	98 015
D. General services	Rb mn	27	51	44	76	57	83	2 032	8 166	57 771	106 130	67 064	72 296
VIII. Gross unit PSE	Rb/t	271	393	329	474	361	638	–9 737	–108 194	–92 627	51 702	21 524	–91 871
IX. Gross percentage PSE	**%**	**98**	**107**	**88**	**94**	**84**	**73**	**–55**	**–150**	**–36**	**6**	**3**	**–12**

e: estimate; p: provisional.
Source: OECD.

Annex Table I.5.*ii.* **Consumer Subsidy Equivalents: Sunflower**

	Units	1986	1987	1988	1989	1990	1991	1992	1993	1994	1995	1996p	1997e
I. Level of consumption	'000 t	3 153	3 070	2 948	3 719	3 322	2 705	3 055	2 285	2 075	2 630	2 025	2 000
II. Consumption price (farm gate)	Rb/t	272	363	375	507	428	837	17 799	72 062	260 457	829 762	791 645	784 303
III. Value of consumption	Rb mn	858	1 114	1 106	1 886	1 422	2 264	54 376	164 662	540 448	2 182 274	1 603 081	1 568 606
IV. Total CSE	Rb mn	–333	–352	88	–352	–339	–810	42 270	284 362	309 011	93 099	164 465	304 506
A. Market transfers	Rb mn	–604	–896	–765	–1 466	–988	–1 517	40 228	284 362	309 011	93 099	164 465	304 506
B. Other transfers	Rb mn	271	544	853	1 115	649	706	2 042	0	0	0	0	0
V. Unit CSE	Rb/t	–106	–115	30	–95	–102	–300	13 836	124 447	148 921	35 399	81 217	152 253
VI. Percentage CSE	**%**	**–39**	**–32**	**8**	**–19**	**–24**	**–36**	**78**	**173**	**57**	**4**	**10**	**19**

e: estimate; p: provisional.
Source: OECD.

RUSSIA

Definitions and notes to Table 5

PSE: Sunflower

I. **Level of production:** total sunflower production, calendar year.[1]

II. **Production price (farm gate):** before 1991, these prices are the average state purchases prices (received by large-scale producers and household plots), as other marketing channels were insignificant. As of 1991, average prices received by large-scale agricultural producers reflecting all marketing channels in the Goskomstat classification (procurement agencies, private traders, barter deals, payments in kind to farm workers, free market...).[1]

III. **Value of production:** (I)*(II).

V. **Direct payments**

VI. **Adjusted value of production:** III + V.

A. **Market price support:** for calculation, see following table.

B, C, D See notes to Table 12.

CSE: Sunflower

I. **Level of consumption:** total domestic use of sunflower (sum of crush, food use, feed, seed, and waste), marketing year.[2]

II. **Consumption price (farm gate):** implicit price measured at the farm gate; equal to the production price minus the sum of unit market price support and unit market transfers transfers [= Pp – (MPSu + MTu)]. As MPSu = –MTu, consumption price = average production price (see following table).

III. **Value of consumption:** (I)*(II).

A. **Market transfers:** for calculation, see following table.

B. **Other transfers:** 1986-1991: consumer subsidies paid to food processing industries, designed to maintain relatively low and stable prices for food. In 1992, consumer subsidies were applied to vegetable oil (on a regional basis); they were abolished in 1993.

Sources:

1. Goskomstat database. 1991-1995 production data published by Goskomstat in *Agriculture of the Russian Federation*, and *Russian Statistical Yearbook*.
2. USDA database (Production, Supply, and Distribution database).

Annex Table I.5.*iii.* **Market Price Support and Market Transfers: Sunflower**

	Units	1986	1987	1988	1989	1990	1991	1992	1993	1994	1995	1996p	1997e
PSE – MARKET PRICE SUPPORT (MPS)													
1. Total consumption	'000 t	3 153	3 070	2 948	3 719	3 322	2 705	3 055	2 285	2 075	2 630	2 025	2 000
2. Total production	'000 t	2 363	3 067	2 958	3 789	3 427	2 896	3 110	2 765	2 553	4 200	2 765	2 824
3. Average price at farm level	**Rb/t**	**272**	**363**	**375**	**507**	**428**	**837**	**17 799**	**72 062**	**260 457**	**829 762**	**791 645**	**784 303**
4. Handling margin	%	20	20	20	20	20	25	40	45	50	50	50	50
5. Adjusted domestic price	Rb/t	326	436	450	608	514	1 046	24 919	104 490	390 686	1 244 643	1 187 468	1 176 455
6. World reference price	ECU/t	164	123	196	205	158	160	174	261	235	218	201	212
7. Official exchange rate	Rb/ECU	0.59	0.69	0.71	0.66	0.99	2.16	249	1 092	2 614	5 953	6 504	6 627
8. World reference price	Rb/t	96	85	139	135	157	345	43 354	284 939	614 067	1 297 741	1 309 293	1 404 834
9. Price difference	Rb/t	230	350	311	473	357	701	–18 435	–180 449	–223 381	–53 098	–121 826	–228 379
10. Percentage MPS	**%**	**70**	**80**	**69**	**78**	**69**	**67**	**–74**	**–173**	**–57**	**–4**	**–10**	**–19**
11. Unit market price support	Rb/t	192	292	259	394	297	561	–13 168	–124 447	–148 921	–35 399	–81 217	–152 253
12. Consumer transfers	Rb mn	453	895	765	1 466	988	1 517	–40 228	–284 362	–309 011	–93 099	–164 465	–304 506
13. Budget transfers	Rb mn	0	0	3	28	31	107	–724	–59 735	–71 184	–55 576	–60 101	–125 456
14. Market price support	Rb mn	453	895	767	1 494	1 019	1 624	–40 952	–344 097	–380 195	–148 675	–224 565	–429 962
CSE – MARKET TRANSFERS (MT)													
15. Unit market transfers	Rb/t	–192	–292	–259	–394	–297	–561	13 168	124 447	148 921	35 399	81 217	152 253
16. Consumer transfers	Rb mn	–453	–895	–765	–1 466	–988	–1 517	40 228	284 362	309 011	93 099	164 465	304 506
17. Budget transfers	Rb mn	–151	–1	0	0	0	0	0	0	0	0	0	0
18. Market transfers	Rb mn	–604	–896	–765	–1 466	–988	–1 517	40 228	284 362	309 011	93 099	164 465	304 506
19. Consumption price (farm level)	Rb/t	272	363	375	507	428	837	17 799	72 062	260 457	829 762	791 645	784 303

e: estimate; p: provisional.

Notes:

5 = 3*(1 + (4/100))
6 = EU import price.
8 = 6*7
9 = 5 – 8
10 = 9/5*100
11 = 10*3/100
12 = If 1 > 2 then 11*2; if 2 > 1 then 11*1
13 = If 1 > 2 then 0; if 2 > 1 then 11*(2 – 1)
14 = 12 + 13 or 11*2
15 = –11
16 = –12
17 = If 1 > 2 then 15*(1 – 2); if 2 > 1 then 0
18 = 16 + 17 or 15*1
19 = Consumption price (farm gate) = average producer price –(MPSu + MTu); as MPSu = –MTu consumption price = average producer price.

Source: OECD.

Annex Table I.6.*i.* **Producer Subsidy Equivalents: Refined Sugar**

	Units	1986	1987	1988	1989	1990	1991	1992	1993	1994	1995	1996p	1997e
I. Level of production	'000 t	3 159	3 860	3 315	3 749	3 207	2 217	2 688	2 987	1 764	2 346	1 455	1 566
II. Production price (farm gate)	Rb/t	619	602	703	728	696	909	24 306	201 420	441 550	1 164 577	2 100 836	1 735 732
III. Value of production	Rb mn	1 956	2 323	2 331	2 729	2 231	2 015	65 326	601 722	778 969	2 731 929	3 056 590	2 718 746
IV. Levies	Rb mn	0	0	0	0	0	0	0	0	0	0	0	0
V. Direct payments	Rb mn	33	31	0	0	0	83	202	1 334	797	13 694	24 635	256
VI. Adjusted value of production	Rb mn	1 989	2 353	2 331	2 729	2 231	2 098	65 527	603 056	779 767	2 745 622	3 081 225	2 719 003
VII. Gross total PSE	Rb mn	2 329	2 720	2 470	2 686	2 024	1 564	10 547	358 436	290 336	1 073 640	2 215 637	1 560 056
A. Market price support	Rb mn	1 760	2 074	2 038	2 255	1 694	1 377	–2 305	222 494	114 428	713 691	1 755 625	1 342 231
B. Direct payments	Rb mn	33	31	0	0	0	83	202	1 334	797	13 694	24 635	256
C. Reduction of input costs	Rb mn	453	509	341	323	243	36	10 252	109 949	107 434	263 058	341 729	128 826
D. General services	Rb mn	83	106	92	108	87	69	2 398	24 659	67 677	83 197	93 648	88 743
VIII. Gross unit PSE	Rb/t	737	705	745	716	631	706	3 924	119 983	164 574	457 675	1 522 837	995 988
IX. Gross percentage PSE	%	**117**	**116**	**106**	**98**	**91**	**75**	**16**	**59**	**37**	**39**	**72**	**57**

e: estimate; p: provisional.
Source: OECD.

Annex Table I.6.*ii*. **Consumer Subsidy Equivalents: Refined sugar**

	Units	1986	1987	1988	1989	1990	1991	1992	1993	1994	1995	1996p	1997e
I. Level of consumption	'000 t	6 850	7 134	7 432	6 707	7 024	5 637	4 478	4 677	4 601	4 711	4 828	4 828
II. Consumption price (farm gate)	Rb/t	619	602	703	728	696	909	24 306	201 420	441 550	1 164 577	2 100 836	1 735 732
III. Value of consumption	Rb mn	4 241	4 293	5 225	4 882	4 886	5 125	108 849	942 063	2 031 529	5 486 087	10 142 835	8 380 114
IV. Total CSE	Rb mn	–2 214	–2 272	–2 025	–1 058	–1 596	–2 786	3 840	–348 339	–298 424	–1 433 189	–5 825 777	–4 137 219
A. Market transfers	Rb mn	–3 817	–3 833	–4 569	–4 035	–3 711	–3 502	3 840	–348 339	–298 424	–1 433 189	–5 825 777	–4 137 219
B. Other transfers	Rb mn	1 603	1 560	2 544	2 977	2 116	717	0	0	0	0	0	0
V. Unit CSE	Rb/t	–323	–319	–272	–158	–227	–494	857	–74 478	–64 862	–304 235	–1 206 665	–856 922
VI. Percentage CSE	**%**	**–52**	**–53**	**–39**	**–22**	**–33**	**–54**	**4**	**–37**	**–15**	**–26**	**–57**	**–49**

e: estimate; p: provisional.
Source: OECD.

RUSSIA

Definitions and notes to Table 6

PSE: Refined sugar

This table sets out the calculation for sugar beet (Table 6A, given below) expressed as white sugar equivalent.

I. **Level of production:** quantity of white sugar obtained from sugar beet production, calendar year.[1]

II. **Production price (farm gate):** implicit price for beet in white sugar equivalent, obtained by dividing the production value of sugar beet (Table 6.iii) by the level of sugar production (I) (see following table).[1]

III. **Value of production:** (I)*(II).

V. **Direct payments**

VI. **Adjusted value of production:** III + V.

A. **Market price support:** for calculation, see following table.

B, C, D See notes to Table 12.

CSE: Refined sugar

I. **Level of consumption:** total domestic use of white sugar defined as production plus imports minus exports minus net change in stocks (balance sheets), calendar year.[2]

II. **Consumption price (farm gate):** implicit price measured at the farm gate; equal to the production price minus the sum of unit market price support and unit market transfers transfers [= Pp – (MPSu + MTu)]. As MPSu = –MTu, consumption price = average production price (see following table).

III. **Value of consumption:** (I)*(II).

A. **Market transfers:** for calculation, see following table.

B. **Other transfers:** 1986-1991: consumer subsidies paid to food processing industries, designed to maintain relatively low and stable prices for food.

Sources:

1. Goskomstat database.
2. Goskomstat. *Consumption of basic food products by population of the Russian Federation.*

Annex Table I.6.*iii.* **Market Price Support and Market Transfers: Sugar beet/Refined sugar**

	Units	1986	1987	1988	1989	1990	1991	1992	1993	1994	1995	1996p	1997e
PSE – MARKET PRICE SUPPORT (MPS)													
1. Total sugar consumption	'000 t	6 850	7 134	7 432	6 707	7 024	5 637	4 478	4 677	4 601	4 711	4 828	4 828
2. Coefficient (beet to sugar)	%	10.8	11.3	10.1	10.0	9.9	9.1	10.5	11.7	12.7	12.3	9.0	11.3
3. Total consumption in terms of beet	'000 t	63 306	63 130	73 586	66 871	70 807	61 746	42 569	39 873	36 371	38 299	53 644	42 663
4. Production of white sugar	'000 t	3 159	3 860	3 315	3 749	3 207	2 217	2 688	2 987	1 764	2 346	1 455	1 566
5. Total beet production	'000 t	29 200	34 156	32 824	37 378	32 327	24 280	25 548	25 468	13 946	19 072	16 166	13 841
6. Production marketed by large-scale farms (beet)	'000 t						18 662	10 976	7 143	1 312	2 071	2 375	
7. Production marketed by households (beet)	'000 t						1	0	0	0	0	0	
8. Production marketed by private farms (beet)	'000 t						0	61	99	5	3	3	
9. Producer prices for large-scale farms (beet)	Rb/t	67	68	71	73	69	83	2 555	23 632	55 877	143 259	189 067	196 427
10. Producer prices for households (beet)	Rb/t						98	3 603	25 385	0	117 154	270 028	
11. Producer prices for private farms (beet)	Rb/t						0	2 911	23 234	50 661	132 996	195 081	
12. Average price at farm level (beet)	**Rb/t**	**67**	**68**	**71**	**73**	**69**	**83**	**2 557**	**23 627**	**55 856**	**143 243**	**189 075**	**196 427**
13. Value of production	Rb mn	1 956	2 323	2 331	2 729	2 231	2 015	65 326	601 722	778 969	2 731 929	3 056 590	2 718 746
14. Processing margin factor		1.8	1.8	1.8	1.8	1.8	1.8	2.1	2.1	2.1	2.1	2.1	2.1
15. World reference price of white sugar	ECU/t	190	168	224	344	304	240	212	244	303	304	289	279
16. Official exchange rate	Rb/ECU	0.59	0.69	0.71	0.66	0.99	2.16	249	1 092	2 614	5 953	6 504	6 627
17. World reference price of white sugar	Rb/t	112	116	159	227	301	518	52 843	266 580	791 045	1 806 718	1 877 759	1 845 501
18. (11) in terms of beet at farm level	Rb/t	7	7	9	13	17	26	2 647	14 890	47 651	105 822	80 475	99 452
19. Price difference (beet)	Rb/t	60	61	62	60	52	57	–90	8 736	8 205	37 421	108 600	96 975
20. Percentage MPS	**%**	**90**	**89**	**87**	**83**	**76**	**68**	**–4**	**37**	**15**	**26**	**57**	**49**
21. Unit market price support	Rb/t	60	61	62	60	52	57	–90	8 736	8 205	37 421	108 600	96 975
22. Consumer transfers	Rb mn	1 760	2 074	2 038	2 255	1 694	1 377	–2 305	222 494	114 428	713 691	1 755 625	1 342 231
23. Budget transfers	Rb mn	0	0	0	0	0	0	0	0	0	0	0	0
24. Market price support	Rb mn	1 760	2 074	2 038	2 255	1 694	1 377	–2 305	222 494	114 428	713 691	1 755 625	1 342 231
CSE – MARKET TRANSFERS (MT)													
25. Unit market transfers	Rb/t	–60	–61	–62	–60	–52	–57	90	–8 736	–8 205	–37 421	–108 600	–96 975
26. Consumer transfers	Rb mn	–1 760	–2 074	–2 038	–2 255	–1 694	–1 377	2 305	–222 494	–114 428	–713 691	–1 755 625	–1 342 231
27. Budget transfers	Rb mn	–2 056	–1 759	–2 531	–1 780	–2 017	–2 125	1 535	–125 845	–183 996	–719 498	–4 070 152	–2 794 988
28. Market transfers	Rb mn	–3 817	–3 833	–4 569	–4 035	–3 711	–3 502	3 840	–348 339	–298 424	–1 433 189	–5 825 777	–4 137 219
29. Consumption price (farm level) (refined sugar)	Rb/t	619	602	703	728	696	909	24 306	201 420	441 550	1 164 577	2 100 836	1 735 732

e: estimate; p: provisional.

Notes:

3 = 1/(2/100)
4 = 5*(2/100)
12 = (6*9 + 7*10 + 8*11)/(6 + 7 + 8)
13 = 5*12
15 = EU export price of white sugar.
17 = 15*16
18 = 17*(2/100)/14
19 = 12 – 18
20 = 19/12*100
21 = 20*12/100
22 = If 3 > 5 then 21*5; if 5 > 3 then 21*3
23 = If 3 > 5 then 0; if 5 > 3 then 21*(5 – 3)
24 = 22 + 23 or 21*5
25 = –21
26 = –22
27 = If 3 > 5 then 25*(3 – 5); if 5 > 3 then 0
28 = 26 + 27 or 25*3
29 = Consumption price (farm gate) = average producer price –(MPSu + MTu); as MPSu = –MTu consumption price = average producer price.

Source: OECD.

RUSSIA

Definitions and notes to Table 6A

PSE: Sugar beet

I. **Level of production:** total production of sugar beet, calendar year.[1]

II. **Production price (farm gate):** before 1991, these prices are the average state purchases prices (received by large-scale producers and household plots), as other marketing channels were insignificant. Between 1991 and 1996, these prices are weighted average producer prices received by large-scale farms, private farms, and households with weights based on their marketed output. In 1997, prices received by large-scale farms were applied, as the price information for households and private farms was not available at the time when 1997 calculations were made (for calculation, see following table).[1]

III. **Value of production:** (I)*(II).

V. **Direct payments**

VI. **Adjusted value of production:** III + V.

A. **Market price support:** for calculation, see following table.

B, C, D See notes to Table 12.

Sources:

1. Goskomstat database.
1991-1995 production data published by Goskomstat in *Agriculture of the Russian Federation*, and *Russian Statistical Yearbook*.

Annex Table I.6.A.*i.* **Producer Subsidy Equivalents: Sugar beet**

	Units	1986	1987	1988	1989	1990	1991	1992	1993	1994	1995	1996p	1997e
I. Level of production	'000 t	29 200	34 156	32 824	37 378	32 327	24 280	25 548	25 468	13 946	19 072	16 166	13 841
II. Production price (farm gate)	Rb/t	67	68	71	73	69	83	2 557	23 627	55 856	143 243	189 075	196 427
III. Value of production	Rb mn	1 956	2 323	2 331	2 729	2 231	2 015	65 326	601 722	778 969	2 731 929	3 056 590	2 718 746
IV. Levies	Rb mn	0	0	0	0	0	0	0	0	0	0	0	0
V. Direct payments	Rb mn	33	31	0	0	0	83	202	1 334	797	13 694	24 635	256
VI. Adjusted value of production	Rb mn	1 989	2 353	2 331	2 729	2 231	2 098	65 527	603 056	779 767	2 745 622	3 081 225	2 719 003
VII. Gross total PSE	Rb mn	2 329	2 720	2 470	2 686	2 024	1 564	10 547	358 436	290 336	1 073 640	2 215 637	1 560 056
A. Market price support	Rb mn	1 760	2 074	2 038	2 255	1 694	1 377	–2 305	222 494	114 428	713 691	1 755 625	1 342 231
B. Direct payments	Rb mn	33	31	0	0	0	83	202	1 334	797	13 694	24 635	256
C. Reduction of input costs	Rb mn	453	509	341	323	243	36	10 252	109 949	107 434	263 058	341 729	128 826
D. General services	Rb mn	83	106	92	108	87	69	2 398	24 659	67 677	83 197	93 648	88 743
VIII. Gross unit PSE	Rb/t	80	80	75	72	63	64	413	14 074	20 819	56 294	137 055	112 713
IX. Gross percentage PSE	**%**	**117**	**116**	**106**	**98**	**91**	**75**	**16**	**59**	**37**	**39**	**72**	**57**

e: estimate; p: provisional.
Source: OECD.

Annex Table I.7.*i.* **Producer Subsidy Equivalents: Milk**

	Units	1986	1987	1988	1989	1990	1991	1992	1993	1994	1995	1996p	1997e
I. Level of production	'000 t	52 217	52 880	54 535	55 742	55 715	51 887	47 236	46 524	42 176	39 241	35 819	34 066
II. Production price (farm gate)	Rb/t	479	484	644	653	654	806	5 671	49 335	157 976	808 912	1 056 774	1 109 559
III. Value of production	Rb mn	25 012	25 594	35 121	36 400	36 438	41 821	267 871	2 295 263	6 662 807	31 742 502	37 852 583	37 798 237
IV. Levies	Rb mn	0	0	0	0	0	0	0	0	0	0	0	0
V. Direct payments	Rb mn	940	899	0	0	0	1 716	68 030	391 251	600 801	2 176 339	2 153 861	1 283 308
VI. Adjusted value of production	Rb mn	25 952	26 493	35 121	36 400	36 438	43 538	335 901	2 686 514	7 263 608	33 918 841	40 006 444	39 081 545
VII. Gross total PSE	Rb mn	29 489	29 025	36 801	37 479	37 161	36 223	–632 692	–1 565 532	–1 966 430	12 389 483	18 074 905	15 872 825
A. Market price support	Rb mn	23 067	22 725	31 134	32 105	31 995	32 337	–772 644	–2 505 523	–4 182 906	6 456 748	10 333 034	11 061 366
B. Direct payments	Rb mn	940	899	0	0	0	1 716	68 030	391 251	600 801	2 176 339	2 153 861	1 283 308
C. Reduction of input costs	Rb mn	4 367	4 184	4 238	3 895	3 679	662	60 135	430 827	929 580	2 514 250	4 087 302	1 904 288
D. General services	Rb mn	1 115	1 217	1 429	1 479	1 487	1 508	11 787	117 913	686 094	1 242 146	1 500 709	1 623 862
VIII. Feed adjustment	Rb mn	–1 974	–1 655	–2 317	–3 007	–3 992	–8 502	113 420	379 537	639 456	1 402 973	157 239	–186 849
IX. Net total PSE	Rb mn	27 514	27 370	34 484	34 472	33 169	27 722	–519 272	–1 185 995	–1 326 974	13 792 456	18 232 144	15 685 976
X. Net unit PSE	Rb/t	527	518	632	618	595	534	–10 993	–25 492	–31 463	351 481	509 008	460 458
XI. Net percentage PSE	%	**106**	**103**	**98**	**95**	**91**	**64**	**–155**	**–44**	**–18**	**41**	**46**	**40**

e: estimate; p: provisional.
Source: OECD.

Annex Table I.7.*ii.* **Consumer Subsidy Equivalents: Milk**

	Units	1986	1987	1988	1989	1990	1991	1992	1993	1994	1995	1996p	1997e
I. Level of consumption	'000 t	58 811	60 023	63 527	65 701	64 608	58 938	49 617	51 145	48 516	44 549	40 954	39 289
II. Consumption price (farm gate)	Rb/t	479	484	644	653	654	806	5 671	49 335	157 976	808 912	1 056 774	1 109 559
III. Value of consumption	Rb mn	28 170	29 051	40 912	42 903	42 254	47 505	281 374	2 523 255	7 664 423	36 036 205	43 279 117	43 593 464
IV. Total CSE	Rb mn	–15 774	–15 287	–21 717	–21 588	–22 256	–16 706	835 993	2 832 238	4 974 164	–7 330 131	–11 814 375	–12 757 295
A. Market transfers	Rb mn	–25 980	–25 795	–36 268	–37 841	–37 102	–36 732	811 592	2 754 400	4 811 720	–7 330 131	–11 814 375	–12 757 295
B. Other transfers	Rb mn	10 205	10 508	14 550	16 254	14 846	20 025	24 401	77 838	162 444	0	0	0
V. Unit CSE	Rb/t	–268	–255	–342	–329	–344	–283	16 849	55 376	102 526	–164 541	–288 479	–324 704
VI. Percentage CSE	%	**–56**	**–53**	**–53**	**–50**	**–53**	**–35**	**297**	**112**	**65**	**–20**	**–27**	**–29**

e: estimate; p: provisional.
Source: OECD.

RUSSIA

Definitions and notes to Table 7

PSE: Milk

I. **Level of production:** total milk production (cow milk-including suckled milk- and also milk from goats, sheep, buffalo, horse and camel), calendar year.[1]

II. **Production price (farm gate):** before 1991, these prices are the average state purchases prices (received by large-scale producers and household plots), as other marketing channels were insignificant. Between 1991 and 1996, these prices are weighted average producer prices received by large-scale farms, private farms, and households with weights based on their marketed output. In 1997, prices received by large-scale farms were applied, as the price information for households and private farms was not available at the time when 1997 calculations were made (for calculation, see following table).[1]

III. **Value of production:** (I)*(II).

V. **Direct payments**

VI. **Adjusted value of production:** III + V.

A. **Market price support:** for calculation, see following table.

B, C, D See notes to Table 12.

CSE: Milk

I. **Level of consumption:** total domestic use of milk and dairy products expressed as milk equivalents, calendar year; coefficient for butter: 20.3; hard cheese: 8.8; processed cheese: 7.7; canned milk: 1; dry milk products: 7.3; whole milk products: 0.92. Total domestic use defined as production plus imports minus exports minus net change in stocks (balance sheets).[2]

II. **Consumption price (farm gate):** implicit price measured at the farm gate; equal to the production price minus the sum of unit market price support and unit market transfers transfers [= Pp – (MPSu + MTu)]. As MPSu = –MTu, consumption price = average production price (see following table).

III. **Value of consumption:** (I)*(II).

A. **Market transfers:** for calculation, see following table.

B. **Other transfers:** 1986-1991: consumer subsidies paid to food processing industries (and also in some cases to agricultural producers delivering milk and cream directly to retailers), designed to maintain relatively low and stable prices for food. In 1992-1994, consumer subsidies were maintained for milk and milk products (on a regional basis). In 1995, they became negligible.

Sources:

1. Goskomstat database. 1991-1995 production data published by Goskomstat in *Agriculture of the Russian Federation*, and *Russian Statistical Yearbook*.

2. Goskomstat database. 1991-1995 data published by Goskomstat in *Consumption of basic food products by population of the Russian Federation*.

Annex Table I.7.*iii.* **Market Price Support and Market Transfers: Milk**

	Units	1986	1987	1988	1989	1990	1991	1992	1993	1994	1995	1996p	1997e
PSE – MARKET PRICE SUPPORT (MPS)													
1. Total consumption	'000 t	58 811	60 023	63 527	65 701	64 608	58 938	49 617	51 145	48 516	44 549	40 954	39 289
2. Total production	'000 t	52 217	52 880	54 535	55 742	55 715	51 887	47 236	46 524	42 176	39 241	35 819	34 066
3. Production marketed by large-scale farms	'000 t						34 594	26 983	25 271	19 690	17 144	14 452	
4. Production marketed by households	'000 t						3 545	3 108	3 295	4 629	3 233	3 268	
5. Production marketed by private farms	'000 t						0	80	121	85	81	141	
6. Producer prices for large-scale farms	Rb/t	479	484	644	653	654	818	5 684	48 644	162 420	757 402	942 127	1 109 559
7. Producer prices for households	Rb/t						689	5 555	54 444	138 476	1 075 503	1 542 805	
8. Producer prices for private farms	Rb/t						0	5 758	54 534	190 557	1 070 538	1 545 548	
9. Average price at farm level	**Rb/t**	**479**	**484**	**644**	**653**	**654**	**806**	**5 671**	**49 335**	**157 976**	**808 912**	**1 056 774**	**1 109 559**
10. Fat content – domestic	%	3.2	3.2	3.2	3.2	3.2	3.2	3.2	3.2	3.2	3.2	3.2	3.2
11. Fat content – New Zealand	%	4.6	4.6	4.7	4.6	4.7	4.7	4.7	4.7	4.8	4.7	4.7	4.71
12. Transport cost milk equivalent	US$/t	15	18	18	17	18	18	17	18	18	17	20	20
13. Price of milk farmgate – New Zealand	US$/t	68	105	151	162	122	129	144	138	147	184	191	170
14. N-Z price + transport adjustment to domestic fat content	US$/t	62	90	122	128	102	105	114	111	117	141	150	136
15. Official exchange rate	Rb/US$	0.60	0.60	0.60	0.60	0.78	1.75	193	932	2 204	4 554	5 124	5 785
16. World reference price	Rb/t	37	54	73	77	80	183	22 028	103 189	257 154	644 371	768 295	784 855
17. Price difference	Rb/t	442	430	571	576	574	623	–16 357	–53 854	–99 177	164 541	288 479	324 704
18. Percentage MPS	**%**	**92**	**89**	**89**	**88**	**88**	**77**	**–288**	**–109**	**–63**	**20**	**27**	**29**
19. Unit market price support	Rb/t	442	430	571	576	574	623	–16 357	–53 854	–99 177	164 541	288 479	324 704
20. Consumer transfers	Rb mn	23 067	22 725	31 134	32 105	31 995	32 337	–772 644	–2 505 523	–4 182 906	6 456 748	10 333 034	11 061 366
21. Budget transfers	Rb mn	0	0	0	0	0	0	0	0	0	0	0	0
22. Market price support	Rb mn	23 067	22 725	31 134	32 105	31 995	32 337	–772 644	–2 505 523	–4 182 906	6 456 748	10 333 034	11 061 366
CSE – MARKET TRANSFERS (MT)													
23. Unit market transfers	Rb/t	–442	–430	–571	–576	–574	–623	16 357	53 854	99 177	–164 541	–288 479	–324 704
24. Consumer transfers	Rb mn	–23 067	–22 725	–31 134	–32 105	–31 995	–32 337	772 644	2 505 523	4 182 906	–6 456 748	–10 333 034	–11 061 366
25. Budget transfers	Rb mn	–2 913	–3 070	–5 134	–5 736	–5 107	–4 394	38 948	248 877	628 814	–873 383	–1 481 340	–1 695 929
26. Market transfers	Rb mn	–25 980	–25 795	–36 268	–37 841	–37 102	–36 732	811 592	2 754 400	4 811 720	–7 330 131	–11 814 375	–12 757 295
27. Consumption price (farm level)	Rb/t	479	484	644	653	654	806	5 671	49 335	157 976	808 912	1 056 774	1 109 559

e: estimate; p: provisional.

Notes:

9 = (3*6 + 4*7 + 5*8)/(3 + 4 + 5)

12 = Transport cost from New Zealand to United Kingdom of butter and SMP adjusted to milk equivalent on the basis of New Zealand product yields *i.e.* one tonne of milk yields 56 kg of butter and 82 kg of skim milk powder.

14 = ((13/11)*10) + 12

16 = 14*15

17 = 9 – 16

18 = 17/9*100

19 = 17; as New Zealand price is at farm gate price difference = MPSu.

20 = If 1 > 2 then 19*2; if 2 > 1 then 19*1.

21 = If 1 > 2 then 0; if 2 > 1 then 19*(2 – 1)

22 = 20 + 21 or 19*2

23 = –19

24 = –20

25 = If 1 > 2 then 23*(1 – 2); if 2 > 1 then 0.

26 = 24 + 25 or 23*1

27 = Consumption price (farm gate) = average producer price –(MPSu + MTu); as MPSu = –MTu consumption price = average producer price.

Source: OECD.

Annex Table I.8.*i.* **Producer Subsidy Equivalents: Beef and veal**

	Units	1986	1987	1988	1989	1990	1991	1992	1993	1994	1995	1996p	1997e
I. Level of production	'000 t	3 756	3 991	4 150	4 256	4 329	3 989	3 632	3 359	3 240	2 733	2 630	2 338
II. Production price (farm gate)	Rb/t	4 480	4 639	5 550	5 697	6 245	8 702	41 556	502 401	1 292 774	4 736 274	7 017 231	6 911 252
III. Value of production	Rb mn	16 827	18 514	23 033	24 246	27 035	34 711	150 933	1 687 566	4 188 588	12 944 236	18 455 318	16 158 507
IV. Levies	Rb mn	0	0	0	0	0	0	0	0	0	0	0	0
V. Direct payments	Rb mn	632	650	0	0	0	1 424	44 797	217 313	448 913	712 732	794 482	444 931
VI. Adjusted value of production	Rb mn	17 459	19 165	23 033	24 246	27 035	36 135	195 730	1 904 879	4 637 502	13 656 969	19 249 800	16 603 438
VII. Gross total PSE	Rb mn	19 534	20 923	24 207	25 019	26 345	30 330	–489 285	–1 521 644	–3 375 667	–4 635 626	702 498	–1 377 430
A. Market price support	Rb mn	15 214	16 366	20 490	21 439	22 512	27 105	–582 276	–2 143 194	–4 840 277	–6 880 174	–2 816 462	–3 330 623
B. Direct payments	Rb mn	632	650	0	0	0	1 424	44 797	217 313	448 913	712 732	794 482	444 931
C. Reduction of input costs	Rb mn	2 938	3 026	2 779	2 595	2 730	549	41 552	317 543	584 383	1 025 283	1 992 796	814 071
D. General services	Rb mn	750	881	937	985	1 103	1 252	6 642	86 694	431 315	506 533	731 682	694 191
VIII. Feed adjustment	Rb mn	–1 671	–1 394	–1 976	–2 567	–3 439	–7 159	93 063	297 228	486 812	1 002 124	106 205	–126 205
IX. Net total PSE	Rb mn	17 863	19 530	22 231	22 451	22 906	23 171	–396 222	–1 224 416	–2 888 855	–3 633 502	808 703	–1 503 635
X. Net unit PSE	Rb/t	4 756	4 893	5 357	5 275	5 291	5 809	–109 092	–364 518	–891 622	–1 329 492	307 492	–643 129
XI. Net percentage PSE	**%**	**102**	**102**	**97**	**93**	**85**	**64**	**–202**	**–64**	**–62**	**–27**	**4**	**–9**

e: estimate; p: provisional.
Source: OECD.

Annex Table I.8.*ii.* **Consumer Subsidy Equivalents: Beef and veal**

	Units	1986	1987	1988	1989	1990	1991	1992	1993	1994	1995	1996p	1997e
I. Level of consumption	'000 t	4 521	4 812	5 224	5 335	4 905	5 060	4 165	3 797	3 791	3 402	3 055	2 795
II. Consumption price (farm gate)	Rb/t	4 480	4 639	5 550	5 697	6 245	8 702	41 556	502 401	1 292 774	4 736 274	7 017 231	6 911 252
III. Value of consumption	Rb mn	20 254	22 323	28 993	30 393	30 632	44 030	173 082	1 907 618	4 900 907	16 112 804	21 437 641	19 316 949
IV. Total CSE	Rb mn	–10 598	–12 272	–13 304	–14 967	–7 577	–17 733	672 948	2 422 658	5 663 423	8 564 344	3 271 594	3 981 648
A. Market transfers	Rb mn	–18 312	–19 733	–25 793	–26 874	–25 508	–34 382	667 726	2 422 658	5 663 423	8 564 344	3 271 594	3 981 648
B. Other transfers	Rb mn	7 715	7 461	12 489	11 907	17 930	16 649	5 222	0	0	0	0	0
V. Unit CSE	Rb/t	–2 344	–2 550	–2 547	–2 805	–1 545	–3 505	161 572	638 045	1 493 913	2 517 444	1 070 898	1 424 561
VI. Percentage CSE	**%**	**–52**	**–55**	**–46**	**–49**	**–25**	**–40**	**389**	**127**	**116**	**53**	**15**	**21**

e: estimate; p: provisional.
Source: OECD.

RUSSIA

Definitions and notes to Table 8

PSE: Beef and veal

I. **Level of production:** total production of beef and veal (but including also buffalo's meat), carcass weight, calendar year.[1]

II. **Production price (farm gate):** before 1991, these prices are the average state purchases prices (received by large-scale producers and household plots), as other marketing channels were insignificant. Between 1991 and 1996, these prices are weighted average producer prices received by large-scale farms, private farms, and households with weights based on their marketed output. In 1997, prices received by large-scale farms were applied, as the price information for households and private farms was not available at the time when 1997 calculations were made (for calculation, see following table).[1]

III. **Value of production:** (I)*(II).

V. **Direct payments**

VI. **Adjusted value of production:** III + V.

A. **Market price support:** for calculation, see following table.

B, C, D See notes to Table 12.

CSE: Beef and veal

I. **Level of consumption:** total domestic use of beef and veal, carcass weight, calendar year.[2]

II. **Consumption price (farm gate):** implicit price measured at the farm gate; equal to the production price minus the sum of unit market price support and unit market transfers transfers [= Pp – (MPSu + MTu)]. As MPSu = –MTu, consumption price = average production price (see following table).

III. **Value of consumption:** (I)*(II).

A. **Market transfers:** for calculation, see following table.

B. **Other transfers:** 1986-1991: consumer subsidies paid to food processing industries, designed to maintain relatively low and stable prices for food. In 1992, consumer subsidies were maintained for meat products on a regional basis -they were allocated between livestock commodities in proportion to their share in total value of purchases of meat products concerned. In 1993, they were abolished.

Sources:

1. Goskomstat database. 1991-1995 production data published by Goskomstat in *Agriculture of the Russian Federation*, and *Russian Statistical Yearbook*.
2. USDA database (Production, Supply, and Distribution database).

Annex Table I.8.*iii.* **Market Price Support and Market Transfers: Beef and veal**

	Units	1986	1987	1988	1989	1990	1991	1992	1993	1994	1995	1996p	1997e
PSE – MARKET PRICE SUPPORT (MPS)													
1. Total consumption (carcass)	'000 t	4 521	4 812	5 224	5 335	4 905	5 060	4 165	3 797	3 791	3 402	3 055	2 795
2. Total production (live)	'000 t	6 381	6 735	7 027	7 214	7 331	6 770	6 210	5 773	5 577	4 783	4 579	4 068
3. Total production (carcass)	'000 t	3 756	3 991	4 150	4 256	4 329	3 989	3 632	3 359	3 240	2 733	2 630	2 338
4. Production marketed by large-scale farms (carcass)	'000 t						3 226	2 722	2 234	1 903	1 423	1 322	
5. Production marketed by households (carcass)	'000 t						314	354	361	589	415	424	
6. Production marketed by private farms (carcass)	'000 t						0	12	19	10	7	13	
7. Producer prices for large-scale farms (carcass)	Rb/t	4 480	4 639	5 550	5 697	6 245	8 588	40 079	471 014	1 246 816	4 279 896	5 722 380	6 911 252
8. Producer prices for households (carcass)	Rb/t						9 871	52 658	688 238	1 434 280	6 285 384	10 904 128	
9. Producer prices for private farms (carcass)	Rb/t						0	49 013	667 618	1 707 676	5 705 657	12 028 680	
10. Average price at farm level (carcass)	**Rb/t**	**4 480**	**4 639**	**5 550**	**5 697**	**6 245**	**8 702**	**41 556**	**502 401**	**1 292 774**	**4 736 274**	**7 017 231**	**6 911 252**
11. Handling margin	%	8	8	8	8	8	15	30	40	50	50	50	50
12. Adjusted domestic price (carcass)	Rb/t	4 838	5 010	5 994	6 153	6 745	10 007	54 023	703 362	1 939 161	7 104 411	10 525 847	10 366 878
13. World reference price (carcass)	US$/t	773	969	1 103	1 187	1 445	1 256	1 363	1 713	1 897	2 389	2 368	2 161
14. Official exchange rate	Rb/US$	0.60	0.60	0.60	0.60	0.78	1.75	193	932	2 204	4 554	5 124	5 785
15. World reference price (carcass)	Rb/t	464	581	662	712	1 128	2 193	262 437	1 596 625	4 180 030	10 880 577	12 132 194	12 503 719
16. Price difference	Rb/t	4 375	4 429	5 332	5 440	5 616	7 814	–208 414	–893 264	–2 240 869	–3 776 166	–1 606 347	–2 136 841
17. Percentage MPS	**%**	**90**	**88**	**89**	**88**	**83**	**78**	**–386**	**–127**	**–116**	**–53**	**–15**	**–21**
18. Unit market price support	Rb/t	4 051	4 101	4 937	5 037	5 200	6 795	–160 318	–638 045	–1 493 913	–2 517 444	–1 070 898	–1 424 561
19. Consumer transfers	Rb mn	15 214	16 366	20 490	21 439	22 512	27 105	–582 276	–2 143 194	–4 840 277	–6 880 174	–2 816 462	–3 330 623
20. Budget transfers	Rb mn	0	0	0	0	0	0	0	0	0	0	0	0
21. Market price support	Rb mn	15 214	16 366	20 490	21 439	22 512	27 105	–582 276	–2 143 194	–4 840 277	–6 880 174	–2 816 462	–3 330 623
CSE – MARKET TRANSFERS (MT)													
22. Unit market transfers	Rb/t	–4 051	–4 101	–4 937	–5 037	–5 200	–6 795	160 318	638 045	1 493 913	2 517 444	1 070 898	1 424 561
23. Consumer transfers	Rb mn	–15 214	–16 366	–20 490	–21 439	–22 512	–27 105	582 276	2 143 194	4 840 277	6 880 174	2 816 462	3 330 623
24. Budget transfers	Rb mn	–3 099	–3 367	–5 303	–5 435	–2 995	–7 277	85 450	279 464	823 146	1 684 170	455 132	651 024
25. Market transfers	Rb mn	–18 312	–19 733	–25 793	–26 874	–25 508	–34 382	667 726	2 422 658	5 663 423	8 564 344	3 271 594	3 981 648
26. Consumption price (farm level)	Rb/t	4 480	4 639	5 550	5 697	6 245	8 702	41 556	502 401	1 292 774	4 736 274	7 017 231	6 911 252

e: estimate; p: provisional.

Notes:

10 = (4*7 + 5*8 + 6*9)/(4 + 5 + 6)
12 = 10*(1 + (11/100))
13 = Hungarian unit export value for carcasses.
This price is expressed in carcass weight and recalculated from the Hungarian carcass coefficient (0.56) using the Russian carcass coefficient.
15 = 13*14
16 = 12 – 15
17 = 16/12*100
18 = 17*10/100
19 = If 1 > 3 then 18*3; if 3 > 1 then 18*1
20 = If 1 > 3 then 0; if 3 > 1 then 18*(3 – 1)
21 = 19 + 20 or 18*3
22 = –18
23 = –19
24 = If 1 > 3 then 22*(1 – 3); if 3 > 1 then 0
25 = 23 + 24 or 22*1
26 = Consumption price (farm gate) = average producer price –(MPSu + MTu);
as MPSu = –MTu consumption price = average producer price.

Source: OECD.

Annex Table I.9.*i.* **Producer Subsidy Equivalents: Pigmeat**

	Units	1986	1987	1988	1989	1990	1991	1992	1993	1994	1995	1996p	1997e
I. Level of production	'000 t	3 093	3 264	3 399	3 499	3 480	3 190	2 784	2 432	2 103	1 865	1 705	1 565
II. Production price (farm gate)	Rb/t	2 924	2 978	3 618	3 728	4 168	6 133	43 402	590 119	1 795 704	6 195 864	8 767 629	9 980 635
III. Value of production	Rb mn	9 044	9 720	12 298	13 044	14 505	19 564	120 832	1 435 169	3 776 364	11 555 287	14 948 808	15 619 694
IV. Levies	Rb mn	0	0	0	0	0	0	0	0	0	0	0	0
V. Direct payments	Rb mn	340	341	0	0	0	803	18 739	105 293	158 616	389 557	573 424	311 621
VI. Adjusted value of production	Rb mn	9 384	10 062	12 298	13 044	14 505	20 367	139 571	1 540 462	3 934 981	11 944 845	15 522 231	15 931 315
VII. Gross total PSE	Rb mn	9 893	10 494	12 645	12 607	13 026	14 732	–481 294	–279 044	645 327	3 157 152	6 028 822	5 091 272
A. Market price support	Rb mn	7 571	8 102	10 661	10 681	10 969	12 914	–529 065	–726 208	–429 026	1 400 146	3 248 571	3 321 683
B. Direct payments	Rb mn	340	341	0	0	0	803	18 739	105 293	158 616	389 557	573 424	311 621
C. Reduction of input costs	Rb mn	1 579	1 589	1 484	1 396	1 465	310	23 715	268 143	526 870	915 267	1 614 164	786 925
D. General services	Rb mn	403	462	500	530	592	705	5 317	73 728	388 866	452 181	592 662	671 043
VIII. Feed adjustment	Rb mn	–2 547	–2 046	–2 771	–3 612	–5 101	–10 729	152 356	443 361	713 978	1 525 360	178 309	–211 887
IX. Net total PSE	Rb mn	7 346	8 449	9 874	8 995	7 925	4 002	–328 938	164 317	1 359 305	4 682 512	6 207 131	4 879 385
X. Net unit PSE	Rb/t	2 375	2 589	2 905	2 571	2 277	1 255	–118 153	67 564	646 365	2 510 730	3 640 546	3 117 818
XI. Net percentage PSE	**%**	**78**	**84**	**80**	**69**	**55**	**20**	**–236**	**11**	**35**	**39**	**40**	**31**

e: estimate; p: provisional.
Source: OECD.

Annex Table I.9.*ii.* **Consumer Subsidy Equivalents: Pigmeat**

	Units	1986	1987	1988	1989	1990	1991	1992	1993	1994	1995	1996p	1997e
I. Level of consumption	'000 t	3 366	3 552	3 648	3 875	3 502	3 487	2 974	2 694	2 466	2 570	2 149	1 969
II. Consumption price (farm gate)	Rb/t	2 924	2 978	3 618	3 728	4 168	6 133	43 402	590 119	1 795 704	6 195 864	8 767 629	9 980 635
III. Value of consumption	Rb mn	9 842	10 578	13 198	14 446	14 596	21 386	129 078	1 589 780	4 428 205	15 923 372	18 841 635	19 651 870
IV. Total CSE	Rb mn	–5 100	–5 774	–6 456	–7 018	–3 926	–7 572	567 854	804 443	503 081	–1 929 424	–4 094 533	–4 179 165
A. Market transfers	Rb mn	–8 239	–8 817	–11 442	–11 829	–11 039	–14 116	565 172	804 443	503 081	–1 929 424	–4 094 533	–4 179 165
B. Other transfers	Rb mn	3 139	3 042	4 986	4 810	7 112	6 544	2 682	0	0	0	0	0
V. Unit CSE	Rb/t	–1 515	–1 626	–1 770	–1 811	–1 121	–2 171	190 940	298 605	204 007	–750 749	–1 905 320	–2 122 481
VI. Percentage CSE	**%**	**–52**	**–55**	**–49**	**–49**	**–27**	**–35**	**440**	**51**	**11**	**–12**	**–22**	**–21**

e: estimate; p: provisional.
Source: OECD.

RUSSIA

Definitions and notes to Table 9

PSE: Pigmeat

I. **Level of production:** total production of pigmeat, carcass weight, calendar year.[1]

II. **Production price (farm gate):** before 1991, these prices are the average state purchases prices (received by large-scale producers and household plots), as other marketing channels were insignificant. Between 1991 and 1996, these prices are weighted average producer prices received by large-scale farms, private farms, and households with weights based on their marketed output. In 1997, prices received by large-scale farms were applied, as the price information for households and private farms was not available at the time when 1997 calculations were made (for calculation, see following table).[1]

III. **Value of production:** (I)*(II).

V. **Direct payments**

VI. **Adjusted value of production:** III + V.

A. **Market price support:** for calculation, see following table.

B, C, D See notes to Table 12.

CSE: Pigmeat

I. **Level of consumption:** total domestic use of pigmeat, carcass weight, calendar year.[2]

II. **Consumption price (farm gate):** implicit price measured at the farm gate; equal to the production price minus the sum of unit market price support and unit market transfers transfers [= Pp – (MPSu + MTu)]. As MPSu = –MTu, consumption price = average production price (see following table).

III. **Value of consumption:** (I)*(II).

A. **Market transfers:** for calculation, see following table.

B. **Other transfers:** 1986-1991: consumer subsidies paid to food processing industries, designed to maintain relatively low and stable prices for food. In 1992, consumer subsidies were maintained for meat products on a regional basis – they were allocated between livestock commodities in proportion to their share in total value of purchases of meat products concerned. In 1993, they were abolished.

Sources:

1. Goskomstat database. 1991-1995 production data published by Goskomstat in *Agriculture of the Russian Federation*, and *Russian Statistical Yearbook*.
2. USDA database (Production, Supply, and Distribution database).

Annex Table I.9.*iii.* **Market Price Support and Market Transfers: Pigmeat**

	Units	1986	1987	1988	1989	1990	1991	1992	1993	1994	1995	1996p	1997e
PSE – MARKET PRICE SUPPORT (MPS)													
1. Total consumption (carcass)	'000 t	3 366	3 552	3 648	3 875	3 502	3 487	2 974	2 694	2 466	2 570	2 149	1 969
2. Total production (live)	'000 t	4 136	4 337	4 556	4 719	4 684	4 296	3 751	3 308	2 876	2 572	2 335	2 148
3. Total production (carcass)	'000 t	3 093	3 264	3 399	3 499	3 480	3 190	2 784	2 432	2 103	1 865	1 705	1 565
4. Production marketed by large-scale farms (carcass)	'000 t						1 669	1 251	924	732	555	496	
5. Production marketed by households (carcass)	'000 t						197	177	145	214	156	159	
6. Production marketed by private farms (carcass)	'000 t						0	4	3	1	2	5	
7. Producer prices for large–scale farms (carcass)	Rb/t	2 924	2 978	3 618	3 728	4 168	6 006	43 337	588 472	1 783 378	5 839 902	7 769 915	9 980 635
8. Producer prices for households (carcass)	Rb/t						7 208	43 810	599 820	1 833 247	7 450 648	11 966 950	
9. Producer prices for private farms (carcass)	Rb/t						0	45 446	628 784	2 462 450	7 241 756	5 822 470	
10. Average price at farm level (carcass)	**Rb/t**	**2 924**	**2 978**	**3 618**	**3 728**	**4 168**	**6 133**	**43 402**	**590 119**	**1 795 704**	**6 195 864**	**8 767 629**	**9 980 635**
11. Handling margin	%	10	10	10	10	10	15	30	40	50	50	50	50
12. Adjusted domestic price (carcass)	Rb/t	3 216	3 276	3 980	4 101	4 585	7 053	56 423	826 166	2 693 555	9 293 797	13 151 444	14 970 953
13. World reference price (carcass)	US$/t	873	909	883	1 238	1 431	1 373	1 576	1 335	1 361	1 794	2 009	2 038
14. Official exchange rate	Rb/US$	0.60	0.60	0.60	0.60	0.78	1.75	193	932	2 204	4 554	5 124	5 785
15. World reference price (carcass)	Rb/t	524	545	530	743	1 118	2 398	303 472	1 244 214	2 999 566	8 167 674	10 293 463	11 787 231
16. Price difference	Rb/t	2 692	2 730	3 450	3 358	3 467	4 655	–247 049	–418 048	–306 010	1 126 123	2 857 980	3 183 722
17. Percentage MPS	**%**	**84**	**83**	**87**	**82**	**76**	**66**	**–438**	**–51**	**–11**	**12**	**22**	**21**
18. Unit market price support	Rb/t	2 448	2 482	3 137	3 053	3 152	4 048	–190 038	–298 605	–204 007	750 749	1 905 320	2 122 481
19. Consumer transfers	Rb mn	7 571	8 102	10 661	10 681	10 969	12 914	–529 065	–726 208	–429 026	1 400 146	3 248 571	3 321 683
20. Budget transfers	Rb mn	0	0	0	0	0	0	0	0	0	0	0	0
21. Market price support	Rb mn	7 571	8 102	10 661	10 681	10 969	12 914	–529 065	–726 208	–429 026	1 400 146	3 248 571	3 321 683
CSE – MARKET TRANSFERS (MT)													
22. Unit market transfers	Rb/t	–2 448	–2 482	–3 137	–3 053	–3 152	–4 048	190 038	298 605	204 007	–750 749	–1 905 320	–2 122 481
23. Consumer transfers	Rb mn	–7 571	–8 102	–10 661	–10 681	–10 969	–12 914	529 065	726 208	429 026	–1 400 146	–3 248 571	–3 321 683
24. Budget transfers	Rb mn	–668	–715	–781	–1 148	–69	–1 202	36 107	78 235	74 054	–529 278	–845 962	–857 482
25. Market transfers	Rb mn	–8 239	–8 817	–11 442	–11 829	–11 039	–14 116	565 172	804 443	503 081	–1 929 424	–4 094 533	–4 179 165
26. Consumption price (farm level)	Rb/t	2 924	2 978	3 618	3 728	4 168	6 133	43 402	590 119	1 795 704	6 195 864	8 767 629	9 980 635

e: estimate; p: provisional.

Notes:

10 = (4*7 + 5*8 + 6*9)/(4 + 5 + 6)
12 = 10*(1 + (11/100))
13 = Hungarian unit export value for carcasses. This price is expressed in carcass weight and recalculated from the Hungarian carcass coefficient (0.79) using the Russian carcass coefficient.
15 = 13*14
16 = 12 – 15
17 = 16/12*100
18 = 17*10/100
19 = If 1 > 3 then 18*3; if 3 > 1 then 18*1
20 = If 1 > 3 then 0; if 3 > 1 then 18*(3 – 1)
21 = 19 + 20 or 18*3
22 = –18
23 = –19
24 = If 1 > 3 then 22*(1 – 3); if 3 > 1 then 0
25= 23 + 24 or 22*1
26 = Consumption price (farm gate) = average producer price –(MPSu + MTu); as MPSu = –MTu consumption price = average producer price.

Source: OECD.

Annex Table I.10.*i.* **Producer Subsidy Equivalents: Poultry**

	Units	1986	1987	1988	1989	1990	1991	1992	1993	1994	1995	1996p	1997e
I. Level of production	'000 t	1 613	1 712	1 776	1 831	1 801	1 751	1 428	1 277	1 068	859	690	632
II. Production price (farm gate)	Rb/t	2 842	2 960	3 142	3 374	3 629	6 179	43 262	621 765	2 025 573	6 348 689	9 462 896	10 474 625
III. Value of production	Rb mn	4 584	5 068	5 580	6 178	6 536	10 819	61 777	793 994	2 163 312	5 453 523	6 529 398	6 619 963
IV. Levies	Rb mn	0	0	0	0	0	0	0	0	0	0	0	0
V. Direct payments	Rb mn	172	178	0	0	0	444	12 958	99 107	190 464	447 906	613 150	391 685
VI. Adjusted value of production	Rb mn	4 756	5 246	5 580	6 178	6 536	11 263	74 735	893 101	2 353 777	5 901 430	7 142 548	7 011 648
VII. Gross total PSE	Rb mn	4 919	5 404	5 616	6 094	5 932	8 594	–173 997	105 541	1 040 158	3 331 040	4 793 494	4 303 739
A. Market price support	Rb mn	3 742	4 157	4 715	5 182	5 006	7 588	–203 676	–183 287	325 109	2 237 766	3 216 437	3 294 136
B. Direct payments	Rb mn	172	178	0	0	0	444	12 958	99 107	190 464	447 906	613 150	391 685
C. Reduction of input costs	Rb mn	800	828	673	661	660	171	14 002	148 932	301 821	431 961	705 041	333 516
D. General services	Rb mn	204	241	227	251	267	390	2 718	40 789	222 764	213 407	258 865	284 402
VIII. Feed adjustment	Rb mn	–658	–507	–674	–911	–2 354	–5 326	75 943	249 591	434 675	943 478	106 009	–125 972
IX. Net total PSE	Rb mn	4 261	4 897	4 942	5 182	3 578	3 267	–98 054	355 132	1 474 833	4 274 518	4 899 504	4 177 767
X. Net unit PSE	Rb/t	2 642	2 861	2 783	2 830	1 987	1 866	–68 665	278 099	1 380 930	4 976 156	7 100 730	6 610 391
XI. Net percentage PSE	%	**90**	**93**	**89**	**84**	**55**	**29**	**–131**	**40**	**63**	**72**	**69**	**60**

e: estimate; p: provisional.
Source: OECD.

Annex Table I.10.*ii.* **Consumer Subsidy Equivalents: Poultry**

	Units	1986	1987	1988	1989	1990	1991	1992	1993	1994	1995	1996p	1997e
I. Level of consumption	'000 t	1 828	1 956	2 029	2 194	2 194	1 857	1 517	1 486	1 588	1 749	1 755	1 899
II. Consumption price (farm gate)	Rb/t	2 842	2 960	3 142	3 374	3 629	6 179	43 262	621 765	2 025 573	6 348 689	9 462 896	10 474 625
III. Value of consumption	Rb mn	5 195	5 790	6 375	7 403	7 962	11 474	65 628	923 943	3 216 610	11 103 856	16 607 382	19 891 313
IV. Total CSE	Rb mn	–1 698	–1 580	–2 451	–3 231	–1 751	–2 867	218 499	213 285	–483 401	–4 556 289	–8 180 939	–9 898 045
A. Market transfers	Rb mn	–4 241	–4 749	–5 387	–6 209	–6 098	–8 048	216 370	213 285	–483 401	–4 556 289	–8 180 939	–9 898 045
B. Other transfers	Rb mn	2 543	3 169	2 936	2 979	4 347	5 181	2 129	0	0	0	0	0
V. Unit CSE	Rb/t	–929	–808	–1 208	–1 472	–798	–1 544	144 033	143 530	–304 409	–2 605 082	–4 661 503	–5 212 241
VI. Percentage CSE	**%**	**–33**	**–27**	**–38**	**–44**	**–22**	**–25**	**333**	**23**	**–15**	**–41**	**–49**	**–50**

e: estimate; p: provisional.
Source: OECD.

RUSSIA

Definitions and notes to Table 10

PSE: Poultry

I. **Level of production:** total production of poultry (chicken, ducks, goose, turkeys), carcass weight, calendar year.[1]

II. **Production price (farm gate):** before 1991, these prices are the average state purchases prices (received by large-scale producers and household plots), as other marketing channels were insignificant. Between 1991 and 1996, these prices are weighted average producer prices received by large-scale farms, private farms, and households with weights based on their marketed output. In 1997, prices received by large-scale farms were applied, as the price information for households and private farms was not available at the time when 1997 calculations were made (for calculation, see following table).[1]

III. **Value of production:** (I)*(II).

V. **Direct payments**

VI. **Adjusted value of production:** III + V.

A. **Market price support:** for calculation, see following table.

B, C, D See notes to Table 12.

CSE: Poultry

I. **Level of consumption:** total domestic use of poultrymeat (all kinds), carcass weight, calendar year.[2]

II. **Consumption price (farm gate):** implicit price measured at the farm gate; equal to the production price minus the sum of unit market price support and unit market transfers transfers [= Pp – (MPSu + MTu)]. As MPSu = –MTu, consumption price = average production price (see following table).

III. **Value of consumption:** (I)*(II).

A. **Market transfers:** for calculation, see following table.

B. **Other transfers:** 1986-1991: consumer subsidies paid to food processing industries, designed to maintain relatively low and stable prices for food (and also in some cases to agricultural producers delivering poultrymeat directly to retailers). In 1992, consumer subsidies were maintained for meat products on a regional basis – they were allocated between livestock commodities in proportion to their share in total value of purchases of meat products concerned. In 1993, they were abolished.

Sources:

1. Goskomstat database.
   1991-1995 production data published by Goskomstat in *Agriculture of the Russian Federation*, and *Russian Statistical Yearbook*.
2. USDA database (Production, Supply, and Distribution database).

Annex Table I.10.*iii.* **Market Price Support and Market Transfers: Poultry**

	Units	1986	1987	1988	1989	1990	1991	1992	1993	1994	1995	1996p	1997e
PSE – MARKET PRICE SUPPORT (MPS)													
1. Total consumption (carcass)	'000 t	1 828	1 956	2 029	2 194	2 194	1 857	1 517	1 486	1 588	1 749	1 755	1 899
2. Total production (live)	'000 t	2 142	2 282	2 442	2 551	2 553	2 499	2 023	1 836	1 547	1 262	1 010	931
3. Total production (carcass)	'000 t	1 613	1 712	1 776	1 831	1 801	1 751	1 428	1 277	1 068	859	690	632
4. Production marketed by large-scale farms (carcass)	'000 t						1 163	961	865	741	557	426	
5. Production marketed by households (carcass)	'000 t						85	105	109	190	123	137	
6. Production marketed by private farms (carcass)	'000 t						0	0	1	1	1	4	
7. Producer prices for large-scale farms (carcass)	Rb/t	2 842	2 960	3 142	3 374	3 629	6 123	43 167	631 526	1 975 193	5 852 416	8 249 102	10 474 625
8. Producer prices for households (carcass)	Rb/t						6 952	44 124	542 816	2 223 001	8 588 026	13 054 875	
9. Producer prices for private farms (carcass)	Rb/t						0	44 652	866 901	1 945 887	8 790 663	15 879 122	
10. Average price at farm level (carcass)	**Rb/t**	**2 842**	**2 960**	**3 142**	**3 374**	**3 629**	**6 179**	**43 262**	**621 765**	**2 025 573**	**6 348 689**	**9 462 896**	**10 474 625**
11. Handling margin	%	10	10	10	10	10	15	30	40	50	50	50	50
12. Adjusted domestic price (carcass)	Rb/t	3 126	3 256	3 456	3 711	3 992	7 106	56 240	870 471	3 038 360	9 523 033	14 194 343	15 711 938
13. World reference price (carcass)	ECU/t	975	846	755	906	943	983	970	981	988	943	1 107	1 191
14. Official exchange rate	Rb/ECU	0.59	0.69	0.71	0.66	0.99	2.16	249	1 092	2 614	5 953	6 504	6 627
15. World reference price (carcass)	Rb/t	574	585	536	598	935	2 122	241 659	1 071 412	2 581 746	5 615 409	7 202 088	7 893 576
16. Price difference	Rb/t	2 552	2 671	2 921	3 113	3 057	4 984	–185 419	–200 941	456 613	3 907 623	6 992 255	7 818 361
17. Percentage MPS	**%**	**82**	**82**	**85**	**84**	**77**	**70**	**–330**	**–23**	**15**	**41**	**49**	**50**
18. Unit market price support	Rb/t	2 320	2 428	2 655	2 830	2 779	4 334	–142 630	–143 530	304 409	2 605 082	4 661 503	5 212 241
19. Consumer transfers	Rb mn	3 742	4 157	4 715	5 182	5 006	7 588	–203 676	–183 287	325 109	2 237 766	3 216 437	3 294 136
20. Budget transfers	Rb mn	0	0	0	0	0	0	0	0	0	0	0	0
21. Market price support	Rb mn	3 742	4 157	4 715	5 182	5 006	7 588	–203 676	–183 287	325 109	2 237 766	3 216 437	3 294 136
CSE – MARKET TRANSFERS (MT)													
22. Unit market transfers	Rb/t	–2 320	–2 428	–2 655	–2 830	–2 779	–4 334	142 630	143 530	–304 409	–2 605 082	–4 661 503	–5 212 241
23. Consumer transfers	Rb mn	–3 742	–4 157	–4 715	–5 182	–5 006	–7 588	203 676	183 287	–325 109	–2 237 766	–3 216 437	–3 294 136
24. Budget transfers	Rb mn	–499	–592	–672	–1 027	–1 092	–459	12 694	29 998	–158 293	–2 318 523	–4 964 501	–6 603 909
25. Market transfers	Rb mn	–4 241	–4 749	–5 387	–6 209	–6 098	–8 048	216 370	213 285	–483 401	–4 556 289	–8 180 939	–9 898 045
26. Consumption price (farm level)	Rb/t	2 842	2 960	3 142	3 374	3 629	6 179	43 262	621 765	2 025 573	6 348 689	9 462 896	10 474 625

e: estimate; p: provisional.

Notes:

10 = (4*7 + 5*8 + 6*9)/(4 + 5 + 6)
12 = 10*(1 + (11/100))
13 = Extra–EU unit export value.
15 = 13*14
16 = 12 – 15
17 = 16/12*100
18 = 17*10/100
19 = If 1 > 3 then 18*3; if 3 > 1 then 18*1
20 = If 1 > 3 then 0; if 3 > 1 then 18*(3 – 1)
21 = 19 + 20 or 18*3
22 = –18
23 = –19
24 = If 1 > 3 then 22*(1 – 3); if 3 > 1 then 0
25 = 23 + 24 or 22*1
26 = Consumption price (farm gate) = average producer price –(MPSu + MTu);
as MPSu = –MTu consumption price = average producer price.

Source: OECD.

Annex Table I.11.*i.* **Producer Subsidy Equivalents: Eggs**

	Units	1986	1987	1988	1989	1990	1991	1992	1993	1994	1995	1996p	1997e
I. Level of production	'000 t	2 566	2 636	2 730	2 724	2 637	2 604	2 383	2 239	2 082	1 879	1 772	1 771
II. Production price (farm gate)	Rb/t	1 800	1 818	1 836	1 800	1 890	4 077	33 224	359 080	1 505 790	4 652 757	6 841 361	7 343 388
III. Value of production	Rb mn	4 620	4 792	5 013	4 902	4 984	10 617	79 189	803 880	3 135 056	8 744 598	12 125 173	13 007 588
IV. Levies	Rb mn	0	0	0	0	0	0	0	0	0	0	0	0
V. Direct payments	Rb mn	174	168	0	0	0	436	11 392	77 996	140 881	378 352	535 778	302 652
VI. Adjusted value of production	Rb mn	4 793	4 960	5 013	4 902	4 984	11 053	90 581	881 876	3 275 937	9 122 950	12 660 951	13 310 240
VII. Gross total PSE	Rb mn	4 913	4 630	4 621	4 506	3 748	7 388	–243 828	–417 302	871 629	4 769 055	6 804 851	6 695 228
A. Market price support	Rb mn	3 727	3 451	3 812	3 782	3 041	6 402	–274 849	–687 181	–29 477	3 355 871	4 479 086	5 178 426
B. Direct payments	Rb mn	174	168	0	0	0	436	11 392	77 996	140 881	378 352	535 778	302 652
C. Reduction of input costs	Rb mn	807	783	605	525	503	168	16 144	150 586	437 396	692 639	1 309 270	655 327
D. General services	Rb mn	206	228	204	199	203	383	3 485	41 297	322 828	342 193	480 716	558 823
VIII. Feed adjustment	Rb mn	–1 066	–863	–1 211	–1 559	–1 758	–3 793	56 411	166 625	279 610	613 119	79 578	–94 564
IX. Net total PSE	Rb mn	3 847	3 767	3 410	2 947	1 990	3 595	–187 417	–250 677	1 151 239	5 382 173	6 884 429	6 600 665
X. Net unit PSE	Rb/t	1 499	1 429	1 249	1 082	755	1 380	–78 633	–111 973	552 949	2 863 704	3 884 387	3 726 382
XI. Net percentage PSE	**%**	**80**	**76**	**68**	**60**	**40**	**33**	**–207**	**–28**	**35**	**59**	**54**	**50**

e: estimate; p: provisional.
Source: OECD.

Annex Table I.11.*ii.* **Consumer Subsidy Equivalents: Eggs**

	Units	1986	1987	1988	1989	1990	1991	1992	1993	1994	1995	1996p	1997e
I. Level of consumption	'000 t	2 718	2 777	2 828	2 807	2 721	2 629	2 374	2 229	2 092	1 889	1 796	1 796
II. Consumption price (farm gate)	Rb/t	1 800	1 818	1 836	1 800	1 890	4 077	33 224	359 080	1 505 790	4 652 757	6 841 361	7 343 388
III. Value of consumption	Rb mn	4 892	5 049	5 193	5 052	5 142	10 718	78 879	800 447	3 149 746	8 787 248	12 285 565	13 187 093
IV. Total CSE	Rb mn	–3 947	–3 636	–3 949	–3 898	–3 138	–6 463	273 773	684 246	29 615	–3 372 238	–4 538 336	–5 249 889
A. Market transfers	Rb mn	–3 947	–3 636	–3 949	–3 898	–3 138	–6 463	273 773	684 246	29 615	–3 372 238	–4 538 336	–5 249 889
B. Other transfers	Rb mn	0	0	0	0	0	0	0	0	0	0	0	0
V. Unit CSE	Rb/t	–1 452	–1 309	–1 396	–1 389	–1 153	–2 458	115 316	306 952	14 158	–1 785 565	–2 527 226	–2 923 462
VI. Percentage CSE	**%**	**–81**	**–72**	**–76**	**–77**	**–61**	**–60**	**347**	**85**	**1**	**–38**	**–37**	**–40**

e: estimate; p: provisional.
Source: OECD.

RUSSIA

Definitions and notes to Table 11

PSE: Eggs

I. **Level of production:** total production of eggs expressed in thousand tonnes using the coefficient 1 kg = 18 eggs, calendar year.[1]

II. **Production price (farm gate):** before 1991, these prices are the average state purchases prices (received by large-scale producers and household plots), as other marketing channels were insignificant. Between 1991 and 1996, these prices are weighted average producer prices received by large-scale farms, private farms, and households with weights based on their marketed output. In 1997, prices received by large-scale farms were applied, as the price information for households and private farms was not available at the time when 1997 calculations were made (for calculation, see following table).[1]

III. **Value of production:** (I)*(II).

V. **Direct payments**

VI. **Adjusted value of production:** III + V.

A. **Market price support:** for calculation, see following table.

B, C, D See notes to Table 12.

CSE: Eggs

I. **Level of consumption:** total domestic use of eggs expressed in thousand tonnes by using the coefficient 1 kg = 18 eggs, calendar year. Total domestic use defined as production plus imports minus exports minus net change in stocks (balance sheets).[2]

II. **Consumption price (farm gate):** implicit price measured at the farm gate; equal to the production price minus the sum of unit market price support and unit market transfers transfers [= Pp – (MPSu + MTu)]. As MPSu = –MTu, consumption price = average production price (see following table).

III. **Value of consumption:** (I)*(II).

A. **Market transfers:** for calculation, see following table.

B. **Other transfers:** no consumer subsidies were applied to eggs.

Sources:

1. Goskomstat database. 1991-1995 production data published by Goskomstat in *Agriculture of the Russian Federation*, and *Russian Statistical Yearbook*.
2. Goskomstat database. 1991-1995 data published by Goskomstat in *Consumption of basic food products by population of the Russian Federation*.

Annex Table I.11.*iii.* **Market Price Support and Market Transfers: Eggs**

	Units	1986	1987	1988	1989	1990	1991	1992	1993	1994	1995	1996p	1997e
PSE – MARKET PRICE SUPPORT (MPS)													
1. Total consumption	'000 t	2 718	2 777	2 828	2 807	2 721	2 629	2 374	2 229	2 092	1 889	1 796	1 796
2. Total production	'000 t	2 566	2 636	2 730	2 724	2 637	2 604	2 383	2 239	2 082	1 879	1 772	1 771
3. Production marketed by large-scale farms	'000 t						2	2	1	1	1	1	
4. Production marketed by households	'000 t						0.1	0.1	0.1	0.1	0.1	0.1	
5. Production marketed by private farms	'000 t						0.0	0.0	0.0	0.0	0.0	0.0	
6. Producer prices for large-scale farms	Rb/t	1 800	1 818	1 836	1 800	1 890	4 122	34 092	365 652	1 519 956	4 594 806	6 742 908	7 343 388
7. Producer prices for households	Rb/t						3 384	22 122	274 734	1 320 246	5 606 874	8 313 138	
8. Producer prices for private farms	Rb/t						0	0	892 440	1 738 890	5 322 636	8 037 720	
9. Average price at farm level	**Rb/t**	**1 800**	**1 818**	**1 836**	**1 800**	**1 890**	**4 077**	**33 224**	**359 080**	**1 505 790**	**4 652 757**	**6 841 361**	**7 343 388**
10. Handling margin	%	7	7	7	7	7	10	20	30	40	40	40	40
11. Adjusted domestic price	Rb/t	1 926	1 945	1 965	1 926	2 022	4 485	39 869	466 804	2 108 107	6 513 860	9 577 906	10 280 743
12. World reference price	ECU/t	632	787	664	666	795	825	715	793	814	674	929	934
13. Official exchange rate	Rb/ECU	0.59	0.69	0.71	0.66	0.99	2.16	249	1 092	2 614	5 953	6 504	6 627
14. World reference price	Rb/t	372	545	471	440	788	1 781	178 248	865 842	2 127 928	4 014 068	6 039 790	6 187 896
15. Price difference	Rb/t	1 554	1 401	1 494	1 486	1 234	2 704	–138 379	–399 038	–19 821	2 499 791	3 538 116	4 092 847
16. Percentage MPS	**%**	**81**	**72**	**76**	**77**	**61**	**60**	**–347**	**–85**	**–1**	**38**	**37**	**40**
17. Unit market price support	Rb/t	1 452	1 309	1 396	1 389	1 153	2 458	–115 316	–306 952	–14 158	1 785 565	2 527 226	2 923 462
18. Consumer transfers	Rb mn	3 727	3 451	3 812	3 782	3 041	6 402	–273 773	–684 246	–29 477	3 355 871	4 479 086	5 178 426
19. Budget transfers	Rb mn	0	0	0	0	0	0	–1 076	–2 935	0	0	0	0
20. Market price support	Rb mn	3 727	3 451	3 812	3 782	3 041	6 402	–274 849	–687 181	–29 477	3 355 871	4 479 086	5 178 426
CSE – MARKET TRANSFERS (MT)													
21. Unit market transfers	Rb/t	–1 452	–1 309	–1 396	–1 389	–1 153	–2 458	115 316	306 952	14 158	–1 785 565	–2 527 226	–2 923 462
22. Consumer transfers	Rb mn	–3 727	–3 451	–3 812	–3 782	–3 041	–6 402	273 773	684 246	29 477	–3 355 871	–4 479 086	–5 178 426
23. Budget transfers	Rb mn	–220	–185	–137	–116	–96	–61	0	0	138	–16 368	–59 249	–71 462
24. Market transfers	Rb mn	–3 947	–3 636	–3 949	–3 898	–3 138	–6 463	273 773	684 246	29 615	–3 372 238	–4 538 336	–5 249 889
25. Consumption price (farm level)	Rb/t	1 800	1 818	1 836	1 800	1 890	4 077	33 224	359 080	1 505 790	4 652 757	6 841 361	7 343 388

e: estimate; p: provisional.

Notes:

9 = (3*6 + 4*7 + 5*8)/(3 + 4 + 5)
11 = 9*(1 + (10/100))
12 = Extra-EU unit export value.
14 = 12*13
15 = 11 – 14
16 = 15/11*100
17 = 16*9/100
18 = If 1 > 2 then 17*2; if 2 > 1 then 17*1
19 = If 1 > 2 then 0; if 2 > 1 then 17*(2 – 1)
20 = 18 + 19 or 17*2
21 = –17
22 = –18
23 = If 1 > 2 then 21*(1 – 2); if 2 > 1 then 0
24 = 22 + 23 or 21*1
25 = Consumption price (farm gate) = average producer price –(MPSu + MTu); as MPSu = –MTu consumption price = average producer price.

Source: OECD.

Annex Table I.12.*i.* **Producer Subsidy Equivalents: Aggregate**

	Units	1986	1987	1988	1989	1990	1991	1992	1993	1994	1995	1996p	1997e
Adjusted value of production	Rb mn	78 143	81 397	97 292	108 827	123 494	179 000	1 951 493	13 027 592	30 418 696	99 511 587	139 669 782	145 893 500
Gross total PSE	Rb mn	84 200	85 641	97 320	105 399	114 954	143 881	–2 531 128	–4 964 938	–5 162 138	15 164 553	43 645 532	38 256 880
A. Market price support	Rb mn	64 258	65 716	81 228	89 089	97 238	127 914	–3 088 397	–8 753 079	–13 597 948	–464 823	19 282 109	22 966 108
B. Direct payments	Rb mn	2 516	2 439	0	0	0	7 056	163 837	972 360	1 595 882	4 229 404	5 031 183	2 739 286
C. Reduction of Input Costs	Rb mn	14 086	13 760	12 156	11 917	12 740	2 819	322 848	2 248 841	4 015 115	7 886 987	14 397 249	6 958 171
D. General Services	Rb mn	3 339	3 726	3 936	4 392	4 976	6 092	70 584	566 941	2 824 813	3 512 986	4 934 991	5 593 315
Total other support C + D	Rb mn	17 425	17 486	16 092	16 309	17 716	8 911	393 433	2 815 782	6 839 927	11 399 972	19 332 240	12 551 486
Feed adjustment	Rb mn	–7 916	–6 464	–8 950	–11 657	–16 645	–35 510	491 193	1 536 343	2 554 532	5 487 054	627 341	–745 476
Net total PSE	Rb mn	76 284	79 176	88 371	93 742	98 310	108 371	–2 039 935	–3 428 595	–2 607 607	20 651 607	44 272 872	37 511 404
Net percentage PSE	**%**	**98**	**97**	**91**	**86**	**80**	**61**	**–105**	**–26**	**–9**	**21**	**32**	**26**

e: estimate; p: provisional.
Source: OECD.

RUSSIA

Definitions and notes to Table 12

PSE: AGGREGATE

Detail of general policy measures

A. **Market price support:** the sum of market price support calculated for each commodity in the commodity PSE tables (PSE tables 1-11).

B. **Direct payments:** the sum of direct payments provided in the commodity PSE tables. Before 1992, premiums and supplementary payments were included in the producer administered prices; therefore direct payments before 1992 include only very few subsidies: premiums for livestock, compensation of insurance payments, and in 1991 subsidies to low-profitable farms (which were previously included in the average producer prices). As of 1992, direct payments consisted mainly of direct product subsidies (mainly for livestock products), compensation payments to farms affected by natural disasters, and subsidies paid in support of private farms. If not commodity specific, they are allocated to PSE commodities in proportion to their share in the total value of agricultural production.

C. **Reduction of input costs:** the sum of subsidies reducing input costs for each commodity in the PSE tables. They include soft credits for agricultural producers (subsidised interest rates, credit subsidies for private farms and agricultural enterprises), debt restructuring and write-offs, long-term credits for capital investment in agriculture. In 1994 long-term state loans for purchases of agricultural machinery were introduced, and in 1995-1996 commodity loans were granted. This category also includes input subsidies to agricultural machinery, fertilisers, electricity, energy, as well as smaller subsidies for fuel, mixed feed, high quality seed, animal breeding. If not commodity specific, they are allocated to PSE commodities in proportion to their share in the total value of agricultural production.

D. **General services:** the sum of subsidies of general services for each commodity in the PSE tables. They mainly include subsidies for research, training and extension services, infrastructure facilities, pest and disease control as well as subsidies for land improvement. They are allocated to PSE commodities in proportion to their share in the total value of agricultural production.

Feed adjustment: amount of Russian production of fodder wheat, maize, rye, barley, and oats used for livestock feeding multiplied by their respective unit market price support levels.

Sources:

Ministry of Agriculture and Food of the Russian Federation, Ministry of Finance of the Russian Federation, Goskomstat, Annual reports of farms. Local budgets.

Annex Table I.12.*ii.* **Consumer Subsidy Equivalents: Aggregate**

	Units	1986	1987	1988	1989	1990	1991	1992	1993	1994	1995	1996p	1997e
Value of consumption	Rb mn	89 980	92 990	120 457	132 754	142 441	214 167	2 011 414	13 540 025	34 601 823	118 951 694	165 957 717	169 372 522
A. Market transfers	Rb mn	–76 796	–77 427	–100 952	–109 211	–112 722	–160 979	3 393 414	9 397 705	15 156 309	–616 792	–30 255 986	–34 309 893
B. Other transfers	Rb mn	28 650	29 307	42 381	44 999	51 677	53 678	73 989	388 225	437 801	0	0	0
Total CSE	Rb mn	–48 146	–48 120	–58 571	–64 212	–61 045	–107 301	3 467 403	9 785 930	15 594 110	–616 792	–30 255 986	–34 309 893
Percentage CSE	**%**	**–54**	**–52**	**–49**	**–48**	**–43**	**–50**	**172**	**72**	**45**	**–1**	**–18**	**–20**

e: estimate; p: provisional.
Source: OECD.

RUSSIA

Definitions and notes to Table 12

PSE: AGGREGATE

I. **Total value of consumption:** sum of value of consumption calculated for each commodity in the commodity CSE tables (CSE tables 1-11).

II. **Total CSE:** (A) + (B).

A. **Market transfers:** sum of market transfers calculated for each commodity in the commodity CSE tables.

B. **Other transfers:** sum of consumer subsidies for each commodity in the commodity CSE tables.

III. **Percentage CSE:** (II)/I.

Sources:

1986-1991: Ministry of Finance. 1992-1995: Annual reports on execution of local budgets.

Annex Table I.13.*i.* **Producer Subsidy Equivalents: by commodity**

	Units	1986	1987	1988	1989	1990	1991	1992	1993	1994	1995	1996p	1997e
Wheat													
Gross total PSE	Rb mn	5 318	4 127	3 852	6 632	11 298	23 987	–384 265	–655 553	–1 238 613	–2 485 388	1 950 122	3 141 640
Gross unit PSE	Rb/t	112	112	97	151	228	617	–8 323	–15 054	–38 551	–82 519	55 850	71 097
Gross percentage PSE	%	90	88	74	76	82	86	–101	–31	–36	–22	9	11
Maize													
Gross total PSE	Rb mn	604	917	1 661	1 878	1 863	2 059	–11 725	–5 860	113 416	296 129	460 460	709 113
Gross unit PSE	Rb/t	353	239	436	403	760	1 046	–5 492	–2 401	127 148	170 385	423 217	265 486
Gross percentage PSE	%	111	106	103	100	104	90	–62	–4	50	31	46	36
Other grains													
Gross total PSE	Rb mn	6 563	6 195	4 474	6 702	12 321	17 155	–94 307	–684 823	–1 305 816	–2 948 080	2 555 230	2 519 880
Gross unit PSE	Rb/t	129	125	105	143	220	397	–1 810	–14 398	–29 813	–103 638	84 574	66 956
Gross percentage PSE	%	98	96	74	76	84	79	–16	–34	–34	–44	16	13
Potatoes (not included in the aggregation)													
Gross total PSE	Rb mn	2 376	5 120	5 367	7 433	7 889	34 132	–236 239	683 939	101 174	–9 081 544	31 408 590	25 490 668
Gross unit PSE	Rb/t	185	135	159	220	256	994	–6 163	18 166	2 991	–227 556	812 599	688 658
Gross percentage PSE	%	100	85	83	91	87	87	–67	32	1	–25	71	67
Sunflower													
Gross total PSE	Rb mn	640	1 205	973	1 797	1 236	1 849	–30 282	–299 156	–236 478	217 150	59 513	–259 443
Gross unit PSE	Rb/t	271	393	329	474	361	638	–9 737	–108 194	–92 627	51 702	21 524	–91 871
Gross percentage PSE	%	98	107	88	94	84	73	–55	–150	–36	6	3	–12
Sugar (refined equivalent)													
Gross total PSE	Rb mn	2 329	2 720	2 470	2 686	2 024	1 564	10 547	358 436	290 336	1 073 640	2 215 637	1 560 056
Gross unit PSE	Rb/t	737	705	745	716	631	706	3 924	119 983	164 574	457 675	1 522 837	995 988
Gross percentage PSE	%	117	116	106	98	91	75	16	59	37	39	72	57
Crops[1]													
Gross total PSE	Rb mn	15 453	15 163	13 430	19 695	28 742	46 614	–510 031	–1 286 956	–2 377 155	–3 846 549	7 240 962	7 671 246
Gross percentage PSE	**%**	**98**	**98**	**83**	**82**	**85**	**82**	**–46**	**–25**	**–27**	**–15**	**16**	**14**

e: estimate; p: provisional.
1. For all crop products the gross and net PSEs are equivalent.
Source: OECD.

Annex Table I.13.*i.* **Producer Subsidy Equivalents: by commodity** *(cont.)*

	Units	1986	1987	1988	1989	1990	1991	1992	1993	1994	1995	1996p	1997e
Milk													
Net total PSE	Rb mn	27 514	27 370	34 484	34 472	33 169	27 722	–519 272	–1 185 995	–1 326 974	13 792 456	18 232 144	15 685 976
Net unit PSE	Rb/t	527	518	632	618	595	534	–10 993	–25 492	–31 463	351 481	509 008	460 458
Net percentage PSE	%	106	103	98	95	91	64	–155	–44	–18	41	46	40
Beef and veal													
Net total PSE	Rb mn	17 863	19 530	22 231	22 451	22 906	23 171	–396 222	–1 224 416	–2 888 855	–3 633 502	808 703	–1 503 635
Net unit PSE	Rb/t	4 756	4 893	5 357	5 275	5 291	5 809	–109 092	–364 518	–891 622	–1 329 492	307 492	–643 129
Net percentage PSE	%	102	102	97	93	85	64	–202	–64	–62	–27	4	–9
Pigmeat													
Net total PSE	Rb mn	7 346	8 449	9 874	8 995	7 925	4 002	–328 938	164 317	1 359 305	4 682 512	6 207 131	4 879 385
Net unit PSE	Rb/t	2 375	2 589	2 905	2 571	2 277	1 255	–118 153	67 564	646 365	2 510 730	3 640 546	3 117 818
Net percentage PSE	%	78	84	80	69	55	20	–236	11	35	39	40	31
Poultry													
Net total PSE	Rb mn	4 261	4 897	4 942	5 182	3 578	3 267	–98 054	355 132	1 474 833	4 274 518	4 899 504	4 177 767
Net unit PSE	Rb/t	2 642	2 861	2 783	2 830	1 987	1 866	–68 665	278 099	1 380 930	4 976 156	7 100 730	6 610 391
Net percentage PSE	%	90	93	89	84	55	29	–131	40	63	72	69	60
Eggs													
Net total PSE	Rb mn	3 847	3 767	3 410	2 947	1 990	3 595	–187 417	–250 677	1 151 239	5 382 173	6 884 429	6 600 665
Net unit PSE	Rb/t	1 499	1 429	1 249	1 082	755	1 380	–78 633	–111 973	552 949	2 863 704	3 884 387	3 726 382
Net percentage PSE	%	80	76	68	60	40	33	–207	–28	35	59	54	50
Livestock products													
Net total PSE	Rb mn	60 831	64 013	74 941	74 048	69 568	61 757	–1 529 904	–2 141 638	–230 452	24 498 157	37 031 911	29 840 158
Net percentage PSE	**%**	**98**	**97**	**92**	**87**	**78**	**50**	**–183**	**–27**	**–1**	**33**	**39**	**32**
All products[1]													
Net total PSE	Rb mn	76 284	79 176	88 371	93 742	98 310	108 371	–2 039 935	–3 428 595	–2 607 607	20 651 607	44 272 872	37 511 404
Net percentage PSE	**%**	**98**	**97**	**91**	**86**	**80**	**61**	**–105**	**–26**	**–9**	**21**	**32**	**26**

e: estimate; p: provisional.
1. For all crop products the gross and net PSEs are equivalent.
Source: OECD.

Annex Table I.13.*ii.* **Consumer Subsidy Equivalents: by commodity**

	Units	1986	1987	1988	1989	1990	1991	1992	1993	1994	1995	1996p	1997e
Wheat													
Total CSE	Rb mn	–2 260	–2 518	–1 333	–3 078	–7 986	–27 171	589 565	1 575 516	2 934 233	5 156 863	1 366 221	–867 836
Unit CSE	Rb/t	–43	–49	–27	–59	–139	–505	10 413	32 190	68 853	130 818	36 130	–22 359
Percentage CSE	%	–35	–40	–21	–30	–50	–73	127	67	65	36	6	–4
Maize													
Total CSE	Rb mn	–2 780	–1 063	–5 115	–4 932	–3 780	–9 468	44 834	95 697	–147 445	–178 518	–378 378	–562 157
Unit CSE	Rb/t	–253	–148	–326	–315	–618	–925	7 215	16 582	–68 452	–99 177	–291 060	–208 206
Percentage CSE	%	–81	–66	–78	–78	–84	–83	82	29	–27	–18	–32	–28
Other grains													
Total CSE	Rb mn	–3 443	–3 367	–2 310	–4 091	–8 695	–15 725	217 829	1 221 824	2 109 854	4 368 693	–225 927	–944 439
Unit CSE	Rb/t	–68	–64	–50	–84	–151	–327	4 115	25 668	51 252	130 401	–7 371	–27 860
Percentage CSE	%	–52	–49	–35	–45	–58	–68	36	61	58	55	–1	–5
Potatoes (not included in the aggregation)													
Total CSE	Rb mn	–5 031	–3 167	–3 922	–5 627	–6 261	–28 958	290 187	–209 084	1 812 877	11 971 302	–25 411 084	–23 205 623
Unit CSE	Rb/t	–123	–82	–107	–166	–194	–856	7 802	–5 444	49 589	320 037	–668 449	–610 433
Percentage CSE	%	–67	–52	–56	–68	–66	–78	85	–9	20	35	–59	–60
Sunflower													
Total CSE	Rb mn	–333	–352	88	–352	–339	–810	42 270	284 362	309 011	93 099	164 465	304 506
Unit CSE	Rb/t	–106	–115	30	–95	–102	–300	13 836	124 447	148 921	35 399	81 217	152 253
Percentage CSE	%	–39	–32	8	–19	–24	–36	78	173	57	4	10	19
Sugar													
Total CSE	Rb mn	–2 214	–2 272	–2 025	–1 058	–1 596	–2 786	3 840	–348 339	–298 424	–1 433 189	–5 825 777	–4 137 219
Unit CSE	Rb/t	–323	–319	–272	–158	–227	–494	857	–74 478	–64 862	–304 235	–1 206 665	–856 922
Percentage CSE	%	–52	–53	–39	–22	–33	–54	4	–37	–15	–26	–57	–49
Crops													
Total CSE	Rb mn	–11 031	–9 572	–10 694	–13 511	–22 397	–55 960	898 337	2 829 060	4 907 228	8 006 947	–4 899 397	–6 207 145
Percentage CSE	**%**	**–51**	**–47**	**–41**	**–41**	**–54**	**–71**	**70**	**49**	**44**	**26**	**–9**	**–12**

e: estimate; p: provisional.
Source: OECD.

Annex Table I.13.*ii.* **Consumer Subsidy Equivalents: by commodity** *(cont.)*

	Units	1986	1987	1988	1989	1990	1991	1992	1993	1994	1995	1996p	1997e
Milk													
Total CSE	Rb mn	–15 774	–15 287	–21 717	–21 588	–22 256	–16 706	835 993	2 832 238	4 974 164	–7 330 131	–11 814 375	–12 757 295
Unit CSE	Rb/t	–268	–255	–342	–329	–344	–283	16 849	55 376	102 526	–164 541	–288 479	–324 704
Percentage CSE	%	–56	–53	–53	–50	–53	–35	297	112	65	–20	–27	–29
Beef and veal													
Total CSE	Rb mn	–10 598	–12 272	–13 304	–14 967	–7 577	–17 733	672 948	2 422 658	5 663 423	8 564 344	3 271 594	3 981 648
Unit CSE	Rb/t	–2 344	–2 550	–2 547	–2 805	–1 545	–3 505	161 572	638 045	1 493 913	2 517 444	1 070 898	1 424 561
Percentage CSE	%	–52	–55	–46	–49	–25	–40	389	127	116	53	15	21
Pigmeat													
Total CSE	Rb mn	–5 100	–5 774	–6 456	–7 018	–3 926	–7 572	567 854	804 443	503 081	–1 929 424	–4 094 533	–4 179 165
Unit CSE	Rb/t	–1 515	–1 626	–1 770	–1 811	–1 121	–2 171	190 940	298 605	204 007	–750 749	–1 905 320	–2 122 481
Percentage CSE	%	–52	–55	–49	–49	–27	–35	440	51	11	–12	–22	–21
Poultry													
Total CSE	Rb mn	–1 698	–1 580	–2 451	–3 231	–1 751	–2 867	218 499	213 285	–483 401	–4 556 289	–8 180 939	–9 898 045
Unit CSE	Rb/t	–929	–808	–1 208	–1 472	–798	–1 544	144 033	143 530	–304 409	–2 605 082	–4 661 503	–5 212 241
Percentage CSE	%	–33	–27	–38	–44	–22	–25	333	23	–15	–41	–49	–50
Eggs													
Total CSE	Rb mn	–3 947	–3 636	–3 949	–3 898	–3 138	–6 463	273 773	684 246	29 615	–3 372 238	–4 538 336	–5 249 889
Unit CSE	Rb/t	–1 452	–1 309	–1 396	–1 389	–1 153	–2 458	115 316	306 952	14 158	–1 785 565	–2 527 226	–2 923 462
Percentage CSE	%	–81	–72	–76	–77	–61	–60	347	85	1	–38	–37	–40
Livestock products													
Total CSE	Rb mn	–37 115	–38 548	–47 877	–50 702	–38 648	–51 341	2 569 067	6 956 870	10 686 882	–8 623 738	–25 356 589	–28 102 748
Percentage CSE	**%**	**–54**	**–53**	**–51**	**–51**	**–38**	**–38**	**353**	**90**	**46**	**–10**	**–23**	**–24**
All products													
Total CSE	Rb mn	–48 146	–48 120	–58 571	–64 212	–61 045	–107 301	3 467 403	9 785 930	15 594 110	–616 792	–30 255 986	–34 309 893
Percentage CSE	**%**	**–54**	**–52**	**–49**	**–48**	**–43**	**–50**	**172**	**72**	**45**	**–1**	**–18**	**–20**

e: estimate; p: provisional.
Source: OECD.

Annex Table I.14. **Composition of Russian assistance, 1986-1997**

	1986	1987	1988	1989	1990	1991	1992	1993	1994	1995	1996p	1997e
TOTAL PSE												
MPS	84	83	92	95	99	118	–151	–255	–521	–2	44	61
DPs	3	3	0	0	0	7	8	28	61	20	11	7
RI	18	17	14	13	13	3	16	66	154	38	33	19
GS	4	5	4	5	5	6	3	17	108	17	11	15
Feed	–10	–8	–10	–12	–17	–33	24	45	98	27	1	–2
Total	100	100	100	100	100	100	–100	–100	–100	100	100	100
TOTAL CSE												
MT	160	161	172	170	185	150	–98	–96	–97	100	100	100
OT	–60	–61	–72	–70	–85	–50	–2	–4	–3	0	0	0
Total	100	100	100	100	100	100	–100	–100	–100	100	100	100

e: estimate; p: provisional.
PSE: Producer subsidy equivalent.
MPS: Market price support.
DPs: Direct payments.
RI: Reduction of input costs.
GS: General services.
CSE: Consumer subsidy equivalent.
MT: Market transfers.
Source: OECD.

Annex Table I.15.*i*. **Shares in total Russian PSE by commodity**

	1986	1987	1988	1989	1990	1991	1992	1993	1994	1995	1996p	1997e
Wheat	7	5	4	7	11	22	19	19	47	–12	4	8
Maize	1	1	2	2	2	2	1	0	–4	1	1	2
Rye	2	3	2	2	4	4	2	6	4	1	3	3
Barley	4	4	2	3	6	7	–1	6	30	–12	0	1
Oats	2	2	1	2	3	5	3	8	16	–3	3	2
Sunflower	1	2	1	2	1	2	1	9	9	1	0	–1
Sugar	3	3	3	3	2	1	–1	–10	–11	5	5	4
Crops	20	19	15	21	29	43	25	38	91	–19	16	20
Milk	36	35	39	37	34	26	25	35	51	67	41	42
Beef and veal	23	25	25	24	23	21	19	36	111	–18	2	–4
Pigmeat	10	11	11	10	8	4	16	–5	–52	23	14	13
Poultry	6	6	6	6	4	3	5	–10	–57	21	11	11
Eggs	5	5	4	3	2	3	9	7	–44	26	16	18
Livestock	80	81	85	79	71	57	75	62	9	119	84	80
Total	100	100	100	100	100	100	100	100	100	100	100	100

e: estimate; p: provisional.
Source: OECD.

Annex Table I.15.*ii.* **Shares in total Russian CSE by commodity**

	1986	1987	1988	1989	1990	1991	1992	1993	1994	1995	1996p	1997e
Wheat	5	5	2	5	13	25	17	16	19	–836	–5	3
Maize	6	2	9	8	6	9	1	1	–1	29	1	2
Rye	2	2	1	1	3	3	2	4	2	11	2	2
Barley	4	4	2	3	8	7	1	5	7	–540	–4	–1
Oats	1	1	1	2	3	4	3	4	4	–179	2	2
Sunflower	1	1	0	1	1	1	1	3	2	–15	–1	–1
Sugar	5	5	3	2	3	3	0	–4	–2	232	19	12
Crops	23	20	18	21	37	52	26	29	31	–1 298	16	18
Milk	33	32	37	34	36	16	24	29	32	1 188	39	37
Beef and veal	22	26	23	23	12	17	19	25	36	–1 389	–11	–12
Pigmeat	11	12	11	11	6	7	16	8	3	313	14	12
Poultry	4	3	4	5	3	3	6	2	–3	739	27	29
Eggs	8	8	7	6	5	6	8	7	0	547	15	15
Livestock	77	80	82	79	63	48	74	71	69	1 398	84	82
Total	100	100	100	100	100	100	100	100	100	100	100	100

e: estimate; p: provisional.
Source: OECD.

Annex Table I.16.*i.* **OECD Producer Subsidy Equivalents: by country**

	Units	1986	1987	1988	1989	1990	1991	1992	1993	1994	1995	1996	1997p
Australia													
Total PSE	A$ mn	1 670	1 445	1 244	1 204	1 933	1 563	1 585	1 506	1 581	1 717	1 438	1 461
Total PSE	US$ mn	1 116	1 011	972	952	1 508	1 217	1 163	1 022	1 154	1 272	1 126	1 094
Total PSE	ECU mn	1 138	877	822	864	1 188	985	899	873	973	973	887	955
Percentage PSE	%	14	10	7	7	13	11	11	10	10	10	8	9
Canada													
Total PSE	C$ mn	8 546	7 964	6 602	6 227	8 204	8 486	7 071	6 167	5 210	5 400	5 178	4 335
Total PSE	US$ mn	6 151	6 005	5 362	5 259	7 030	7 404	5 849	4 780	3 815	3 934	3 797	3 135
Total PSE	ECU mn	6 268	5 209	4 537	4 776	5 537	5 990	4 519	4 080	3 216	3 010	2 991	2 737
Percentage PSE	%	47	44	37	35	45	44	37	31	25	22	22	20
Czech Republic													
Total PSE	CZK mn	46 479	40 627	37 577	48 516	50 626	42 240	23 045	21 677	16 476	11 732	12 878	10 329
Total PSE	US$ mn	3 567	3 003	2 686	3 379	3 446	2 729	1 039	832	586	442	474	327
Total PSE	ECU mn	3 635	2 605	2 273	3 068	2 715	2 208	803	710	494	338	374	286
Percentage PSE	%	70	61	55	55	54	51	30	27	21	15	14	11
EU[1]													
Total PSE	ECU mn	64 323	61 849	58 420	53 536	64 501	66 889	64 365	67 110	65 453	70 183	64 742	63 451
Total PSE	US$ mn	63 124	71 304	69 037	58 947	81 886	82 681	83 310	78 611	77 633	91 742	82 181	72 682
Percentage PSE	%	50	49	46	40	47	47	47	49	48	49	43	42
Hungary													
Total PSE	Ft mn	94 040	90 816	88 450	83 206	84 750	46 563	55 697	72 450	116 903	98 541	102 099	122 747
Total PSE	US$ mn	2 390	2 194	1981	1 647	1 398	623	705	788	1 112	784	669	659
Total PSE	ECU mn	2 435	1 903	1 677	1 496	1 101	504	545	673	938	600	527	575
Percentage PSE	%	48	44	40	31	27	15	20	24	31	21	15	16
Iceland													
Total PSE	I.kr mn	5 331	6 608	7 284	8 570	9 630	10 601	9 885	8 859	8 291	8 737	7 770	8 220
Total PSE	US$ mn	130	171	169	150	165	179	172	131	118	135	117	116
Total PSE	ECU mn	132	148	143	136	130	145	133	112	100	103	92	101
Percentage PSE	%	79	83	84	82	84	86	83	76	72	75	69	69
Japan													
Total PSE	¥ bn	5 577	4 968	4 558	4 463	4 293	4 107	4 624	3 944	4 774	4 571	4 327	3 981
Total PSE	US$ mn	33 093	34 354	35 575	32 348	29 647	30 533	36 505	35 478	46 695	48 597	39 761	33 184
Total PSE	ECU mn	33 722	29 799	30 103	29 379	23 353	24 701	28 203	30 288	39 368	37 177	31 324	28 970
Percentage PSE	%	74	74	72	69	66	66	74	74	75	76	71	69
Mexico													
Total PSE	MN$ mn	–144	838	2 076	3 296	10 368	12 040	15 129	17 505	17 592	–71	9 325	19 277
Total PSE	US$ mn	–226	591	910	1 321	3 650	3 983	4 888	5 619	5 192	–11	1 227	2 431
Total PSE	ECU mn	–230	513	770	1 200	2 875	3 223	3 777	4 797	4 377	–8	967	2 122
Percentage PSE	%	–3	7	10	12	29	29	32	36	33	0	8	16

p: provisional.
1. EU-12 for 1986-1994, EU-15 from 1995; as from 1990 includes ex-GDR.
Source: OECD Secretariat.

Annex Table I.16.*i.* **OECD Producer Subsidy Equivalents: by country** *(cont.)*

	Units	1986	1987	1988	1989	1990	1991	1992	1993	1994	1995	1996	1997p
New Zealand													
Total PSE	NZ$ mn	1 773	656	400	341	265	187	169	178	200	223	204	215
Total PSE	US$ mn	925	387	261	204	158	108	91	96	119	146	140	143
Total PSE	ECU mn	943	336	221	185	125	87	70	82	100	112	111	125
Percentage PSE	%	33	13	7	6	5	3	3	3	3	3	3	3
Norway													
Total PSE	NKr mn	15 122	16 669	16 821	16 725	19 164	19 875	19 244	18 671	17 078	16 301	16 303	16 833
Total PSE	US$ mn	2 046	2 474	2 581	2 423	3 062	3 065	3 097	2 632	2 420	2 572	2 525	2 385
Total PSE	ECU mn	2 085	2 146	2 184	2 200	2 412	2 480	2 393	2 247	2 041	1 968	1 989	2 082
Percentage PSE	%	74	73	73	71	74	76	76	74	73	72	70	71
Poland													
Total PSE	NZl mn	80	64	112	77	–907	55	2 249	2 207	3 568	4 569	7 163	7 119
Total PSE	US$ mn	3 572	2 301	2 508	450	–955	52	1 650	1 217	1 570	1 884	2 657	2 171
Total PSE	ECU mn	3 640	1 996	2 122	409	–752	42	1 275	1 039	1 324	1 441	2 094	1 895
Percentage PSE	%	42	28	27	5	–15	1	20	15	20	19	23	22
Switzerland													
Total PSE	SF mn	6 675	6 864	7 169	6 953	7 500	7 686	7 139	7 553	7 408	6 895	6 710	6 624
Total PSE	US$ mn	3 712	4 605	4 900	4 251	5 400	5 361	5 079	5 113	5 419	5 833	5 428	4 572
Total PSE	ECU mn	3 782	3 994	4 146	3 861	4 253	4 337	3 924	4 365	4 569	4 462	4 276	3 991
Percentage PSE	%	79	79	78	73	79	79	78	80	81	79	77	76
Turkey													
Total PSE	TL bn	864	1 494	2 446	4 133	8 124	17 783	27 024	42 752	54 627	154 324	256 800	781 178
Total PSE	US$ mn	1 291	1 748	1 721	1 950	3 117	4 266	3 939	3 899	1 834	3 374	3 159	5 161
Total PSE	ECU mn	1 316	1 516	1 456	1 771	2 455	3 451	3 043	3 329	1 547	2 581	2 489	4 506
Percentage PSE	%	24	29	26	28	31	41	37	35	24	30	25	38
United States													
Total PSE	US$ mn	36 664	35 699	25 233	23 488	28 117	24 841	26 082	27 657	25 307	17 344	22 614	22 791
Total PSE	ECU mn	37 361	30 965	21 352	21 332	22 148	20 097	20 151	23 610	21 336	13 268	17 815	19 897
Percentage PSE	%	35	32	23	20	23	21	21	23	19	13	15	16
OECD 24[2]													
Total PSE	US$ mn	155 347	166 057	154 365	138 461	171 141	170 032	174 679	167 174	172 766	174 949	160 847	145 264
Total PSE	ECU mn	158 298	144 038	130 623	125 750	134 807	137 556	134 957	142 716	145 659	133 836	126 715	126 816
Percentage PSE	**%**	**47**	**46**	**41**	**37**	**41**	**42**	**42**	**42**	**41**	**40**	**35**	**35**
OECD 28[3]													
Total PSE	US$ mn	164 649	174 145	162 450	145 258	178 679	177 420	182 962	175 630	181 226	178 048	165 875	150 852
Total PSE	ECU mn	167 778	151 053	137 465	131 923	140 746	143 533	141 356	149 935	152 791	136 207	130 676	131 694
Percentage PSE	**%**	**47**	**45**	**40**	**36**	**40**	**41**	**41**	**41**	**41**	**38**	**34**	**34**

p: provisional.
2. Excludes Korea, the Czech Republic, Hungary, Mexico and Poland.
3. Excludes Korea.
Source: OECD Secretariat.

Annex Table I.16.*ii.* **OECD Consumer Subsidy Equivalents: by country**

	Units	1986	1987	1988	1989	1990	1991	1992	1993	1994	1995	1996	1997p
Australia													
Total CSE	A$ mn	–565	–410	–346	–353	–591	–538	–485	–421	–433	–436	–307	–403
Total CSE	US$ mn	–377	–287	–270	–279	–461	–419	–356	–286	–316	–323	–240	–302
Total CSE	ECU mn	–385	–249	–228	–254	–363	–339	–275	–244	–267	–247	–189	–264
Percentage CSE	%	–12	–8	–6	–6	–11	–9	–8	–7	–6	–6	–5	–6
Canada													
Total CSE	C$ mn	–3 285	–3 344	–2 551	–2 566	–3 212	–3 111	–2 939	–3 078	–2 717	–1 999	–1 971	–2 210
Total CSE	US$ mn	–2 365	–2 522	–2 072	–2 167	–2 752	–2 715	–2 431	–2 386	–1 989	–1 456	–1 445	–1 598
Total CSE	ECU mn	–2 410	–2 187	–1 753	–1 968	–2 168	–2 196	–1 879	–2 037	–1 677	–1 114	–1 139	–1 395
Percentage CSE	%	–27	–26	–19	–19	–26	–25	–23	–21	–17	–12	–12	–14
Czech Republic													
Total CSE	CKr mn	–20 793	–16 963	–12 164	–8 858	–20 074	–32 770	–17 752	–16 662	–12 169	–2 386	–2 759	–3 687
Total CSE	US$ mn	–1 596	–1 254	–870	–617	–1 366	–2 117	–801	–640	–433	–90	–102	–117
Total CSE	ECU mn	–1 626	–1 087	–736	–560	–1 076	–1 713	–619	–546	–365	–69	–80	–102
Percentage CSE	%	–41	–33	–23	–14	–29	–51	–27	–24	–18	–3	–3	–4
EU[1]													
Total CSE	ECU mn	–53 957	–53 317	–47 592	–40 943	–48 628	–51 761	–48 819	–45 570	–42 405	–40 466	–29 467	–30 673
Total CSE	US$ mn	–52 950	–61 468	–56 241	–45 082	–61 734	–63 981	–63 187	–53 380	–50 296	–52 897	–37 404	–35 135
Percentage CSE	%	–45	–46	–40	–33	–41	–42	–40	–39	–36	–33	–24	–25
Hungary													
Total CSE	Ft mn	–49 213	–43 080	–36 094	–28 279	–58 617	–18 757	–27 412	–55 972	–74 968	–37 409	–40 409	–47 923
Total CSE	US$ mn	–1 251	–1 041	–809	–560	–967	–251	–347	–609	–713	–298	–265	–257
Total CSE	ECU mn	–1 274	–903	–684	–508	–761	–203	–268	–520	–601	–228	–209	–224
Percentage CSE	%	–31	–26	–21	–14	–21	–8	–10	–22	–24	–11	–8	–9
Iceland													
Total CSE	I.kr mn	–3 353	–4 514	–3 332	–3 457	–3 842	–4 398	–2 643	–3 064	–3 283	–3 381	–2 867	–2 865
Total CSE	US$ mn	–82	–117	–77	–61	–66	–74	–46	–45	–47	–52	–43	–40
Total CSE	ECU mn	–83	–101	–65	–55	–52	–60	–35	–39	–40	–40	–34	–35
Percentage CSE	%	–55	–57	–38	–32	–33	–36	–25	–37	–40	–40	–34	–34
Japan													
Total CSE	¥ bn	–5 134	–5 022	–4 855	–4 525	–4 384	–4 473	–4 844	–4 644	–4 372	–4 544	–4 364	–4 280
Total CSE	US$ mn	–30 463	–34 722	–37 893	–32 795	–30 277	–33 254	–38 241	–41 771	–42 763	–48 310	–40 105	–35 681
Total CSE	ECU mn	–31 042	–30 118	–32 065	–29 785	–23 849	–26 903	–29 545	–35 660	–36 054	–36 957	–31 595	–31 150
Percentage CSE	%	–58	–58	–55	–50	–47	–48	–53	–51	–50	–51	–47	–46
Mexico													
Total CSE	M$ mn	1 506	811	1 142	840	–2 047	–7 235	–8 324	–10 652	–5 057	11 600	10 330	–113
Total CSE	US$ mn	2 355	572	501	337	–721	–2 394	–2 689	–3 419	–1 492	1 806	1 359	–14
Total CSE	ECU mn	2 400	497	424	306	–568	–1 937	–2 078	–2 919	–1 258	1 382	1 071	–12
Percentage CSE	%	29	6	4	3	–5	–16	–16	–20	–9	14	8	0

p: provisional.
1. EU-12 for 1986-1994, EU-15 from 1995; as from 1990 includes ex-GDR.
Source: OECD Secretariat.

Annex Table I.16.*ii.* **OECD Consumer Subsidy Equivalents: by country** *(cont.)*

	Units	1986	1987	1988	1989	1990	1991	1992	1993	1994	1995	1996	1997p
New Zealand													
Total CSE	NZ$ mn	–117	–104	–88	–88	–71	–53	–51	–69	–91	–109	–92	–98
Total CSE	US$ mn	–61	–61	–57	–52	–43	–31	–27	–37	–54	–72	–63	–65
Total CSE	ECU mn	–62	–53	–49	–47	–34	–25	–21	–32	–45	–55	–50	–57
Percentage CSE	%	–10	–9	–6	–5	–5	–4	–3	–4	–6	–6	–5	–6
Norway													
Total CSE	NKr mn	–8 263	–9 222	–8 553	–8 197	–9 927	–9 843	–9 760	–9 218	–8 403	–7 754	–7 080	–7 787
Total CSE	US$ mn	–1 118	–1 369	–1 312	–1 187	–1 586	–1 518	–1 571	–1 299	–1 191	–1 224	–1 096	–1 103
Total CSE	ECU mn	–1 139	–1 187	–1 111	–1 078	–1 249	–1 228	–1 213	–1 109	–1 004	–936	–864	–963
Percentage CSE	%	–63	–64	–59	–56	–64	–64	–63	–60	–56	–53	–48	–50
Poland													
Total CSE	NZl mn	–29	–17	60	414	1 636	575	–1 903	–2 213	–3 091	–3 710	–7 478	–6 363
Total CSE	US$ mn	–1 299	–624	1 332	2 421	1 722	544	–1 396	–1 220	–1 360	–1 530	–2 774	–1 940
Total CSE	ECU mn	–1 324	–541	1 127	2 199	1 357	440	–1 079	–1 042	–1 147	–1 170	–2 185	–1 694
Percentage CSE	%	–16	–8	14	29	28	8	–17	–14	–17	–16	–23	–20
Switzerland													
Total CSE	SF mn	–5 316	–5 696	–5 906	–5 223	–5 482	–5 522	–4 822	–5 181	–4 939	–4 432	–3 901	–3 766
Total CSE	US$ mn	–2 956	–3 821	–4 037	–3 193	–3 947	–3 852	–3 430	–3 507	–3 613	–3 749	–3 156	–2 599
Total CSE	ECU mn	–3 012	–3 314	–3 416	–2 900	–3 109	–3 116	–2 650	–2 994	–3 046	–2 868	–2 486	–2 269
Percentage CSE	%	–64	–65	–65	–58	–61	–62	–58	–61	–61	–59	–54	–53
Turkey													
Total CSE	TL bn	–613	–1 112	–1 471	–2 855	–6 980	–16 495	–25 112	–37 474	–33 248	–117 890	–201 164	–776 024
Total CSE	US$ mn	–917	–1 301	–1 035	–1 347	–2 678	–3 957	–3 660	–3 418	–1 117	–2 577	–2 475	–5 127
Total CSE	ECU mn	–934	–1 129	–876	–1 223	–2 109	–3 201	–2 828	–2 918	–941	–1 972	–1 950	–4 476
Percentage CSE	%	–17	–21	–16	–18	–27	–37	–32	–30	–13	–20	–18	–34
United States													
Total CSE	US$ mn	–13 931	–13 224	–7 855	–8 482	–12 363	–10 825	–10 724	–11 989	–10 769	–7 517	–9 992	–9 159
Total CSE	ECU mn	–14 196	–11 470	–6 647	–7 704	–9 738	–8 757	–8 285	–10 235	–9 080	–5 750	–7 872	–7 996
Percentage CSE	%	–17	–14	–8	–8	–12	–10	–10	–11	–10	–7	–8	–8
OECD 24[2]													
Total CSE	US$ mn	–111 550	–126 295	–118 346	–101 569	–124 661	–128 958	–131 302	–124 135	–118 395	–118 177	–96 020	–90 811
Total CSE	ECU mn	–113 670	–109 548	–100 144	–92 245	–98 195	–104 327	–101 444	–105 974	–99 819	–90 405	–75 644	–79 278
Percentage CSE	**%**	**–39**	**–39**	**–34**	**–29**	**–33**	**–34**	**–34**	**–33**	**–31**	**–29**	**–24**	**–24**
OECD 28[3]													
Total CSE	US$ mn	–113 341	–128 640	–118 191	–99 988	–125 992	–133 176	–136 536	–130 024	–122 393	–118 287	–97 801	–93 140
Total CSE	ECU mn	–115 494	–111 583	–100 014	–90 809	–99 244	–107 739	–105 487	–111 001	–103 190	–90 490	–77 048	–81 311
Percentage CSE	**%**	**–36**	**–37**	**–31**	**–26**	**–31**	**–33**	**–33**	**–32**	**–30**	**–27**	**–22**	**–23**

p: provisional.
2. Excludes Korea, the Czech Republic, Hungary, Mexico and Poland.
3. Excludes Korea.
Source: OECD Secretariat.

Annex Table I.17.*i.* **Selected CEECs: Producer Subsidy Equivalents 1986-1997**

	Units	1986	1987	1988	1989	1990	1991	1992	1993	1994	1995	1996	1997p
Estonia													
Total PSE	LC mn	1 068	1 074	1 062	1 113	1 195	1 796	–1 909	–871	–190	94	289	386
Total PSE	US$ mn	1 548	1 705	1 770	1 767	2 060	1 030	–151	–66	–15	8	24	28
Total PSE	ECU mn	1 578	1 479	1 498	1 605	1 622	833	–117	–56	–12	6	19	24
Percentage PSE	%	**79**	**80**	**80**	**80**	**72**	**57**	**–91**	**–30**	**–6**	**3**	**7**	**9**
Latvia													
Total PSE	LC mn	2 131	2 084	2 551	2 440	2 182	4 471	–36 699	–88	20	19	18	19
Total PSE	US$ mn	3 088	3 308	4 252	3 873	3 762	7 708	–270	–130	37	36	32	33
Total PSE	ECU mn	3 147	2 869	3 598	3 518	2 963	6 236	–208	–111	31	27	25	29
Percentage PSE	%	**87**	**85**	**87**	**83**	**77**	**83**	**–93**	**–38**	**9**	**8**	**7**	**8**
Lithuania													
Total PSE	LC mn	3 341	2 830	3 301	3 133	3 283	–24 511	–80 674	–840	–191	225	705	823
Total PSE	US$ mn	4 842	4 492	5 502	4 973	5 661	–704	–490	–195	–48	56	176	206
Total PSE	ECU mn	4 934	3 896	4 656	4 516	4 459	–570	–379	–167	–40	43	139	180
Percentage PSE	%	**94**	**79**	**83**	**78**	**71**	**–259**	**–113**	**–33**	**–8**	**6**	**14**	**18**
Czech Republic													
Total PSE	CZK mn	46 479	40 627	37 577	48 516	50 626	42 240	23 045	21 677	16 476	11 732	12 878	10 329
Total PSE	US$ mn	3 567	3 003	2 686	3 379	3 446	2 729	1 039	832	586	442	474	327
Total PSE	ECU mn	3 635	2 605	2 273	3 068	2 715	2 208	803	710	494	338	374	286
Percentage PSE	%	**70**	**61**	**55**	**55**	**54**	**51**	**30**	**27**	**21**	**15**	**14**	**11**
Hungary													
Total PSE	Ft mn	94 040	90 816	88 450	83 206	84 750	46 563	55 697	72 450	116 903	98 541	102 099	122 747
Total PSE	US$ mn	2 390	2 194	1 981	1 647	1 398	623	705	788	1 112	784	669	659
Total PSE	ECU mn	2 435	1 903	1 677	1 496	1 101	504	545	673	938	600	527	575
Percentage PSE	%	**48**	**44**	**40**	**31**	**27**	**15**	**20**	**24**	**31**	**21**	**15**	**16**
Poland													
Total PSE	NZl mn	80	64	112	77	–907	55	2 249	2 207	3 568	4 569	7 163	7 119
Total PSE	US$ mn	3 572	2 301	2 508	450	–955	52	1 650	1 217	1 570	1 884	2 657	2 171
Total PSE	ECU mn	3 640	1 996	2 122	409	–752	42	1 275	1 039	1 324	1 441	2 094	1 895
Percentage PSE	%	**42**	**28**	**27**	**5**	**–15**	**1**	**20**	**15**	**20**	**19**	**23**	**22**
Slovakia													
Total PSE	SKK mn	23 477	22 055	20 813	27 242	28 934	19 082	15 626	14 942	14 431	12 040	9 974	13 540
Total PSE	US$ mn	1 354	1 298	1 260	1 706	1 669	879	613	520	450	405	325	403
Total PSE	ECU mn	1 380	1 126	1 066	1 550	1 315	711	474	444	380	310	256	352
Percentage PSE	%	**63**	**57**	**52**	**56**	**57**	**45**	**40**	**35**	**31**	**25**	**19**	**25**
EU[1]													
Total PSE	ECU mn	64 323	61 849	58 420	53 536	64 501	66 889	64 365	67 110	65 453	70 183	64 742	63 451
Total PSE	US$ mn	63 124	71 304	69 037	58 947	81 886	82 681	83 310	78 611	77 633	91 742	82 181	72 682
Percentage PSE	%	**50**	**49**	**46**	**40**	**47**	**47**	**47**	**49**	**48**	**49**	**43**	**42**
OECD[2]													
Total PSE	US$ mn	155 347	166 057	154 365	138 461	171 141	170 032	174 679	167 174	172 766	174 949	160 847	145 264
Total PSE	ECU mn	158 298	144 038	130 623	125 750	134 807	137 556	134 957	142 716	145 659	133 836	126 715	126 816
Percentage PSE	%	**47**	**46**	**41**	**37**	**41**	**42**	**42**	**42**	**41**	**40**	**35**	**35**

p: provisional.
1. EU-12 for 1986-1994 EU-15 from 1995; as from 1990 includes ex-GDR.
2. OECD does not include Czech Republic Hungary Poland Mexico and Korea.
Source: OECD Secretariat.

Annex Table I.17.*ii.* **Selected CEECs: Consumer Subsidy Equivalents 1986-1997**

	Units	1986	1987	1988	1989	1990	1991	1992	1993	1994	1995	1996	1997p
Estonia													
Total CSE	LC mn	–224	–151	–139	–17	–333	–1 996	1 930	875	232	4	–152	–290
Total CSE	US$ mn	–325	–240	–231	–26	–574	–1 144	153	66	18	0	–13	–21
Total CSE	ECU mn	–331	–208	–195	–24	–452	–926	118	56	15	0	–10	–18
Percentage CSE	%	**–21**	**–15**	**–14**	**–2**	**–26**	**–76**	**86**	**34**	**8**	**0**	**–3**	**–6**
Latvia													
Total CSE	LC mn	–630	–635	–836	–613	–516	–4 593	38 012	76	–26	–6	–10	–20
Total CSE	US$ mn	–913	–1 008	–1 393	–973	–889	–7 919	279	113	–46	–11	–18	–35
Total CSE	ECU mn	–931	–874	–1 179	–884	–700	–6 406	216	97	–39	–9	–14	–30
Percentage CSE	%	**–36**	**–35**	**–43**	**–31**	**–24**	**–86**	**91**	**32**	**–10**	**–2**	**–3**	**–7**
Lithuania													
Total CSE	LC mn	–309	–292	–617	–335	–892	16 962	63 249	721	325	–133	–426	–434
Total CSE	US$ mn	–448	–464	–1 029	–532	–1 538	487	384	168	81	–33	–106	–109
Total CSE	ECU mn	–456	–402	–870	–483	–1 211	394	297	143	68	–25	–84	–95
Percentage CSE	%	**–14**	**–13**	**–24**	**–14**	**–29**	**285**	**123**	**39**	**16**	**–5**	**–12**	**–13**
Czech Republic													
Total CSE	CZK mn	–20 793	–16 963	–12 164	–8 858	–20 074	–32 770	–17 752	–16 662	–12 169	–2 386	–2 759	–3 687
Total CSE	US$ mn	–1 596	–1 254	–870	–617	–1 366	–2 117	–801	–640	–433	–90	–102	–117
Total CSE	ECU mn	–1 626	–1 087	–736	–560	–1 076	–1 713	–619	–546	–365	–69	–80	–102
Percentage CSE	%	**–41**	**–33**	**–23**	**–14**	**–29**	**–51**	**–27**	**–24**	**–18**	**–3**	**–3**	**–4**
Hungary													
Total CSE	Ft mn	–49 213	–43 080	–36 094	–28 279	–58 617	–18 757	–27 412	–55 972	–74 968	–37 409	–40 409	–47 923
Total CSE	US$ mn	–1 251	–1 041	–809	–560	–967	–251	–347	–609	–713	–298	–265	–257
Total CSE	ECU mn	–1 274	–903	–684	–508	–761	–203	–268	–520	–601	–228	–209	–224
Percentage CSE	%	**–31**	**–26**	**–21**	**–14**	**–21**	**–8**	**–10**	**–22**	**–24**	**–11**	**–8**	**–9**
Poland													
Total CSE	NZl mn	–29	–17	60	414	1 636	575	–1 903	–2 213	–3 091	–3 710	–7 478	–6 363
Total CSE	US$ mn	–1 299	–624	1 332	2 421	1 722	544	–1 396	–1 220	–1 360	–1 530	–2 774	–1 940
Total CSE	ECU mn	–1 324	–541	1 127	2 199	1 357	440	–1 079	–1 042	–1 147	–1 170	–2 185	–1 694
Percentage CSE	%	**–16**	**–8**	**14**	**29**	**28**	**8**	**–17**	**–14**	**–17**	**–16**	**–23**	**–20**
Slovakia													
Total CSE	SKK mn	–8 043	–6 725	–4 679	–3 940	–7 722	–9 681	–4 127	–7 680	–9 286	–2 959	74	–2 537
Total CSE	US$ mn	–464	–396	–283	–247	–446	–446	–162	–267	–290	–99	2	–75
Total CSE	ECU mn	–473	–343	–240	–224	–351	–361	–125	–228	–244	–76	2	–66
Percentage CSE	%	**–30**	**–24**	**–16**	**–11**	**–21**	**–28**	**–13**	**–22**	**–21**	**–7**	**0**	**–5**
EU[1]													
Total CSE	ECU mn	–53 957	–53 317	–47 592	–40 943	–48 628	–51 761	–48 819	–45 570	–42 405	–40 466	–29 467	–30 673
Total CSE	US$ mn	–52 950	–61 468	–56 241	–45 082	–61 734	–63 981	–63 187	–53 380	–50 296	–52 897	–37 404	–35 135
Percentage CSE	%	**–45**	**–46**	**–40**	**–33**	**–41**	**–42**	**–40**	**–39**	**–36**	**–33**	**–24**	**–25**
OECD[2]													
Total CSE	US$ mn	–111 550	–126 295	–118 346	–101 569	–124 661	–128 958	–131 302	–124 135	–118 395	–118 177	–96 020	–90 811
Total CSE	ECU mn	–113 670	–109 548	–100 144	–92 245	–98 195	–104 327	–101 444	–105 974	–99 819	–90 405	–75 644	–79 278
Percentage CSE	%	**–39**	**–39**	**–34**	**–29**	**–33**	**–34**	**–34**	**–33**	**–31**	**–29**	**–24**	**–24**

p: provisional.
1. EU-12 for 1986-1994 EU-15 from 1995; as from 1990 includes ex-GDR.
2. OECD does not include Czech Republic Hungary Poland Mexico and Korea.
Source: OECD Secretariat.

Annex Table I.18.*i.* **OECD Producer Subsidy Equivalents: by commodity**

	Units	1986	1987	1988	1989	1990	1991	1992	1993	1994	1995	1996	1997p
Wheat													
Gross total PSE	US$ mn	18 820	20 395	16 378	10 151	17 604	22 827	17 007	18 798	16 530	14 215	13 385	14 298
Gross unit PSE	US$/t	86	97	83	48	70	98	72	81	76	63	53	61
Gross percentage PSE	%	54	58	45	28	42	54	41	45	42	29	25	32
Maize													
Gross total PSE	US$ mn	13 432	15 046	9 046	8 221	9 826	9 806	11 826	10 152	10 294	6 211	5 937	7 750
Gross unit PSE	US$/t	51	63	50	33	39	39	39	45	32	25	19	24
Gross percentage PSE	%	47	48	35	27	31	30	33	33	27	17	15	20
Other grains													
Gross total PSE	US$ mn	10 491	11 790	8 060	6 267	9 969	10 216	8 706	10 454	10 521	9 689	8 717	7 829
Gross unit PSE	US$/t	78	92	65	53	77	81	73	91	98	89	65	62
Gross percentage PSE	%	59	62	40	33	46	50	46	55	56	44	34	37
Rice													
Gross total PSE	US$ mn	22 768	23 067	22 086	20 730	20 183	19 591	24 145	22 273	31 911	32 762	26 919	21 934
Gross unit PSE	US$/t	1 097	1 168	1 085	996	954	958	1 085	1 204	1 308	1 485	1 191	974
Gross percentage PSE	%	91	90	85	82	82	81	88	92	88	91	82	80
Oilseeds													
Gross total PSE	US$ mn	5 138	6 742	5 545	5 529	7 277	6 329	6 006	5 798	5 345	5 561	4 000	5 001
Gross unit PSE	US$/t	73	90	88	75	98	81	75	80	57	65	46	55
Gross percentage PSE	%	33	33	27	28	34	31	28	27	22	22	15	20
Sugar (Refined Equivalent)													
Gross total PSE	US$ mn	5 725	6 254	5 423	4 767	5 538	6 790	7 505	6 610	4 664	4 631	5 466	5 676
Gross unit PSE	US$/t	177	191	165	144	158	199	208	174	130	128	142	144
Gross percentage PSE	%	65	65	53	42	46	59	60	56	42	40	46	49
Crops													
Gross total PSE	US$ mn	76 372	83 294	66 537	55 665	70 396	75 559	75 195	74 085	79 266	73 068	64 424	62 486
Gross percentage PSE	**%**	**58**	**59**	**48**	**39**	**46**	**50**	**48**	**50**	**47**	**41**	**34**	**37**

p: provisional.
Source: OECD Secretariat.

Annex Table I.18.*i.* **OECD Producer Subsidy Equivalents: by commodity** *(cont.)*

	Units	1986	1987	1988	1989	1990	1991	1992	1993	1994	1995	1996	1997p
Milk													
Net total PSE	US$ mn	51 351	48 187	45 567	44 167	60 001	54 656	55 576	51 879	50 730	49 774	49 495	44 513
Net unit PSE	US$/t	196	188	178	173	226	209	215	200	194	189	187	168
Net percentage PSE	%	74	64	56	54	66	63	62	61	59	54	52	52
Beef and Veal													
Net total PSE	US$ mn	21 375	22 807	26 451	24 530	27 061	26 336	28 525	28 911	29 679	32 251	31 671	27 350
Net unit PSE	US$/t	806	868	1 029	984	1 054	991	1 086	1 137	1 152	1 218	1 179	1 020
Net percentage PSE	%	33	31	33	31	32	31	33	35	35	37	36	33
Pigmeat													
Net total PSE	US$ mn	6 527	7 764	9 909	8 395	8 083	8 768	9 759	8 784	9 882	9 326	8 406	7 441
Net unit PSE	US$/t	240	278	340	290	269	291	317	280	317	296	265	236
Net percentage PSE	%	15	17	21	17	14	16	18	19	21	21	17	16
Poultry													
Net total PSE	US$ mn	2 995	5 289	5 533	4 295	5 151	4 164	4 861	4 531	4 941	4 952	4 989	3 867
Net unit PSE	US$/t	167	272	275	208	235	183	205	185	191	181	176	131
Net percentage PSE	%	14	23	22	16	18	15	16	15	16	16	15	12
Sheepmeat													
Net total PSE	US$ mn	3 783	4 671	5 468	5 233	5 633	5 696	5 928	4 423	4 068	5 284	3 785	3 099
Net unit PSE	US$/t	1 313	1 622	1 878	1 790	1 849	1 821	1 971	1 493	1 393	1 871	1 337	1 078
Net percentage PSE	%	54	58	62	59	59	58	56	45	44	51	37	33
Wool													
Net total PSE	US$ mn	344	398	347	315	652	474	426	372	308	255	196	172
Net unit PSE	US$/t	277	313	265	221	473	403	375	334	303	266	204	182
Net percentage PSE	%	12	8	6	6	16	16	18	17	10	10	7	7
Eggs													
Net total PSE	US$ mn	1 901	1 735	2 638	2 659	1 703	1 768	2 693	2 644	2 351	3 138	2 909	1 925
Net unit PSE	US$/t	133	120	180	187	119	120	182	178	155	206	188	122
Net percentage PSE	%	13	12	18	17	10	10	16	16	14	17	14	10
Livestock Products													
Net total PSE	US$ mn	88 277	90 851	95 913	89 593	108 283	101 861	107 767	101 545	101 960	104 980	101 451	88 365
Net percentage PSE	**%**	**40**	**37**	**36**	**34**	**37**	**36**	**37**	**37**	**37**	**36**	**34**	**32**
All Products													
Net total PSE	US$ mn	164 649	174 145	162 450	145 258	178 679	177 420	182 962	175 630	181 226	178 048	165 875	150 852
Net percentage PSE	**%**	**47**	**45**	**40**	**36**	**40**	**41**	**41**	**41**	**41**	**38**	**34**	**34**

p: provisional.
Source: OECD Secretariat.

Annex Table I.18.*ii*. **OECD Consumer Subsidy Equivalents: by commodity**

	Units	1986	1987	1988	1989	1990	1991	1992	1993	1994	1995	1996	1997p
Wheat													
Total CSE	US$ mn	–8 685	–11 226	–9 171	–4 234	–7 462	–12 375	–8 262	–7 244	–5 719	–2 393	–538	–2 485
Unit CSE	US$/t	–59	–77	–65	–30	–49	–81	–54	–47	–35	–15	–3	–15
Percentage CSE	%	–39	–46	–35	–16	–28	–44	–30	–29	–22	–8	–2	–9
Maize													
Total CSE	US$ mn	–2 473	–4 409	–2 540	–1 832	–3 658	–4 455	–4 329	–3 623	–1 752	–1 467	1 263	–137
Unit CSE	US$/t	–11	–19	–12	–8	–16	–19	–17	–15	–7	–6	5	0
Percentage CSE	%	–13	–19	–10	–7	–13	–16	–16	–13	–6	–4	4	0
Other grains													
Total CSE	US$ mn	–7 099	–8 426	–5 063	–3 608	–6 466	–7 055	–6 397	–5 424	–4 528	–2 480	–1 029	–1 312
Unit CSE	US$/t	–62	–72	–45	–33	–57	–64	–58	–52	–44	–24	–9	–12
Percentage CSE	%	–49	–52	–29	–21	–35	–40	–37	–38	–33	–16	–6	–10
Rice													
Total CSE	US$ mn	–16 932	–19 650	–20 415	–17 977	–17 379	–18 451	–20 605	–23 509	–22 440	–26 171	–22 131	–19 740
Unit CSE	US$/t	–971	–1 109	–1 153	–983	–941	–975	–1 077	–1 220	–1 212	–1 363	–1 116	–962
Percentage CSE	%	–80	–81	–76	–73	–74	–73	–78	–79	–76	–78	–71	–70
Oilseeds													
Total CSE	US$ mn	–488	–559	–341	–505	–642	–707	–607	–345	–389	–326	–527	–449
Unit CSE	US$/t	–7	–7	–5	–7	–8	–9	–7	–4	–4	–4	–6	–5
Percentage CSE	%	–4	–3	–2	–3	–3	–3	–3	–1	–1	–1	–2	–2
Sugar (refined equivalent)													
Total CSE	US$ mn	–5 928	–6 749	–5 616	–4 776	–5 482	–7 123	–7 377	–6 442	–5 104	–5 040	–5 506	–5 732
Unit CSE	US$/t	–194	–219	–181	–149	–166	–217	–222	–193	–154	–147	–160	–165
Percentage CSE	%	–61	–62	–48	–39	–42	–57	–57	–54	–41	–37	–44	–48
Crops													
Total CSE	US$ mn	–41 604	–51 018	–43 146	–32 932	–41 090	–50 166	–47 577	–46 588	–39 931	–37 877	–28 466	–29 856
Percentage CSE	**%**	**–41**	**–44**	**–33**	**–26**	**–32**	**–38**	**–35**	**–34**	**–29**	**–24**	**–18**	**–21**

p: provisional.
Source: OECD Secretariat.

Annex Table I.18.*ii.* **OECD Consumer Subsidy Equivalents: by commodity** *(cont.)*

	Units	1986	1987	1988	1989	1990	1991	1992	1993	1994	1995	1996	1997p
Milk													
Total CSE	US$ mn	–37 167	–34 668	–29 584	–28 850	–43 084	–40 079	–41 683	–38 869	–37 639	–34 049	–34 008	–31 391
Unit CSE	US$/t	–162	–151	–127	–128	–185	–173	–179	–167	–161	–147	–146	–135
Percentage CSE	%	–64	–53	–41	–41	–57	–54	–53	–52	–50	–43	–42	–43
Beef and veal													
Total CSE	US$ mn	–17 128	–18 592	–21 433	–20 195	–20 254	–19 158	–22 408	–24 494	–23 750	–27 194	–22 654	–21 088
Unit CSE	US$/t	–678	–734	–856	–820	–836	–778	–904	–995	–951	–1 050	–884	–803
Percentage CSE	%	–27	–25	–27	–25	–24	–23	–27	–28	–27	–29	–26	–25
Pigmeat													
Total CSE	US$ mn	–9 792	–12 633	–12 553	–8 276	–10 642	–13 539	–13 641	–11 628	–12 692	–10 460	–6 200	–6 311
Unit CSE	US$/t	–359	–452	–432	–285	–369	–450	–442	–378	–415	–339	–201	–207
Percentage CSE	%	–23	–28	–28	–18	–20	–25	–25	–25	–28	–23	–13	–14
Poultry													
Total CSE	US$ mn	–2 717	–5 363	–4 787	–3 509	–4 980	–4 575	–4 926	–4 145	–4 095	–3 828	–3 238	–2 474
Unit CSE	US$/t	–155	–286	–245	–174	–236	–207	–213	–176	–167	–152	–123	–91
Percentage CSE	%	–13	–25	–20	–14	–19	–17	–17	–14	–14	–13	–10	–8
Sheepmeat													
Total CSE	US$ mn	–2 778	–4 043	–4 074	–3 774	–3 988	–3 380	–3 365	–1 665	–2 124	–2 108	–1 226	–718
Unit CSE	US$/t	–1 183	–1 685	–1 681	–1 509	–1 548	–1 270	–1 272	–662	–860	–896	–547	–307
Percentage CSE	%	–45	–53	–53	–49	–49	–43	–40	–22	–28	–28	–15	–10
Wool													
Total CSE	US$ mn	–11	–13	–12	–13	–15	–3	–1	–1	1	0	–2	1
Unit CSE	US$/t	–14	–15	–15	–17	–19	–3	–1	–2	1	0	–3	1
Percentage CSE	%	–1	0	0	0	0	0	0	0	0	0	0	0
Eggs													
Total CSE	US$ mn	–2 143	–2 311	–2 602	–2 438	–1 940	–2 276	–2 935	–2 633	–2 163	–2 772	–2 008	–1 301
Unit CSE	US$/t	–159	–169	–189	–178	–141	–161	–206	–184	–149	–190	–136	–87
Percentage CSE	%	–16	–16	–19	–16	–11	–13	–19	–17	–13	–16	–10	–7
Livestock products													
Total CSE	US$ mn	–71 737	–77 622	–75 045	–67 056	–84 902	–83 010	–88 959	–83 436	–82 462	–80 411	–69 335	–63 284
Percentage CSE	**%**	**–35**	**–34**	**–31**	**–27**	**–32**	**–31**	**–32**	**–32**	**–31**	**–29**	**–25**	**–24**
All products													
Total CSE	US$ mn	–113 341	–128 640	–118 191	–99 988	–125 992	–133 176	–136 536	–130 024	–122 393	–118 287	–97 801	–93 140
Percentage CSE	**%**	**–37**	**–37**	**–32**	**–27**	**–32**	**–33**	**–33**	**–33**	**–30**	**–27**	**–22**	**–23**

p: provisional.
Source: OECD Secretariat.

Annex Table I.19. **Feed adjustment**

	Units	1986	1987	1988	1989	1990	1991	1992	1993	1994	1995	1996p	1997e
Unit Market Price Support													
Wheat	Rb/t	77	77	73	119	187	553	–9 919	–26 891	–63 371	–130 818	–36 130	22 359
Maize	Rb/t	262	177	357	339	651	943	–7 215	–16 582	68 452	99 177	291 060	208 206
Rye	Rb/t	118	133	108	113	187	344	–5 460	–34 166	–40 279	11 681	122 057	105 111
Barley	Rb/t	87	83	73	105	182	321	–1 548	–16 534	–46 767	–184 900	–66 190	–17 302
Oats	Rb/t	81	64	55	131	173	427	–8 159	–36 136	–64 530	–112 295	67 767	59 475
Feed usage													
Milk production	'000 t	52 217	52 880	54 535	55 742	55 715	51 887	47 236	46 524	42 176	39 241	35 819	34 066
Wheat	'000 t	6 785	6 958	6 129	6 871	7 535	6 799	7 222	6 186	4 922	4 436	3 570	3 664
Maize	'000 t	2 402	1 657	3 527	3 532	1 148	2 067	1 132	1 109	314	239	89	185
Rye	'000 t	1 769	1 689	1 700	1 599	2 167	1 428	1 747	1 102	207	136	177	208
Barley	'000 t	4 993	5 379	4 321	4 757	5 756	4 640	5 030	5 057	4 593	3 353	2 778	3 103
Oats	'000 t	2 237	2 418	2 050	2 373	2 242	1 905	1 998	2 035	1 951	2 030	1 596	1 653
Beef and veal production	'000 t	3 756	3 991	4 150	4 256	4 329	3 989	3 632	3 359	3 240	2 733	2 630	2 338
Wheat	'000 t	5 741	5 860	5 228	5 866	6 491	5 726	5 925	4 844	3 747	3 169	2 411	2 475
Maize	'000 t	2 033	1 395	3 009	3 016	989	1 741	929	869	239	171	60	125
Rye	'000 t	1 497	1 422	1 450	1 365	1 867	1 203	1 433	863	158	97	119	140
Barley	'000 t	4 225	4 530	3 686	4 061	4 959	3 908	4 127	3 960	3 497	2 395	1 876	2 096
Oats	'000 t	1 893	2 036	1 749	2 026	1 932	1 604	1 639	1 594	1 485	1 450	1 078	1 117
Pigmeat production	'000 t	3 093	3 264	3 399	3 499	3 480	3 190	2 784	2 432	2 103	1 865	1 705	1 565
Wheat	'000 t	8 751	8 600	7 330	8 252	9 628	8 581	9 701	7 226	5 496	4 823	4 048	4 155
Maize	'000 t	3 099	2 048	4 218	4 242	1 467	2 609	1 521	1 296	351	260	101	209
Rye	'000 t	2 282	2 087	2 033	1 921	2 769	1 803	2 346	1 288	231	148	200	235
Barley	'000 t	6 440	6 648	5 168	5 713	7 355	5 856	6 757	5 907	5 128	3 646	3 150	3 518
Oats	'000 t	2 886	2 989	2 452	2 851	2 865	2 404	2 684	2 377	2 178	2 207	1 810	1 875
Poultry production	'000 t	1 613	1 712	1 776	1 831	1 801	1 751	1 428	1 277	1 068	859	690	632
Wheat	'000 t	2 261	2 131	1 782	2 082	4 443	4 260	4 835	4 068	3 346	2 983	2 407	2 470
Maize	'000 t	800	507	1 025	1 071	677	1 295	758	730	214	161	60	125
Rye	'000 t	589	517	494	485	1 278	895	1 170	725	141	92	119	140
Barley	'000 t	1 664	1 647	1 256	1 442	3 394	2 907	3 368	3 325	3 122	2 255	1 873	2 092
Oats	'000 t	746	740	596	719	1 322	1 193	1 338	1 338	1 326	1 365	1 076	1 115
Egg production	'000 t	2 566	2 636	2 730	2 724	2 637	2 604	2 383	2 239	2 082	1 879	1 772	1 771
Wheat	'000 t	3 663	3 630	3 204	3 562	3 317	3 034	3 592	2 716	2 152	1 939	1 807	1 854
Maize	'000 t	1 297	864	1 844	1 831	505	922	563	487	137	104	45	93
Rye	'000 t	955	881	889	829	954	637	869	484	91	60	89	105
Barley	'000 t	2 695	2 806	2 259	2 466	2 534	2 070	2 502	2 220	2 008	1 465	1 406	1 570
Oats	'000 t	1 208	1 262	1 072	1 231	987	850	994	893	853	887	808	837
Total feed adjustment													
Milk	Rb mn	–1 974	–1 655	–2 317	–3 007	–3 992	–8 502	113 420	379 537	639 456	1 402 973	157 239	–186 849
Beef and veal	Rb mn	–1 671	–1 394	–1 976	–2 567	–3 439	–7 159	93 063	297 228	486 812	1 002 124	106 205	–126 205
Pigmeat	Rb mn	–2 547	–2 046	–2 771	–3 612	–5 101	–10 729	152 356	443 361	713 978	1 525 360	178 309	–211 887
Poultry	Rb mn	–658	–507	–674	–911	–2 354	–5 326	75 943	249 591	434 675	943 478	106 009	–125 972
Eggs	Rb mn	–1 066	–863	–1 211	–1 559	–1 758	–3 793	56 411	166 625	279 610	613 119	79 578	–94 564
Total	**Rb mn**	**–7 916**	**–6 464**	**–8 950**	**–11 657**	**–16 645**	**–35 510**	**491 193**	**1 536 343**	**2 554 532**	**5 487 054**	**627 341**	**–745 476**

e: estimate; p: provisional.
Source: OECD.

Annex Table I.20. **Official and adjusted exchange rate**

		1986	1987	1988	1989	1990	1991	1992	1993	1994	1995	1996p	1997e
1. Official rate (r)	Rb/US$	0.60	0.60	0.60	0.60	0.78	1.75	192.50	932	2 204	4 554	5 124	5 785
2. Adjusted exchange rate	Rb/US$	1.24	1.075	0.963	0.955	1.112	2.58	42.663	435.913	1 874	4 559	5 121	n.a.
Ratio r'/r		2.07	1.79	1.61	1.59	1.42	1.48	0.22	0.47	0.85	1.00	1.00	n.a.
R/US$		0.60	0.60	0.60	0.60	0.78	1.75	192.50	932.00	2 204.00	4 554.00	5 124.00	5 785.00
ECU/US$		1.02	0.87	0.85	0.91	0.79	0.81	0.77	0.85	0.84	0.77	0.79	0.873
R/ECU		0.59	0.69	0.71	0.66	0.99	2.16	249.16	1 091.72	2 614.16	5 952.94	6 504.19	6 626.58

e: estimate; p: provisional.
n.a.: not available.
Sources: Official rate: Goskomstat in 1986-1989 and 1992-1997.
1990: weighted average between the official exchange rate for the last 10 months and the commercial exchange rate (1.7462 Rb/US$) for November and December 1990.
1991: commercial exchange rate based on foreign trade of Russia in Roubles and US dollars.
Adjusted exchange rate: World bank. The adjusted exchange rate is the World Bank's "Atlas conversion factor".

Annex Table I.21.*i.* **Producer Subsidy Equivalents: by commodity (adjusted)**[1]

	Units	1986	1987	1988	1989	1990	1991	1992	1993	1994	1995	1996p
Wheat												
Gross total PSE	Rb mn	3 024	2 730	2 465	4 588	9 354	21 457	266 148	1 086 147	–423 461	–2 501 792	1 963 942
Gross unit PSE	Rb/t	64	74	62	104	189	552	5 765	24 942	–13 180	–83 064	56 246
Gross percentage PSE	%	51	58	48	53	68	77	70	51	–12	–23	9
Maize												
Gross total PSE	Rb mn	512	779	1 514	1 698	1 777	1 898	14 953	90 792	137 770	295 291	460 860
Gross unit PSE	Rb/t	299	203	397	364	725	964	7 004	37 195	154 451	169 903	423 585
Gross percentage PSE	%	94	90	94	90	99	83	79	63	61	31	46
Other grains												
Gross total PSE	Rb mn	4 451	4 652	2 862	4 632	10 409	14 401	527 714	1 002 912	–403 704	–2 959 631	2 564 541
Gross unit PSE	Rb/t	87	94	67	99	186	333	10 126	21 085	–9 217	–104 044	84 882
Gross percentage PSE	%	66	72	48	53	71	66	89	49	–10	–44	16
Potatoes (not included in the aggregation)												
Gross total PSE	Rb mn	1 678	3 132	3 974	6 234	6 805	30 664	271 167	1 723 471	1 604 506	–9 135 441	31 419 142
Gross unit PSE	Rb/t	131	82	118	185	221	893	7 075	45 776	47 431	–228 907	812 872
Gross percentage PSE	%	70	52	61	76	75	78	77	80	19	–25	71
Sunflower												
Gross total PSE	Rb mn	437	1 032	766	1 544	1 047	1 467	44 681	–9 942	–79 943	213 160	60 926
Gross unit PSE	Rb/t	185	337	259	408	305	506	14 367	–3 596	–31 314	50 752	22 035
Gross percentage PSE	%	67	92	69	80	71	58	80	–5	–12	6	3
Sugar (refined equivalent)												
Gross total PSE	Rb mn	2 120	2 523	2 293	2 406	1 797	1 259	63 189	560 293	389 866	1 071 424	2 216 399
Gross unit PSE	Rb/t	671	654	692	642	560	568	23 511	187 552	220 991	456 730	1 523 361
Gross percentage PSE	%	107	107	98	88	81	60	96	93	50	39	72
Crops[2]												
Gross total PSE	Rb mn	10 543	11 716	9 901	14 869	24 384	40 481	916 684	2 730 202	–379 472	–3 881 548	7 266 668
Gross percentage PSE	**%**	**67**	**76**	**61**	**62**	**72**	**71**	**82**	**53**	**–4**	**–16**	**16**

p: provisional.
1. 1997 calculations with adjusted exchange rate were not made, as Atlas Conversion Factor was not available.
2. For all crop products, the gross and net PSEs are equivalent.
Source: OECD.

Annex Table I.21.*i.* **Producer Subsidy Equivalents: by commodity (adjusted)** *(cont.)*[1]

	Units	1986	1987	1988	1989	1990	1991	1992	1993	1994	1995	1996p
Milk												
Net total PSE	Rb mn	26 264	25 718	32 728	32 763	31 961	24 327	69 872	792 930	24 823	13 769 523	18 245 406
Net unit PSE	Rb/t	503	486	600	588	574	469	1 479	17 043	589	350 896	509 378
Net percentage PSE	%	101	97	93	90	88	56	21	30	0	41	46
Beef and Veal												
Net total PSE	Rb mn	16 840	18 350	21 252	21 500	21 572	20 493	–6 653	363 194	–1 744 101	–3 651 818	819 232
Net unit PSE	Rb/t	4 483	4 598	5 121	5 052	4 983	5 137	–1 832	108 126	–538 303	–1 336 194	311 495
Net percentage PSE	%	96	96	92	89	80	57	–3	19	–38	–27	4
Pigmeat												
Net total PSE	Rb mn	6 838	7 932	9 669	8 596	7 290	2 257	–119 628	641 394	1 684 791	4 676 613	6 210 749
Net unit PSE	Rb/t	2 211	2 430	2 845	2 457	2 095	708	–42 970	263 731	801 137	2 507 567	3 642 668
Net percentage PSE	%	73	79	79	66	50	11	–86	42	43	39	40
Poultry												
Net total PSE	Rb mn	3 638	4 366	4 610	4 845	3 328	2 435	–39 252	496 237	1 564 845	4 274 235	4 899 522
Net unit PSE	Rb/t	2 255	2 550	2 596	2 646	1 848	1 391	–27 487	388 596	1 465 211	4 975 827	7 100 756
Net percentage PSE	%	76	83	83	78	51	22	–53	56	66	72	69
Eggs												
Net total PSE	Rb mn	3 340	3 027	3 026	2 716	1 464	2 088	–21 645	289 915	1 506 004	5 378 367	6 887 463
Net unit PSE	Rb/t	1 301	1 148	1 108	997	555	802	–9 081	129 500	723 345	2 861 679	3 886 099
Net percentage PSE	%	70	61	60	55	29	19	–24	33	46	59	54
Livestock products												
Net total PSE	Rb mn	56 918	59 392	71 285	70 421	65 615	51 600	–117 306	2 583 671	3 036 362	24 446 920	37 062 372
Net percentage PSE	**%**	**91**	**90**	**88**	**83**	**73**	**42**	**–14**	**33**	**14**	**33**	**39**
All products[2]												
Net total PSE	Rb mn	67 461	71 109	81 186	85 290	89 999	92 081	799 378	5 313 873	2 656 890	20 565 372	44 329 039
Net percentage PSE	**%**	**86**	**87**	**83**	**78**	**73**	**51**	**41**	**41**	**9**	**21**	**32**

p: provisional.
1. 1997 calculations with adjusted exchange rate were not made, as Atlas Conversion Factor was not available.
2. For all crop products the gross and net PSEs are equivalent.
Source: OECD.

Annex Table I.21.*ii.* **Consumer Subsidy Equivalents: by commodity (adjusted)**[1]

	Units	1986	1987	1988	1989	1990	1991	1992	1993	1994	1995	1996p
Wheat												
Total CSE	Rb mn	282	–587	366	–665	–5 743	–23 670	–208 070	–382 082	1 853 013	5 178 333	1 351 255
Unit CSE	Rb/t	5	–12	7	–13	–100	–440	–3 675	–7 806	43 482	131 363	35 734
Percentage CSE	%	4	–9	6	–6	–36	–64	–45	–16	41	36	6
Maize												
Total CSE	Rb mn	–2 188	–805	–4 512	–4 329	–3 566	–8 630	–32 811	–132 806	–206 255	–177 650	–378 857
Unit CSE	Rb/t	–199	–112	–288	–276	–583	–843	–5 280	–23 013	–95 755	–98 695	–291 428
Percentage CSE	%	–64	–50	–68	–68	–79	–76	–60	–40	–38	–18	–32
Other grains												
Total CSE	Rb mn	–1 382	–1 738	–572	–1 942	–6 743	–12 689	–412 938	–462 250	1 259 214	4 382 220	–235 379
Unit CSE	Rb/t	–27	–33	–12	–40	–117	–264	–7 801	–9 711	30 589	130 805	–7 679
Percentage CSE	%	–21	–25	–9	–21	–45	–55	–69	–23	35	55	–1
Potatoes (not included in the aggregation)												
Total CSE	Rb mn	–2 809	–1 141	–2 413	–4 421	–5 130	–25 541	–202 177	–1 269 465	188 218	12 021 819	–25 421 462
Unit CSE	Rb/t	–69	–29	–66	–130	–159	–755	–5 436	–33 055	5 148	321 387	–668 722
Percentage CSE	%	–38	–19	–34	–54	–54	–68	–59	–58	2	35	–59
Sunflower												
Total CSE	Rb mn	–63	–179	295	–104	–155	–453	–31 367	45 355	181 784	95 597	163 430
Unit CSE	Rb/t	–20	–58	100	–28	–47	–168	–10 268	19 849	87 607	36 349	80 706
Percentage CSE	%	–7	–16	27	–6	–11	–20	–58	28	34	4	10
Sugar												
Total CSE	Rb mn	–1 760	–1 908	–1 628	–557	–1 098	–2 011	–83 874	–664 368	–557 996	–1 428 739	–5 828 305
Unit CSE	Rb/t	–257	–267	–219	–83	–156	–357	–18 729	–142 047	–121 280	–303 290	–1 207 188
Percentage CSE	%	–42	–44	–31	–11	–22	–39	–77	–71	–27	–26	–57
Crops												
Total CSE	Rb mn	–5 111	–5 217	–6 052	–7 596	–17 305	–47 454	–769 061	–1 596 151	2 529 759	8 049 761	–4 927 856
Percentage CSE	**%**	**–24**	**–26**	**–23**	**–23**	**–41**	**–60**	**–60**	**–28**	**23**	**26**	**–9**

p: provisional.
1. 1997 calculations with adjusted exchange rate were not made, as Atlas Conversion Factor was not available.
Source: OECD.

Annex Table I.21.*ii.* **Consumer Subsidy Equivalents: by commodity (adjusted)** *(cont.)*[1]

	Units	1986	1987	1988	1989	1990	1991	1992	1993	1994	1995	1996p
Milk												
Total CSE	Rb mn	–13 437	–12 709	–18 907	–18 593	–20 072	–11 562	–14 743	23 036	3 105 573	–7 298 614	–11 832 797
Unit CSE	Rb/t	–228	–212	–298	–283	–311	–196	–297	450	64 011	–163 833	–288 929
Percentage CSE	%	–48	–44	–46	–43	–48	–24	–5	1	41	–20	–27
Beef and Veal												
Total CSE	Rb mn	–8 526	–10 221	–11 368	–12 885	–5 406	–13 126	18 485	117 730	4 081 170	8 591 438	3 257 127
Unit CSE	Rb/t	–1 886	–2 124	–2 176	–2 415	–1 102	–2 594	4 438	31 006	1 076 542	2 525 408	1 066 163
Percentage CSE	%	–42	–46	–39	–42	–18	–30	11	6	83	53	15
Pigmeat												
Total CSE	Rb mn	–3 389	–4 380	–5 393	–5 469	–2 418	–4 101	27 468	–469 959	–235 493	–1 914 059	–4 103 167
Unit CSE	Rb/t	–1 007	–1 233	–1 478	–1 411	–691	–1 176	9 236	–174 447	–95 496	–744 770	–1 909 338
Percentage CSE	%	–34	–41	–41	–38	–17	–19	21	–30	–5	–12	–22
Poultry												
Total CSE	Rb mn	–680	–756	–1 853	–2 524	–961	–1 231	–1 001	–392 041	–892 763	–4 549 100	–8 185 872
Unit CSE	Rb/t	–372	–387	–913	–1 151	–438	–663	–660	–263 823	–562 193	–2 600 972	–4 664 315
Percentage CSE	%	–13	–13	–29	–34	–12	–11	–2	–42	–28	–41	–49
Eggs												
Total CSE	Rb mn	–2 938	–2 517	–3 197	–3 215	–2 288	–4 431	–722	–106 030	–446 568	–3 366 293	–4 542 872
Unit CSE	Rb/t	–1 081	–906	–1 130	–1 145	–841	–1 685	–304	–47 565	–213 489	–1 782 417	–2 529 752
Percentage CSE	%	–60	–50	–62	–64	–44	–41	–1	–13	–14	–38	–37
Livestock products												
Total CSE	Rb mn	–28 971	–30 583	–40 719	–42 687	–31 145	–34 451	29 486	–827 264	5 611 919	–8 536 628	–25 407 581
Percentage CSE	**%**	**–42**	**–42**	**–43**	**–43**	**–31**	**–25**	**4**	**–11**	**24**	**–10**	**–23**
All products												
Total CSE	Rb mn	–34 082	–35 800	–46 770	–50 283	–48 450	–81 904	–739 575	–2 423 415	8 141 678	–486 867	–30 335 437
Percentage CSE	**%**	**–38**	**–38**	**–39**	**–38**	**–34**	**–38**	**–37**	**–18**	**24**	**0**	**–18**

p: provisional.
1. 1997 calculations with adjusted exchange rate were not made, as Atlas Conversion Factor was not available.
Source: OECD.

Annex Table I.22. **Reference prices**

Commodity	Reference price country	Currency	1986	1987	1988	1989	1990	1991	1992	1993	1994	1995	1996p	1997e
Wheat	EU	ECU/t	92.4	83.0	97.3	142.6	111.9	78.9	101.7	99.8	97.2	125.0	155.9	135.7
Maize	EU	ECU/t	103.0	78.5	107.6	118.2	100.0	99.3	90.2	98.8	104.6	110.7	144.9	118.3
Rye	Germany	ECU/t	75.8	63.1	88.9	160.3	110.1	46.2	82.6	105.3	81.4	75.3	117.7	109.3
Feed Barley	EU	ECU/t	71.3	53.6	91.5	117.0	86.1	76.9	79.9	72.1	70.8	98.0	131.3	117.0
Oats	EU	ECU/t	94.0	104.5	152.6	145.1	105.9	110.3	105.6	113.4	97.9	92.9	105.1	102.6
Potatoes	Germany	US$/t	84.9	110.1	113.9	100.0	106.1	121.2	88.3	55.7	134.6	270.1	91.0	70.9
Sunflower	EU	ECU/t	163.8	123.3	195.6	204.7	158.0	160.0	174.0	261.0	234.9	218.0	201.3	212.0
Refined sugar	EU	ECU/t	189.6	167.8	224.1	343.9	303.5	240.0	212.1	244.2	302.6	303.5	288.7	278.5
Milk	NZ	US$/t	67.6	105.2	151.1	161.7	122.3	128.6	143.8	137.8	147.4	184.0	190.9	170.4
Beef and veal (CW)	Hungary	US$/t	773.1	968.9	1 102.6	1 187.4	1 444.6	1 255.7	1 363.3	1 713.1	1 896.6	2 389.2	2 367.7	2 161.4
Pigmeat (CW)	Hungary	US$/t	873.2	909.0	882.7	1 238.4	1 430.9	1 373.0	1 576.5	1 335.0	1 361.0	1 793.5	2 008.9	2 037.6
Poultry	EU	ECU/t	975.3	846.1	755.4	905.7	942.6	983.0	969.9	981.4	987.6	943.3	1 107.3	1 191.2
Eggs	EU	ECU/t	632.1	787.3	663.6	666.0	795.2	824.9	715.4	793.1	814.0	674.3	928.6	933.8

e: estimate; p: provisional.
Source: OECD.

RUSSIA

Definitions and notes to Table 22

REFERENCE PRICES

Common wheat: EU export price of commercial quality wheat, fob Rouen (including handling and trading margin).[1]

Maize: EU import price of US yellow corn number 3, cif Rotterdam (including handling and trading margin).[1]

Rye: German rye unit export value to non EU-members countries.[2]

Barley: EU export price for feed barley, fob French ports (including handling and trading margin).[1]

Oats: EU cif Rotterdam, yearly average. As from 1992, fob Sweden price (including handling and trading margin).[1]

Potatoes: German farm gate price of main crop potatoes.[3]

Sunflower: EU (average France and Spain) import price, cif Rotterdam.[4]

Refined sugar: EU export price of white sugar, Bourse de Paris, fob EU.[5]

Milk: New Zealand farm gate price.[6]

Beef and veal: Hungarian unit export value for carcasses. This price is expressed in carcass weight and recalculated from the Hungarian carcass coefficient (0.56) using the Russian carcass coefficient.[7]

Pigmeat: Hungarian unit export value for carcasses. This price is expressed in carcass weight and recalculated from Hungarian carcass coefficient (0.79) using the Russian carcass coefficient.[7]

Poultry: Extra-EU unit export value, mainly frozen chicken 70%.[8]

Eggs: Extra-EU unit export value – extra EU trade NIMEXE.[8]

Sources:

1. International Grain Council.
2. Eurostat trade data (NIMEXE).
3. German Ministry of Food, Agriculture and Forestry.
4. Commission of the European Communities.
5. F.O Licht, **Internal Sugar Report**, several issues.
6. Communication from Delegation of New Zealand authorities.
7. Hungarian Trade Statistics.
8. Eurostat, **External Trade – Exports**, various years.

Annex II

MAIN POLITICAL AND AGRICULTURAL POLICY EVENTS, 1985-1997

Year	Date	Political Event	Agricultural Policy Event
1985	11 March	Mikhail Gorbachev becomes General Secretary of the Communist Party of the SU.	
	19 November	SU "Law on Individual Labour Activity" allows individual entrepreneurial activity.	
1987	30 June	SU "Law on State Enterprises" turns managers into quasi-owners of enterprises and allows direct contracting with customers/suppliers.	
1988	May	SU "Law on Co-operation in the USSR" allows setting up of non-state enterprises as co-operatives.	Extension of managerial independence of *kolkhozes* to *sovkhozes* and possibility of establishment of non-state agricultural enterprises as co-operatives.
1989	April	SU decree allows leasing of state enterprises by their workers.	
	23 November		SU "Law on Tenancy" introduces regulation on land leasing by individuals or co-operatives.
1990	28 February		SU "Principles of Land Legislation" establishes a framework for withdrawal from *Sovkhozes/ Kolkhozes* and introduces a new title for private land use: lifetime inheritable proprietorship.
	6 March		SU "Law on Property Relations" transfers ownership of farm assets to Sovkhozes.
	13 March	SU Congress of People's Deputies amends Constitution to eliminate Communist Party's monopoly on power and create post of President.	
	14 March	SU Congress of People's Deputies elects Mikhail Gorbachev as President.	
	June	Russia's Supreme Soviet declares its sovereignty and precedence of its laws over those of SU.	
	9 August	SU Council of Ministers legalises private ownership of businesses.	
	11 September	RF Supreme Soviet approves 500-day plan for transition to a market economy.	

Year	Date	Political Event	Agricultural Policy Event
	22 November		RF Congress of People's Deputies adopts "Law on the Peasant Farm", providing a legal base for private farming.
	23 November		RF Congress of People's Deputies adopts "Law on Agrarian Reform."
	25 December	RF "Law on Enterprises and Enterprise Activity" introduces various legal forms for private firms.	
1991	22 February		RF Government "Resolution on Allotment of Land Plots for Small-Scale Production" allows urban population to expand their collective orchards and vegetable gardens in the countryside.
	12 June	Boris Yeltsin is elected President of the Russian Federation with 57.3 per cent of the popular vote.	
	1 July	Member states sign the protocol for dissolution of the Warsaw Pact.	
	3 July	RF Supreme Soviet adopts "Law on Privatisation of State and Municipal Enterprises."	
	19-21 August	Coup attempted by communist hardliners.	
	24 August	Gorbachev suspends Communist Party and resigns as General Secretary.	
	August-September	Most Soviet Republics declare independence.	
	6 November	Yeltsin bans Communist Party of Soviet Union and of Russia.	
	8 December	Yeltsin and Heads of Ukraine and Belarus decide to dismantle the Soviet Union.	
	22 December	Eleven former Soviet Republics form Commonwealth of Independent States (CIS).	
	27 December	RF Supreme Soviet unanimously adopts provisional privatisation programme.	RF Presidential Decree "On Urgent Measures for Implementing Land Reform" launches agrarian reform programme.
	29 December		RF Governmental Resolution "On the Procedure for the Reorganisation of *Kolkhozes* and *Sovkhozes*" initiates farm restructuring.
1992	2 January	Retail prices for most consumer products are freed and economic reforms launched.	
	March		Subsidies for livestock production are introduced to stop the decline in output. RF Governmental decision, allowing *kolkhozes* and *sovkhozes* to retain their previous legal status if so desired.

Year	Date	Political Event	Agricultural Policy Event
	1 July	RF Presidential Decree mandating corporatisation of all state-owned enterprises by 1 November 1992.	
	4 September		RF Governmental resolution on the procedure for privatisation and reorganisation of agro-industry, specifying the course of privatisation of agricultural up- and downstream enterprises.
	December	Voucher privatisation programme for mid-sized and large enterprises is launched.	
1993	25 April	Yeltsin's economic reform programme is narrowly approved in national referendum.	
	21 September	Yeltsin dissolves Congress of People's Deputies and the Supreme Soviet.	
	3 October	Military units storm the Russian Parliament.	
	27 October		RF Governmental Resolution "On the Regulation of Land Relations and the Development of Agrarian Reform."
	12 December	Constitution of the Russian Federation is adopted in a referendum.	Constitutional guarantees for private ownership of land.
	24 December		RF Presidential Decree "On the Liberalisation of the Grain Market in Russia." The old marketing system based on K*hleboproduct* is privatised and demonopolised.
1994	7 April	RF Council of Ministers adopts cash-based privatisation policies to sell state-owned enterprises.	
	15 April		RF Governmental Resolution "On the Experience of Agrarian Reform in the Nizhny-Novgorod O*blast*," recommending the extension of the NN experiment to all other regions.
	July		First introduction of import levies on agro-food products at Russian borders.
	1 July	Formal end to voucher privatisation programme is announced.	
	27 July		RF Governmental Resolution "On Reorganisation of Agricultural Enterprises based on the Experience in the Nizhny-Novgorod O*blast*."
	12 December	Russia moves military troops into the break-away Republic of Chechnya.	
1995	January		Establishment of the Federal Food Co-operation to co-ordinate procurements of agro-food products for federal stocks.

Year	Date	Political Event	Agricultural Policy Event
	20 January	Prime Ministers of Russia, Belarus, and Kazakhstan sign a treaty establishing a customs union.	
	1 February		RF Governmental Resolution "On the Method of Exercising Rights to Land and Property by Owners."
	March		Introduction of commodity credit programmes benefiting agricultural producers.
	7 March	RF Council of Ministers Resolution "On Methods of Transferring Social Assets and Utilities From Federal to Regional and Local Ownership."	
	17 December	Elections to the Russian Parliament.	
	26 December	RF "Law on Corporations," establishing a legal framework for joint-stock companies is adopted.	
1996	15 February	RF Presidential Decree allowing the Government to confiscate property from enterprises with tax arrears.	
	7 March		RF Presidential Decree "On Guarantees of Constitutional Rights of Citizens to Land," requiring users of land to register the use rights with owners.
	22 May		RF Parliament adopts Land Code, banning private ownership of land and thereby contradicting the Presidential Decree of 7 March.
	3 July	Yeltsin is re-elected President of the Russian Federation with 54 per cent of the vote.	
	30 August	A treaty ending the civil war in the break-away Republic of Chechnya is signed.	
	3 December	Nation-wide strikes in protest of wage arrears bring work in 154 out of 189 coal mines to a halt.	
	December		*Agroprombank*, the main agricultural finance institution, is sold through open tender. Abolishing of the commodity credit sytem for agriculture and establishment of a new special soft credit fund, benefiting agro-food producers.
1997	1 March	RF Government Resolution on tax and payment arrears, effecting write-offs of mutual arrears of enterprises and the state budget.	
	22 May	President Yeltsin and President Lukashenko sign a charter on the Union of Russia and Belarus.	

Year	Date	Political Event	Agricultural Policy Event
	14 July		RF law "On State Regulation of Agricultural Production," providing for state intervention in the case of market imbalance.
	3 July		RF Federation Council approves the Land Code, after having rejected its first version in June 1996.
	21 July		RF President vetoes the new version of the Land Code, which prohibits sale and mortgage of farm land.
	October		RF Governmental Resolution to replace the Federal Food Corporation by a Federal Agency for the Regulation of the Food Market.
	23 October		CIS states, except Uzbekistan and Azerbaijan, reach agreement on the creation of a common agricultural market.

Source: Brooks/Lerman (1994), Blasi *et al.* (1997), OECD (1997*b*), OECD Secretariat.

Annex III

REGIONAL ANALYSIS OF AGRICULTURAL POLICY IN THE RUSSIAN FEDERATION

The breakdown of the highly centralised system of Soviet decision-making has led to the devolution of powers and responsibilities to subnational entities within Russia. The Federation Treaty of 1992 and the Russian constitution of 1993 defined the new division of legislative authorities between the federal and regional levels of state (Part I of the Review). Sub-federal entities gained greater independence from the centre with extended political powers and fiscal resources. Given the huge natural and social diversity within the RF, more decentralised decision-making processes seem in principle to be desirable. However, poor delimitation of federal and regional responsibilities and incomplete enforcement of federal legislation have created problems for securing the unity of the internal market.

In the field of agriculture, unevenly exercised regional responsibility for agricultural policy or poor enforcement of federal policies by sub-federal authorities have meant that progress towards structural and institutional reform in agriculture have varied widely across regions in recent years. Regional impediments to trade, unequal levels of agricultural subsidisation, and differing structural policies endanger the gains from free inter-regional exchange of goods, as well as the benefits from regional specialisation according to comparative advantage. This annex is intended to illustrate the regional diversity in policies towards agriculture, primarily by comparing recent developments in the four Russian regions of Nizhny-Novgorod, Novosibirsk, Rostov, and Tatarstan.

The regions have in part been chosen so as to represent major agricultural production areas that face differing agronomic conditions within the vast Russian state. Nizhny-Novgorod, Rostov, and Tatarstan are located in the European part of the RF, while Novosibirsk lies in Asia. Soil quality in Rostov is very high, with black-soils prevailing in large parts of the region. In contrast, soils in Nizhny-Novgorod are on average of rather poor quality. The climate in Rostov allows for extensive sunflower

Annex Table III.1. **Basic agricultural indicators for selected regions of the Russian Federation, 1995**

	Agricultural output in Mln. Rb (% of RF)	Agricultural employment in '000 (% of RF)	Agricultural land in '000 Ha (% of RF)
Nizhny-Novgorod	4 004 516	167	3 106
	(1.9)	(2.1)	(1.4)
Novosibirsk	5 062 102	156	8 439
	(2.4)	(2)	(3.8)
Rostov	6 877 866	312	8 558
	(3.3)	(4)	(3.9)
Tatarstan	8 180 276	244	4 562
	(3.9)	(3.1)	(2.1)
Russian Federation	209 364 222	7 884	220 958
	(100)	(100)	(100)

Source: Goskomstat.

Annex Table III.2. **Production of major agricultural products in selected regions of the Russian Federation, 1995**

(in '000 tons, production shares in per cent in parenthesis)

	Grains	Sunflower seeds	Potatoes	Milk	Meat	Eggs (million)	*Memo. Item:* Population (million)
Nizhny-Novgorod	943	–	943	1 020	116	972	3.7
	(1.6)		(1.5)	(2.6)	(2)	(2.9)	(2.5)
Novosibirsk	2 505	9	2 565	1 095	156	722	2.7
	(4.2)	(0.2)	(4)	(2.8)	(2.7)	(2.1)	(1.8)
Rostov	3 518	1 063	3 518	1 058	186	714	4.4
	(5.8)	(25.3)	(5.5)	(2.7)	(3.2)	(2.1)	(3)
Tatarstan	2 939	–	2 939	1 615	217	1 083	3.8
	(4.9)		(4.6)	(4.1)	(3.7)	(3.2)	(2.6)
Russian Federation	60 301	4 200	63 400	39 306	5 796	33 830	147.5
	(100)	(100)	(100)	(100)	(100)	(100)	(100)

Source: Goskomstat.

production, while Novosibirsk, for example, faces short, dry summers and long, harsh winters that pose considerable risk for any cropping activity.

But the main consideration for selection of the sample of regions was their differing attitudes towards policy change and transition. In Nizhny-Novgorod, a reform minded regional government initiated institutional and structural changes that often went far beyond those in other regions. On the other hand, semi-independent Tatarstan has maintained many of the structures and policies from the Soviet era. Agricultural reform in Rostov has been helped by entrepreneurial memories and experience of some groups within its diverse rural population, while reforms in Novosibirsk have found few supporters in the ethnically and culturally rather homogenous agricultural workforce.

So both agronomic and socio-political criteria were used to choose the sample of provinces. Annex Table III.1 provides a comparison of some basic agricultural indicators for the four regions.

All regions are major agricultural production areas, contributing at least 1.9 per cent each to total Russian agro-food output. The most significant regions among the four provinces in terms of agricultural output are Tatarstan and Rostov with shares in national production of 3.9 per cent and 3.3 per cent, respectively. Together with the *oblast* of Krasnodar, these two regions are the most important agricultural production centres in Russia.

In 1995, employment in agriculture ranged from 12.5 per cent of total employment in Nizhny-Novgorod to 18.6 per cent in Tatarstan. This compares with a share of 14.7 per cent for the whole of the RF. In terms of agricultural area, the four selected regions are of similar size as medium-sized countries in Western Europe. The smallest of the selected regions, Nizhny-Novgorod, has only about 15 per cent less agricultural land than Austria, for example. Agriculture in Novosibirsk and Rostov is relatively land intensive, while more labour intensive farming practises dominate in Nizhny-Novgorod and Tatarstan. Annex Table III.2 shows the differing agricultural output patterns in the regions.

The *oblast* of Nizhny-Novgorod, located to the East of Moscow, is a net-importer of agro-food products. The province is highly industrialised and its agricultural sector is largely specialised in livestock production. Novosibirsk in south-western Siberia is a region where agricultural production is extensive with a high land/labour ratio and climatic conditions that limit the scope of feasible crops. In contrast, agronomic and climatic conditions in the black soil region of Rostov, which is located in the Northern Caucasus area, are among the most favourable for crop production in the RF. Grain and especially sunflower cultivation feature prominently there. In Tatarstan, a republic in the Eastern part of European Russia, both crop and livestock production account for a relatively large share of all-Russian output. The intensive farming practises are fostered by high protection and support granted to farmers by the regional government (see below).

A. FISCAL DECENTRALISATION

During transition, fiscal means and together with them responsibilities for agricultural support policy were devolved from the federal to regional levels of government. The law on the fundamentals of budget organisation and the budgetary process of October 1991 and the law on the basic principles of taxation of December 1991 have provided the basis for fiscal relations between federal, regional, and local entities. Sub-federal administrations gained fiscal independence by being assigned their own tax bases and the right to decide freely on how to spend their incomes. Subsequently, revenue-sharing arrangements between the three levels of government were worked out. Besides this division of tax receipts, the Federal Government has made both "general transfers" and "special purpose transfers" to many regions. These payments have been aimed at reducing regional income disparities and at fostering particular regional policy objectives of the Federal Government, respectively.[1]

The tax-sharing and transfer arrangements do not apply to all regions to the same extent. The political weakness of the central government in the early 1990s encouraged a number of provinces to seek and obtain greater political and fiscal independence. Initially, concessions by the Federal Government were thereby largely informal. But later, explicit power-sharing treaties between the Federal Government and various subjects of the Federation were negotiated. The first of these bilateral agreements was signed by the Russian government and the Republic of Tatarstan in February 1994. The treaty in connection with several subsequent co-operation agreements allows the Tatarstan government to retain the overwhelming share of tax revenues collected in the region, while foregoing funding from many federal programmes. Similar, yet in most cases less far-reaching power-sharing treaties have subsequently been signed between the Federal Government and about a third of all subjects of the Russian Federation. The resulting fiscal decentralisation has led to increasing budgetary disparities among Russian regions. Those provinces with a strong tax base, like Tatarstan, have tended to press for increased fiscal independence and reductions of their contributions to the federal budget. This has left the central government with fewer resources to finance programmes that would benefit the poorer regions in the country.

In parallel to fiscal decentralisation, a number of other functions formerly carried out at the central level have been shifted to regional budgets.[2] Expenditures related to food price subsidisation and agricultural support policy have been among those responsibilities that have been placed upon regions. Initially, the Federal Government provided regional administrations with tied financial transfers to cover outlays related to specific agro-food programmes, particularly livestock production subsidies. The regions then disbursed the funds. But as the financial situation of the federal budget worsened, the central government found itself unable to honour all its promises to farmers for agricultural support payments in full. Regional governments frequently stepped in and paid farming subsidies out of their own budgets. As a result, the origin of agro-food subsidies shifted from federal to regional budgets over time. In Rostov, for example, the share of regional in total budgetary outlays for agriculture increased from about 29 to 63 per cent between 1992 and 1995 (Annex Box III.1).

One example of the fiscal decentralisation of agricultural policy is the commodity credit programme. This policy to provide short-term finance to farmers in order to enable them to buy production inputs like fertilisers and pesticides has been conducted under the authority of the Federal Government. But since 1996, regional administrations were required to guarantee the loans extended to farmers in their regions. Moreover, those provinces that subsequently did not honour their guarantees and fell short of repaying the credits to the federal budget in full because of liquidity problems were encouraged to issue short- to medium-term bonds drawn on regional budgets in order to raise the necessary capital. A total of 37 regions signed contracts with the federal agricultural ministry to restructure their debts into bonds and the first public auction of regional "farm bonds" took place in June 1997.[3] The receipts from the bond issues were directed into a new soft-credit fund that supplies finance only to farmers in those regions that have no outstanding commodity credit debts to the federal budgets. Hence, regional governments have had to take over responsibility for credit programmes that benefit agricultural producers in their regions.

Annex Box III.1. **Decentralisation of agricultural support payments**

Support for agriculture from the federal budget declined markedly during transition. Transfers from regional budgetary sources fell also, but to a lesser extent. Hence, regional budgets gained in relative importance for funding agricultural support payments. This development occurred throughout Russia. Yet, consistently collected data on budgetary expenditures of federal and regional governments are available for only three Russian *oblasts*, namely Orel, Pskov, and Rostov, all located in the European part of the Russian Federation. The following figure illustrates the shift from federal to regional sources of agricultural support payments for these regions.

◆ Box Annex III.1. ***Regional share of budgetary support to agricultural producers, 1992-95***

In per cent of total budgetary support

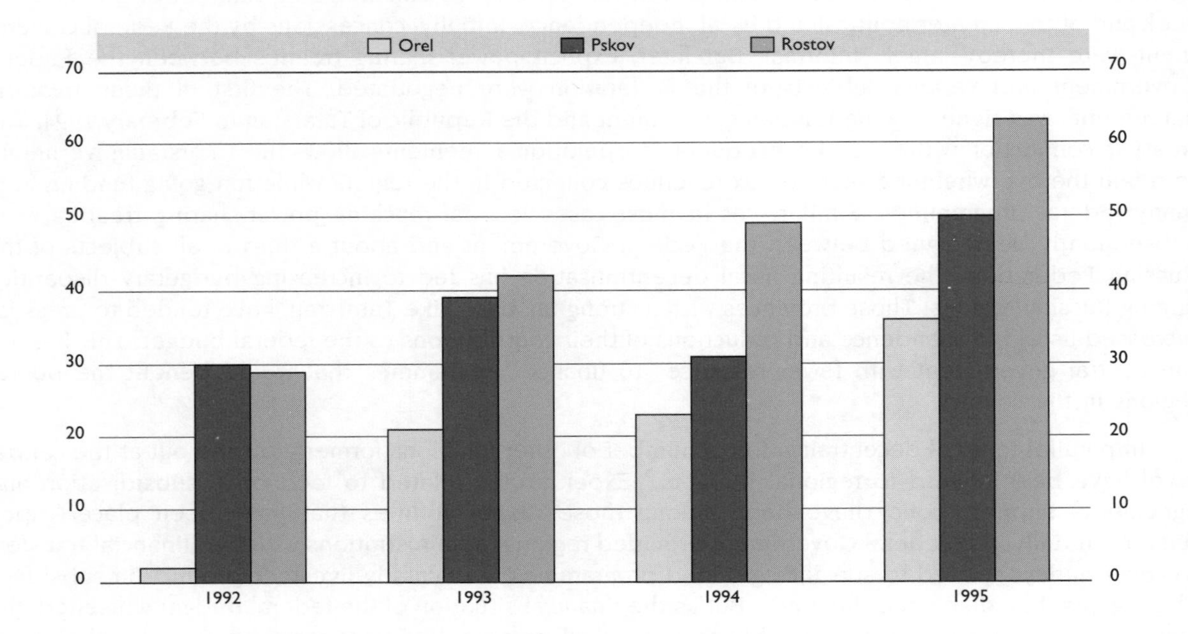

Source: Melyukhina *et al.* (1997).

The share of total budgetary support to agriculture derived from regional sources of funding increased in all three provinces during transition. In 1995, between 35 and 65 per cent of total subsidies for agricultural producers originated from regional budgets, compared with only 2 to 30 per cent in 1992. The differences in the sources of agricultural funding between regions are to a large extent related to the financial resources available to regional government and to the importance attributed to the agricultural sector.

In a federal state, the division of responsibilities for agricultural policies between national and sub-national entities can take different forms. In OECD countries, for example, various arrangements between central and regional levels of government exist (Annex Box III.2). What seems to emerge as a major problem in the Russian Federation, however, is the degree of policy diversity across regions that is a result of differing political objectives as well as of differing budgetary strength of the provinces in

Annex Box III.2. **Delimitation of responsibilities for agricultural policies in Canada, Germany, and the USA**

The division of responsibilities for agricultural policies between central and regional authorities differs considerably among OECD countries. This box aims to illustrate the policy-formulation diversity by describing the institutional framework in Canada, Germany, and the USA.

In **Canada**, both Federal and Provincial levels of government participate in agricultural policy-making. The Federal authorities are mainly responsible for policies affecting international agro-food marketing and trade and for establishing framework programmes for farm income support and stabilisation. Provincial governments can supplement these programmes from their own budgetary resources and thereby shape them to some extent according to their own priorities. Responsibilities for standardisation, licensing, and product promotion are also shared between Federal and Provincial governments, with some marketing boards, such as the Canadian Dairy Commission and the Canadian Wheat Board established under federal legislation, while most others operate at the provincial level. Policies concerning agricultural credit and finance, rural development, and extension services have generally been the preserve of the Provinces. In 1997, about 27 per cent of all support to agricultural producers originated from provincial programmes.

In **Germany**, responsibilities for agricultural policies are shared between the European, the federal and the state (*Länder*) levels of government. Agricultural trade, price, and farm income policies are determined by the European Council of Ministers and implemented by the European Commission in conjunction with national and sub-national authorities. Some policies are set up as framework programmes and leave room for national governments to top up European Union support. Social policy has remained under the exclusive authority of the Federal Government in Germany, while structural policies, aimed at improving agricultural productivity through credit and investment programmes, and rural development initiatives are jointly formulated and financed by the Federal and *Länder* governments. Agricultural extension and education services are provided under the auspices of the *Länder* administrations. About 12 per cent of all support to the agricultural sector came from the German Federal and *Länder* governments in 1997 and 88 per cent from EU sources.

In the **USA**, legislative authority for agricultural policies with respect to price and income support, farm credit and investment programmes, establishment of food quality and safety standards, and agro-food trade rests with the US Congress. The Federal Government is charged with the implementation of the policies. The responsibilities of State assemblies and governments are limited to the conduct of rural development policies, the provision of extension services, the organisation of marketing events, such as trade fairs and product promotion programmes, and some regulatory activities, such as inspection and licensing of agro-food products. In 1997, about 9 per cent of total agricultural support was due to State programmes.

Despite the differing distribution of responsibilities for particular aspects of agricultural policy, national (and supra-national) governments have remained the main force to ensure the unity of the domestic agro-food market as well as the main source of support for farmers in OECD countries. In fact, even in Canada, where agricultural policy is rather decentralised, the Federal Government still accounts for about three quarters of all support to farmers.

the absence of a strong federal policy framework. This regional diversity can pose a serious threat to the unity of the agro-food market and can inhibit the internal specialisation of Russian agriculture according to comparative advantage.

B. STATE PROCUREMENT AND PRICE POLICIES

The traditional state procurement system that set specific delivery targets for farms and allocated produce to downstream enterprises was dismantled at the beginning of the transition period. Yet remnants of the old system remain in the form of so called "food stocks". Federal stocks, acquired through downstream enterprises that are registered as governmental procurement agencies, are thereby purchasing food at administratively determined prices for the national reserve, for the military, and for food-deficit areas in the Russian north. In addition, regional stocks have been established that are run and financed by regional administrations and are supposed to guarantee regional food security. However, there is no definition in Russian federal law of what constitutes food security. In consequence, the volume of procurements and the principles of formation of regional stocks have varied throughout the Russian Federation.

In some provinces regional governments have limited their procurement activities to the supply of food to a core of public institutions, such as schools and hospitals. In Rostov, for example, the share of both grain and sunflower-seed output delivered to state procurement agencies declined from about 30 per cent in 1992 to merely 4 per cent in 1995. This switch of distribution channels went along with a fundamental reorientation of agricultural marketing during transition. Traditionally, Rostov had been a major supplier of agro-food products to Moscow. After trade liberalisation, however, Western imports became available on metropolitan markets and crowded out supplies from provincial sources. On the other hand, a quickly emerging infrastructure for foreign trade allowed Rostov to increase agro-food exports. Helped by its access to the Black Sea, shipping agricultural commodities abroad has become cheaper than transporting them to other regions within Russia. Sales of sunflower seeds to Turkish merchants, for example, expanded markedly during the early 1990s.

In other regions, however, state marketing channels have maintained a dominant role. Procurement for regional stocks, in particular, has often become an actively used policy instrument both to support agricultural producers and/or to subsidise consumers. The share of total supplies that regional authorities have thereby been able to control is closely related to their degree of administrative command over distribution channels as well as to the prices offered for deliveries to regional food stocks. The latter are generally of considerable importance in those provinces where privatisation and restructuring of farms and downstream enterprises has been slow and where regional governments have offered high prices for supplies.

In Tatarstan, for example, almost the entire sugarbeet, oilseed, and milk output, about two-thirds of grain and meat supplies, and about half of all potato and vegetable production was classified as having been delivered to federal and regional food stocks in 1995. These high delivery shares were made possible by the continuing tight grip of public authorities on farming and downstream enterprises (see below) as well as by the extraordinarily high prices that were offered to producers for deliveries to food stocks. In 1996, the administered price for regional state procurement of wheat, for example, was R 1.12 million per tonne, which was more than twice as high as the state support price of R 0.53 million in Rostov.[4]

Support to the agro-food sector has in general been substantial in Tatarstan. In addition to offering high state procurement prices, the Tatar authorities have pursued an input subsidisation policy by which they reduced farm input costs by up to 50 per cent. These massive input and output subsidies have isolated agricultural producers in Tatarstan from the adverse price developments in other parts of the RF and provided them with incentives to maintain their pre-reform level of output. In fact, production of some commodities, such as grain, even increased during the early 1990s, while output in most other regions in Russia fell quite markedly (Annex Figure III.1).

At the retail level, the Tatar government has maintained price controls for bread and some milk products to benefit consumers.[5] However, paying high support prices to farmers while maintaining low retail prices has required considerable budgetary transfers, which have been financed mostly from regional sources. In 1995, total budgetary support to agriculture amounted to about 39 per cent of the value of agricultural output. Maintaining a Soviet-style system of regulation and subsidisation has, therefore, proven to be very costly.

Other remnants of pre-reform price policies, have survived in some regions. The practice of differentiating administrated producer prices across "production cost-zones", for example, was common in the Soviet Union, but was abandoned at the national level in the wake of price liberalisation. However, some regions with rather traditionalist governments, such as Novosibirsk, still use "cost-plus pricing" schemes. Subsidies for milk and meat production in this *oblast* vary with the location of agricultural producers. Enforcing such spatial price differentials requires tight administrative control over commodity flows. Once private, independent economic agents account for a larger share of

◆ Annex Figure III.1. ***Change in grain production between 1986-90 and 1994-96***
In per cent

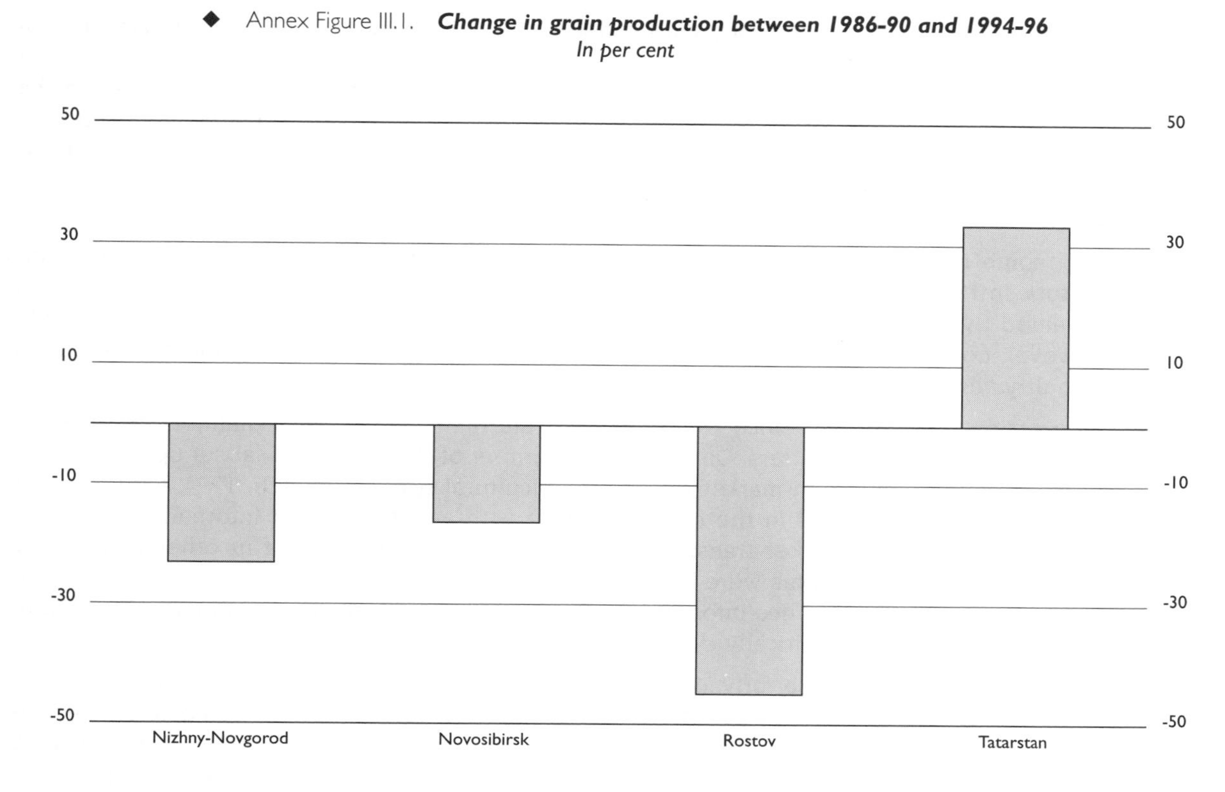

Source: Goskomstat.

agricultural marketing, it will become difficult to sustain zonal-pricing regimes. But more importantly, cost-oriented pricing inhibits the intra-regional specialisation according to comparative advantage and reduces the incentives to adopt more efficient farming practices. It is, therefore, questionable whether the continuation of zonal-pricing is in the best medium to long-run interest of agricultural producers in Novosibirsk.

In addition to regulating producer prices, many regions maintained retail price or mark-up ceilings after price controls had been revoked at the federal level in early 1992.[6] Regional authorities had been entitled by the Federal Government to determine both the scope of products that were covered by price controls and to set the price ceilings. The extent of price regulation in individual regions has thereby varied with the amount of financial resources available to subsidise consumer prices and with the attitude taken towards price liberalisation. In many regions, state-controlled food prices in retail stores have been considerably below prices on free city-markets, although part of the difference might be due to quality differentials.[7]

C. LACK OF MARKET INTEGRATION

Different regional price policies are only sustainable if traders do not or are not able to take advantage of arbitrage opportunities. And indeed, there are strong indicators that the agro-food market in Russia is not integrated, because of inter-regional trade barriers, lack of market price information, and imperfect competition in the downstream sector.

Articles 8 and 27 of the Russian constitution guarantee the unity of economic space, *i.e.* the free movement of goods, services, capital, and labour within the country. Article 74 explicitly prohibits regional barriers to free internal trade or factor movement. Any exemption requires federal legislation.

These common market principles are poorly enforced, though. Many regions have adopted administrative measures to limit the free flow of goods, services, or production factors. Some provinces have, for example, prohibited the export of particular goods from their territories. The local administration in the Tambov *oblast*, for example, has forbidden farmers to "export" meat outside the region in order to ensure supplies for the ailing local processing plant. Other regions have charged levies on trans-regional shipments.[8] These measures have limited the ability of traders to move goods to those locations where the price is highest.

Moreover, some regions have introduced region-specific food quality standards to protect their agro-food sectors. In 1997, the governor of Belgorod, for example, ordered that "imported" poultry had to be accompanied by a certificate showing the quantity of psychotropic (mind-altering) substances in the meat. However, no laboratory is able to conduct such tests, so that the regulation effectively results in a ban of poultry "imports" into the *oblast*.[9]

Also, information on agro-food price differences for alternative marketing channels or different regions is not readily available to traders. Only a limited number of *oblasts* have so far set up agencies to collect and distribute information on market prices for agricultural commodities. In 1994, a first market information system was established in the *oblast* of Kaluga with assistance from international donors. Similar institutions to increase market transparency were subsequently created in other regions. By mid-1997, market information systems were operating in about 30 provinces. In the majority of other regions, though, the only source of price information remains to be the local agricultural administration, which normally quotes only the politically determined prices for state procurement.

In addition, some of the formally privatised trading and processing enterprises are still majority-owned by state-shareholders. These firms often continue to operate in much the same way as under state socialism. Managers frequently have close links with the political establishment and are to a significant extent subject to control by local and regional authorities. The considerable importance of state control over enterprises in the agricultural downstream sector has meant that the marketing of agro-food products has often been dominated by political objectives. Self-sufficiency targets of many regional administrations have thereby favoured local sales over trans-regional shipments of agro-food products. Opportunities to profit from inter-regional price differences have often not been pursued.

Another characteristic of the Russian agricultural marketing system is a widespread lack of competition in handling, trading, and processing of agricultural products. Large enterprises continue to dominate agro-food marketing on local and regional markets. Following price deregulation, these marketing firms have often used their dominant positions to extract monopoly rents. Potential competitors were frequently denied entry into local markets by semi-legal or illegal means ranging from supply discrimination to physical threat.[10] Excessive marketing margins have been the result.

Furthermore, the abolition of subsidies and price regulations since 1992 has led to a relative increase of fuel prices compared to agricultural producer prices, making transport activities relatively more expensive. In addition, lack of investment in transport facilities has resulted in deterioration of the quality of the transportation network. Thus, the costs of transporting agro-food products have risen during transition.[11] This cost increase has certainly contributed to a reduction in transport volume. The quantity of grain that was shipped by truck or train over long-distance within the Russian Federation fell from 70.4 million tonnes in 1992 to 40.7 million tonnes in 1994, *i.e.* by 42 per cent.[12] During the same time period, total grain output fell from 106.9 million tonnes to 81.3 million tonnes, *i.e.* by 24 per cent. Hence, the share of total grain production that was transported over long-distances declined.

The reduced transport volume could just be the result of adjustments from excessive levels of transport activity during the pre-reform period. But inter-regional trade barriers, poor marketing infrastructure, imperfect competition in agricultural downstream sectors, and increased transport costs suggest that agro-food markets in the Russian Federation might have been getting more fragmented during transition. Several studies have used statistical methods to test whether the Russian agro-food market is integrated or indeed regionally segregated. The tests are based on the analysis of price changes over time in different regions. In an integrated market where regions are linked via trading activities, price changes in different provinces should be similar in direction and magnitude. Price

Annex Table III. **Summary of results from market integration tests**

Study	Data	Main results
Gardner/Brooks, 1994	Weekly retail store and city market prices for 19 food commodities in 17 cities in the Central region, Feb. 92-Apr. 93.	Persistence of substantial regional price differences and lack of market integration.
De Masi/Koen, 1996	Weekly retail store and city market prices for 19 food commodities in 99 cities dispersed throughout Russia, Feb. 92-Dec. 94.	Large, persistent price variations across cities.
Goodwin *et al.*, 1996	Weekly retail store and city market prices for 4 food commodities in 5 cities dispersed throughout Russia, June 93-Dec. 94.	Some markets integrated, others not. In some cases long time lags until price changes are transmitted to other regions.
Berkowitz *et al.*, 1996	Weekly retail store and city market prices for 5 food commodities in 25 cities in the Central and Volga regions, Feb. 92-Feb. 95.	Indications that agro-food markets in macro-regions, at least, are integrated.
Loy/Wehrheim, 1996	Weekly retail store and city market prices for 10 commodities in 5 regions within European Russia, Jan. 93-Dec. 95.	Markets in general poorly integrated. Some spatial price linkages exist for high value-added products.

movements in non-integrated markets, on the other hand, would express no systematic relation. Annex Table III.3 summarises the results from several market-integration studies.

The evidence from most of the studies confirms the impression that Russian agro-food markets are poorly integrated. Especially during the first months after price liberalisation started, substantial price differences between regions persisted, indicating a lack of marketing efficiency. During this period many regional authorities retained price controls for agro-food products and traders were just beginning to explore their new roles in a market economy. Over time, the distortionary interventions of regional authorities have receded, better information on inter-regional price differences has become available, and increasing numbers of intermediaries have made use of arbitrage opportunities. As a result, the efficiency of market allocations in the Russian Federation has improved. Markets for some high value-added products within the same macro-region, at least, seem to be becoming more and more integrated.

D. LAND POLICY AND FARM RESTRUCTURING

According to Article 9 of the Russian constitution, land can be in private as well as in state, municipal, or other ownership. However, controversial discussions between the Federal Government, the Federal Parliament, and the Federation Council about the implementation of private land ownership and, in particular, the owner's rights to sell or mortgage land have dragged on over months and years without reaching a conclusion (Part III of the Review as well as Annex II). Meanwhile, a large number of regions have tried to resolve the legal vacuum by adopting their own "land codes". Yet, the way this was done has only added to the confusion about private property rights to land, rather than provided a solid legal basis for private land ownership.

In November 1997, the regional parliament in Saratov adopted a law that allows for the free sale, purchase, and mortgaging of farm land. More than thirty other Russian regions have shown an interest in passing similar legislation and some regional parliaments, such as the one in the Samara *oblast*, have subsequently adopted Saratov-type land laws.[13] On the other hand, resistance by some communist-led regional governments to private ownership of land have obstructed the creation of a country-wide land market. In a direct reaction to the adoption of the land law in Saratov, 13 regions with communist dominated regional parliaments in the "red belt", stretching from Tula to Krasnodar, passed legislation explicitly prohibiting market transactions in land.

Attitudes towards the restructuring of large-scale farming enterprises have also varied among Russian regions. In Nizhny-Novgorod, a reform-minded, progressive regional government initiated the restructuring of large-scale agricultural enterprises faster and more thoroughly than in most other regions of the RF. In addition to the all-Russian re-registration of large-scale farms, a restructuring programme was designed with the help of the International Finance Corporation (IFC) to clearly assign property rights to land and non-land assets held by collective farms and to transform the latter into smaller, more efficient production units (Box 3 in Part III of the Review for details of the programme). By mid-1997, 206 *kolkhozes* and *sovkhozes*, *i.e.* about 30 per cent of all collective farms, had undergone comprehensive reorganisation and been broken up into 807 new enterprises of varying legal status.

Results from IFC farmworker surveys indicate that both work motivation and discipline in reorganised agricultural enterprises is higher than in un-restructured farms. In reorganised farms, a majority of employees reported that salaries from work on the farm were their main source of income, while this was the case for only about 20 per cent of farmworkers on unrestructured farms. Hence, employees of reorganised farms seem to earn higher wages and spend less time working on their own household plots than their colleagues on farms that have not undergone thorough restructuring yet.

Based on the Nizhny Novgorod experience, some other regions have subsequently launched comprehensive farm re-organisation programmes.[14] The *oblast* of Rostov was among the first. A consulting unit was set up in Rostov-upon-Don, the *oblast*'s capital, to administer and co-ordinate the reforms and by the end of 1996, six agricultural enterprises had been transformed into 22 new production units. About 3 000 stakeholders and 38 000 hectares of land were affected by the restructuring.

On the other hand, agricultural restructuring in other parts of Russia has been slow. Novosibirsk followed the federal restructuring model in the sense that a majority of large-scale agricultural enterprises changed their legal status from collective farms into co-operatives, partnerships, or stock holding companies (Part III of the Review). Yet, the only substantive modification triggered by this re-registration concerned remuneration payments to farm workers, which were now to be based both on labour services provided as well as on the amount of farm shares held. Actual farming operations, however, changed little. Moreover, a relatively large fraction of about 30 per cent of all collective farms in Novosibirsk decided not to change at all, but to maintain their status as *kolkhozes* or *sovkhozes*.

Farm restructuring in Tatarstan was similarly cautious. By October 1996, only 129 out of 401 *sovkhozes* had been "privatised" by distributing entitlements on their assets. Even though, the state administration maintained a majority of shares in all of the "privatised" state farms. Reforms of *kolkhozes* in the region were similarly slow. Until Fall 1996, nearly 30 per cent of them had not yet undergone the process of re-registration and land-share allocation mandated by the Federal Government in 1992. Moreover, little if anything has changed in the way the large-scale farming enterprises operate. The attitude towards land reform and farm restructuring started to change at the end of 1996 and by the beginning of 1998 Tatarstan became one of the most progressive Russian regions in this respect. In April 1998, partly inspired by Saratov's example, the local parliament of Tatarstan adopted a Land Code which allows the free sale and purchase of land, including to foreign individuals and companies.

Unequal developments in different Russian regions occurred also with respect to the emergence of private farms. In Rostov, for example, individual farms emerged relatively late, but their number grew rather quickly from 1992. By the end of 1995, about 16 000 individual farms, cultivating more than 8 per cent of agricultural area, had been registered (Annex Table III.4). Most of them concentrate on crop rather than livestock production. Conditions for taking up private farming have generally been quite favourable in Rostov. Some groups within the regional population, notably Cossacks who foster memories of historical independence, have shown considerable entrepreneurial energy and established themselves as private farmers. Moreover, private farming candidates faced fewer problems to obtain land and credit for setting up their own operations than in many other regions. As a result, individual farms are on average bigger than in the whole of the RF and have proven to be more resilient to the unstable economic environment during transition. While 25 per cent of all farmers in the RF that had started on their own since 1991 had given up farming again by the end of 1995, the corresponding figure for Rostov was a mere 8 per cent.

In other regions, the development of private agriculture has been slower. In Novosibirsk, for example, only 5 108 private farmers were registered by the end of 1995, cultivating less than 5 per cent

Annex Table III.4. **Structure of individual farms in selected regions of the Russian Federation, 1996**[1]

	Number of individual farmers	Average size of individual farms (ha)	Share of agricultural land used by individual farmers (per cent)
Nizhny-Novgorod	3 625	25	2.9
Novosibirsk	5 108	67	4.7
Rostov	16 102	48	8.3
Tatarstan	849	31	0.6
Russian Federation	278 613	43	5.4

1. at the beginning of the year.
Source: Goskomstat.

of total agricultural area, even though agricultural land has been readily available in the sparsely populated region. The sceptical attitude of farm enterprise managers towards privatisation and restructuring, the lack of entrepreneurial experience and initiative of the relatively homogenous farming population, and the high weather risk associated with agricultural production in the region are considered to have been major factors inhibiting the creation of more private farms.

Also, while reorganisation of large farms has been relatively swift in Nizhny-Novgorod and is continuing, the development of individual farming has been less dynamic. At the end of 1995, private farms used only 2.9 per cent of total agricultural land in the *oblast*. This share falls well short of the all-Russian average of 5.4 per cent. One factor that has inhibited the development of a more vibrant private farming sector has been the scarcity of high-quality land. Nizhny-Novgorod is a relatively densely populated region, so that competition for farm land is intense. As a result, individual farms are small compared to other regions of the RF. In 1995, average size of individual farms was 25 hectares, which compares to an average of 43 hectares in the RF. Moreover, soil quality in the *oblast* is generally poor, so that yield and profit expectations per acreage unit are relatively low. Hence, establishing a small family-farm is in many cases not a desirable choice for people in rural areas.

In Tatarstan, the traditionalist policy approach with high support payments targeted at large-scale farms has inhibited the emergence of a significant private farming sector altogether. By the end of 1995, only 849 individual farms were operated in the republic, using merely 0.6 per cent of total agricultural land.

E. SUMMARY AND CONCLUSIONS

This annex provided an overview of agricultural policies at the regional level in Russia. It investigated regional price and support policies, impediments to inter-regional trade, and structural policies at the farm level.

The decentralisation of agricultural policy has allowed regional governments to tailor sectoral support measures to a large extent to their own priorities. Diverging objectives for farm income support and food security have thereby led to differing targets for regional food procurement and agricultural producer price support. Those regions with a strong tax base have generally provided more extensive financial transfers to agricultural producers and have paid higher agricultural support prices than poorer provinces.

In order to sustain the inter-regional differences in agro-food prices, many regions have restricted trade flows of agricultural or food products into or out of their territories. These trade barriers, which are in themselves unconstitutional, together with imperfect competition in the downstream sector and a poorly developed market infrastructure have resulted in a fragmented agro-food market in Russia.

The implementation of the farm restructuring programme initiated by the Federal Government in 1992 has varied considerably across regions. Some provinces have reorganised large-scale farms only formally, while others have made decisive efforts to move to farming structures that, by increasing the farm operator's flexibility and self-responsibility, allow for more efficient agricultural production. Also, different attitudes towards private enterprise and land ownership have contributed to the uneven development of the private farming sector in Russia.

Unequal levels of farm support across regions, inter-regional trade barriers, and differing structural policies inhibit the process of specialisation according to comparative advantage in Russia. Farm operators do not engage in those activities that would be most suitable for their particular location, but rather in those that receive the highest support from their respective regional governments. The allocation of resources within the Russian farming sector, hence, is currently sub-optimal.

Moreover, it seems on first sight that regional administrations that have maintained Soviet style agro-food policies, like large-scale state procurement, input subsidisation, and zonal pricing, have been rather successful in terms of preventing a substantial decline in agricultural output in their regions. However, such policies have proven to be very costly for taxpayers and consumers. Also, the experience from other transition countries shows that postponing necessary systemic changes in the transformation from a centrally planned to a market based economy only makes the ultimate adjustments all the more painful later on.

The delimitation of responsibilities for agricultural policy between federal and regional authorities varies considerably in those OECD countries that are organised as federal states. But what is common in all these countries is that the federal level of government retains the key responsibilities for agricultural price and trade policies and establishes a uniform policy framework throughout the country. Moreover, the federal governments in connection with the judicial authorities take responsibility for the implementation of federal law. Such an approach would be beneficial for Russia's agro-food policy also.

NOTES

1. See Le Houeroue/Rutkowski (1996), Chapter 2 of World Bank (1996*b*), and Annex III of OECD (1997*b*) for more elaborate discussions of fiscal federalism in the Russian Federation.
2. The process had already started in 1989, but became explicit policy from 1992.
3. See, for example, Interfax (1997), "Russian Government to Auction First Regional Bonds for Farm Fund," *Food and Agriculture Report*, Issue 23, and Interfax (1997), "Demand Outstrips Supply in First Farm Bond Auction," *Food and Agriculture Report*, Issue 26.
4. See Interfax (1996), "Russian Government Buys 7.5 Million Tonnes of Grain So Far," *Food and Agriculture Report*, Issue 43.
5. In addition, Tatarstan authorities have imposed bans for some agro-food products to be exported from the region in order to ensure ample supplies on the domestic market.
6. See Melyukhina/Wehrheim (1996) for the incidence of retail price regulation in different regions.
7. See Berkowitz *et al.* (1996) for an analysis of state-store versus city-market price relationships.
8. See USDA (1997), "Russian Federation: Processed Meat," Foreign Attache Report No. RS7040, Washington, D.C. The trade restrictions are enforced by police personnel that control cross-regional transport links. In addition, organised crime is reported to frequently interfere with cross-country traffic in order to extract tolls.
9. See USDA (1997), "Russian Federation: Poultry Annual Report," Foreign Attache Report No. RS7036, Washington, D.C.
10. See, for example, "Russia's Breadbasket Hit by Mafia Marketing," The Moscow Times, June 23, 1994.
11. High fuel costs affect both road and rail transport. In addition, railway freight charges are artificially inflated by the need to cross-subsidise passenger transport, since about 50 million Russians are entitled to reduced-fare or free railway transport.
12. See Kopsidis (1997) for a discussion of grain transport within Russia.
13. See, for example, Interfax (1997), "Regional Governors Split on Eve of Land Round Table Session," *Food and Agriculture Report*, Issue No. 49, and Agra Europe (1998), "Veto on Russian Land Code Remains as Regions Take Initiative," *East Europe Agriculture and Food*, February Issue.
14. See Agra Europe (1997), "Nizhny Novgorod Land Reform Programme Spreading," *East Europe Agriculture and Food*, September Issue.

OECD PUBLICATIONS, 2, rue André-Pascal, 75775 PARIS CEDEX 16
PRINTED IN FRANCE
(14 98 04 1 P) ISBN 92-64-16072-8 – No. 50209 1998